Computer Integrated Manufacturing

Volume IV of a four-volume set. Other volumes include:

Computer Integrated Manufacturing, Volume I:
  Revolution in Progress
  R.U. Ayres

Computer Integrated Manufacturing, Volume II:
  The Past, the Present, and the Future
  R.U. Ayres, W. Haywood, M.E. Merchant, J. Ranta,
  and H.-J. Warnecke, eds.

Computer Integrated Manufacturing, Volume III:
  Models, Case Studies, and Forecasts of Diffusion
  R.U. Ayres, W. Haywood, and I. Tchijov, eds.

Computer Integrated Manufacturing

Volume IV: Economic and Social Impacts

Edited by

R.U. Ayres, R. Dobrinsky, W. Haywood, K. Uno and E. Zuscovitch

International Institute for Applied Systems Analysis
Laxenburg, Austria

CHAPMAN & HALL

London · New York · Tokyo · Melbourne · Madras

Published by Chapman & Hall, 2–6 Boundary Row, London SE1 8HN

Chapman & Hall, 2–6 Boundary Row, London SE1 8HN, UK

Van Nostrand Reinhold Inc., 115 5th Avenue, New York NY10003, USA

Chapman & Hall Japan, Thomson Publishing Japan, Hirakawacho Nemoto Building, 7F, 1–7–11 Hirakawa-cho, Chiyoda-ku, Tokyo 102, Japan

Chapman & Hall Australia, Thomas Nelson Australia, 102 Dodds Street, South Melbourne, Victoria 3205, Australia

Chapman & Hall India, R. Seshadri, 32 Second Main Road, CIT East, Madras 600 035, India

First edition 1992
© 1992 International Institute for Applied Systems Analysis

Printed in Great Britain by The University Press, Cambridge

ISBN 0 412 40470 2   0 442 31367 5 (USA)

A catalogue record for this book is available from the British Library

Library of Congress Cataloging-in-Publication data available

♾ Printed on permanent acid-free text paper, manufactured in accordance with the proposed ANSI/NISO Z 39.48–199X and ANSI Z 39.48–1984

Contents

Preface

The purpose of the Computer Integrated Manufacturing (CIM) Project at the International Institute for Applied Systems Analysis (IIASA) is to close the widening gap between the pace of technological, economic, and social events, on the one hand, and the progress of understanding those events, on the other.

The Project, which is currently approaching completion, is not the first at IIASA on the application of computers to manufacturing. In fact, one of the first conferences ever held at IIASA (October 1–3, 1973) concerned the automated control of industrial systems. It was cochaired by Prof. A.Cheliustkin of the Institute for Control Problems, Moscow, and Prof. I. Lefkowitz of Case Western Reserve University, Cleveland. Topics proposed for discussion at that conference included automation of quality control, economies of scale, increased demand for flexibility, automation of multiproduct plants with multipurpose equipment, and adaptation of control systems to changing industrial environments. All of these topics are relevant today.

The IIASA study has attempted, first, to define the existing world situation with regard to the underlying technologies of CIM, and the degrees to which technologies such as NC/CNC machine tools, robotics, and CAD/CAM are currently being used in metal products manufacturing. We have concentrated our attention primarily on this subset of the overall manufacturing sector because of two simple facts: *machines* are metal-products and manufacturing (not to mention transportation, mining, agriculture, utilities, and even households) depends on machines. In other words, the metal-products sectors, and particularly the *machine-building* subsectors, have a unique and vital role in the economic system. It is this sector that produces the capital goods on which all production depends. Above all, it is the machine-building sector that reproduces itself.

The methodology adopted in the study is eclectic. It is multiperspective and multidisciplinary, as well as multinational. It incorporates elements of both "bottom-up" and "top-down" approaches. Finally, it incorporates both historical analysis and "model" forecasts of the future, together with scenario analyses.

The bottom-up part of the study began with a review of the extensive international literature. We have compiled, from (mainly) published sources, a large worldwide data base on so-called flexible manufacturing systems (FMSs). The data base contains 880 entries. This data base is incomplete in many respects, but it is large enough to carry out a variety of statistical analyses and comparisons, both cross-sectional and longitudinal.

In addition, with help from local collaborating institutions, a number of in-depth case studies at the firm level have been carried out in several countries, with both centrally planned and market economies. The results of this work, when complete, will provide perhaps the first truly international comparison of manufacturing technology and manufacturing management of its kind.

The top-down elements of the study consist of two parts. One is the development of a generalized theory of manufacturing as a process of adding information to materials. In particular, one can regard manufacturing as a process of adding morphological information (i.e., shape and form) to materials. Insights derived from this perspective are utilized throughout the Project, although the theory itself is somewhat too technical to be included among the main reports of Project results. (It is to be published separately as a scientific monograph entitled *Information, Evolution, and Economics*.)

The other part of the work that can be called top-down is analysis using various economic models. One category of economic models that we have used is the static or quasi-static input–output (I–O) type of model to explore international impacts of CIM adoption at the sectoral level. Another type of model used for comparing economic growth scenarios on an international basis is a set of linked macro models, previously developed by Wilhelm Krelle at the University of Bonn in collaboration with IIASA (the Bonn–IIASA model). Still a third type is the dynamic simulation model with which we are exploring interactions among such variables as R&D expenditure, product improvement, process improvement, capital flexibility, cost reduction, price reduction, demand, capital investment, profitability, and so on.

This is Volume IV of the Computer Integrated Manufacturing Series. In this volume, *Economic and Social Impacts*, we focus on the effects of CIM on society and the economy in both quantitative and qualitative terms. Volume I, *Revolution in Progress*, is a summary volume and describes the outputs of the bottom-up part of the study. The second volume, *The Past, the Present, and the Future*, examines past technologies and forecasts the future of CIM. Volume III, *Models, Case Studies, and Forecasts of Diffusion*, presents a descriptive and interpretive analysis of the processes of diffusion. The series concludes with a discussion of CIM implementation strategies and related managerial issues at the level of the firms in Volume V, *Fewer and Faster: A Story of Technological, Organizational, and Management Innovation in the Manufacturing Enterprise*.

Robert U. Ayres

THE INTERNATIONAL INSTITUTE FOR APPLIED SYSTEMS ANALYSIS

is a nongovernmental research institution, bringing together scientists from around the world to work on problems of common concern. Situated in Laxenburg, Austria, IIASA was founded in October 1972 by the academies of science and equivalent organizations of twelve countries. Its founders gave IIASA a unique position outside national, disciplinary, and institutional boundaries so that it might take the broadest possible view in pursuing its objectives:

To promote international cooperation in solving problems arising from social, economic, technological, and environmental change

To create a network of institutions in the national member organization countries and elsewhere for joint scientific research

To develop and formalize systems analysis and the sciences contributing to it, and promote the use of analytical techniques needed to evaluate and address complex problems

To inform policy advisors and decision makers about the potential application of the Institute's work to such problems

The Institute now has national member organizations in the following countries:

Austria
The Austrian Academy of Sciences

Bulgaria
The National Committee for Applied Systems Analysis and Management

Canada
The Canadian Committee for IIASA

Czech and Slovak Federal Republic
The Committee for IIASA of the Czech and Slovak Federal Republic

Finland
The Finnish Committee for IIASA

France
The French Association for the Development of Systems Analysis

Germany
Association for the Advancement of IIASA

Hungary
The Hungarian Committee for Applied Systems Analysis

Italy
The National Research Council (CNR) and the National Commission for Nuclear and Alternative Energy Sources (ENEA)

Japan
The Japan Committee for IIASA

Netherlands
The Netherlands Organization for Scientific Research (NWO)

Poland
The Polish Academy of Sciences

Sweden
The Swedish Council for Planning and Coordination of Research (FRN)

Union of Soviet Socialist Republics
The Academy of Sciences of the Union of Soviet Socialist Republics

United States of America
The American Academy of Arts and Sciences

Part I

Overview

Chapter 1

Introduction

Robert U. Ayres and Ehud Zuscovitch

1.1 Motivation of the Study

CIM is an abbreviation for computer integrated manufacturing. It is an acronym that has become fairly well known in recent years in manufacturing and related engineering circles, although most non-engineers (including economists and social scientists) have never heard of the term. Questions naturally arise: Why should a phenomenon as arcane as CIM have any economic or social consequences worth studying? What could they possibly be?

The short answer to the first question is that we think a new industrial revolution is under way. The first industrial revolution which began in the UK in the eighteenth century is usually characterized as the substitution of steam power for human and animal muscles. Steam power – especially as applied in cotton mills, railways, and steamships – gave Great Britain a major economic headstart on the rest of the world and the wherewithal to become the center of a large empire. It also gave rise to the factory system, Luddites, and Marxism.

The "second" industrial revolution was the decentralization of mechanical power from large immobile central steam engines to small mobile internal-combustion engines or even more dispersed electric motor drives. This revolution occurred about a century ago. The exact dates are unimportant.

Among its consequences were the development of the auto industry, mass production, suburban sprawl, and the consumer society.

The present revolution is the substitution of electronic sensors for human eyes and ears, and computers for human brains, at least in a certain category of routine on-line manufacturing operations. Its consequences may be as far-reaching (and as unexpected) as the consequences of the "first" industrial revolution, which was just beginning about two centuries ago. The second question we raise is, What will those consequences be?

The industrial importance of steam power was, of course, obvious to the farsighted entrepreneurs who actively developed and invested in it, from Roebuck, Boulton, and the Wilkinsons to the Stephensons. But the longer-run implications of its use – from "satanic mills," child labor, and Marxism to acid rain and climate change – were not foreseen at all. Nor did the early steam engineers concern themselves with such questions. They simply sensed an economic opportunity and exploited it.[1] Only in later generations did historians begin to piece together an understanding of the causal connection between the eighteenth-century technological changes and the nineteenth- and twentieth-century economic, social, and environmental consequences.

The first industrial revolution began in one country (Great Britain) and spread gradually to France, Germany, the United States, and beyond.

The second industrial revolution began in Germany and the USA. The internal-combustion engine and the automobile were essentially German inventions (Otto, Diesel) and German innovations (Daimler, Benz), although large-scale manufacturing leadership was later taken over in the USA by Ford. On the other hand, the electric-power industry, as such, was initially an American innovation (Edison, Westinghouse), although Siemens in Germany and other European firms pioneered in some areas of electrotechnology while Parsons (UK) innovated the steam turbine.

It is clear that the rise of Germany and the USA to world industrial preeminence in the early part of this century owes much to their leading role in these critical technologies, just as Britain's productivity leadership in the last century was due to its early domination of the technology of steam power, iron making, and machine tools. It is already an obvious conjecture that the countries that gain technological leadership in the next industrial revolution will obtain economic benefits from that leadership for decades thereafter.

Yet, in a superficial sense at least, today's technology spreads as far and as fast as modern communications networks allow. While manufacturing technology is not as easy to imitate as product technology, the free movement

of capital goods, engineers, and scientists across national borders guarantees
that diffusion will be rapid. The scope and speed of the change in physical
production technology that is now under way mean, in turn, that there
will be less time for society to react. Hence there is less time for decision
makers – whether in government or industry – to think about appropriate
governmental responses to crises – such as large-scale closures or relocations
of manufacturing facilities – before the imperative to act becomes irresistible.

Moreover, the industrial and political leaders of most countries are more
sensitive to public opinion than ever before – another consequence of the
new communications technologies. It is not a moment too soon for those in
positions of responsibility to anticipate some of the challenges that lie ahead.
It is by no means too soon to consider the policy options that are likely to
be available – or unavailable – under various plausible future scenarios. (In
fact, it may be more important for decision makers, and their advisers, to
be aware of the attractive dead ends and seductive potential traps than
anything else.)

1.2   Methodological Comments

Rosenberg once wrote that automobile technology did not yet "exist" in
the true sense in the early years when a few rich young sportsmen in open
phaetons were enjoying speed (and loud noises) for their own sake and for
the fun of terrifying the horses on the country roads. Indeed the automobile
as a social and economic system really came into existence when cars were
enclosed, hand cranking was eliminated, roads were paved, gasoline stations
were constructed, and automobiles became a practical means of point-to-
point transportation for ordinary people (Rosenberg, 1976). This, of course,
made possible the new suburban style of living for middle-class households
that, in turn, changed the urban landscape. The new form of transport
created a strong decentralizing force in society, with a new division of labor
and markets. Finally, the automobile manufacturing industry and its satel-
lites (e.g., petroleum refiners and gasoline distributors, materials and parts
suppliers, and subcontractors) became an essential part of the industrial
structure.

Kuznets (1977) describes well the sequence of industrial, economic, and
social transformations that characterize a major technological innovation:

> But the effects of technological innovations were not only on capital forma-
> tion and factor productivity. They were also on the organization of economic

production or management units, in the pressure for the modern type of corporation; and they had a ramifying effect on industrial organization through the use of the discriminating power of monopoly. They affected conditions of work, with changes in labor force status, employment requirements, educational levels, and the effective lifespan of the work population; and they affected conditions of life, through furthering urbanization and modifying patterns of consumption and other elements in the modes of living associated with rising economic standards. The various institutional adjustments, and shifts in conditions of work and life, required for an effective channeling of the continuous stream of technological innovations, were neither easy, nor costless. The gap between the stock of knowledge and inventions as the necessary condition, and the institutional and social adjustments that would convert the former into a sufficient condition, is wide.

For any technological revolution, with or without an important *a priori* "advertisement" effect, it takes time for the downstream influences to change living standards, work organization, industrial structure, demographic patterns, and consumption habits.

With computer technology we face this problem again. Already most office tasks – even word processing – depend on computers. Now information-intensive technologies are increasingly integrated into the materials-handling and production processes. New computerized production tools include computerized numerical control (CNC), computer aided design/computer aided manufacturing (CAD/CAM), robots, and flexible manufacturing system (FMS). Up to now most of these have been introduced on an *ad hoc* or localized basis. We are now approaching the stage of "putting it all together" into an integrated system: this is what is meant by computer integrated manufacturing or CIM. We can now begin for the first time to distinguish some *global* features of the industrial transformation that is taking place.

The first category of new features is directly linked to the permanent cost–benefit evaluation of production, i.e., labor saving, enhancement of capital goods, inventory reductions. As these new (and qualitatively different) forms of automation penetrate to other realms, becoming more and more indirect, even within the very firms that have pioneered the integration, we are able to say less and less, with assurance, about the nature of the industrial transformation. This is simply because we are in the middle (if not at the beginning) of the journey. Since we are on a transition path to a new industrial structure and to a new overall economic and social framework, most of the components of this transformation are not yet adapted to what will progressively become the principles of a self-organized and hopefully self-sustained economic system.

Nevertheless, there are forecasting principles and systems approaches that can help us to conceptualize the main outlines of the new system. Among the possible analytical approaches for dealing with paradigmatic changes at the macroeconomic level is the French "École de Régulation." This approach, pioneered by Boyer and his colleagues, can help us to organize the main features of self-organizing techno-economic systems (see Chapter 15 by Boyer). It is quite probable indeed that after the basically "competitive" era, which lasted until World War I, and the "oligopolistic" mode of regulation, which prevailed afterward, we enter something closer to what economists term "monopolistic competition." Flexible manufacturing, unlike standardized old-style mass production, tolerates and stimulates greater product variety and complexity, enhanced "networking" of cooperating firms, and so forth. This encourages, in turn, greater participation from and creativity of employees. It assigns completely different roles to individual characteristics (and cultures) within the mode of production.

All this is somewhat new. The new, as Schumpeter (1934) put it, is just a "figment of our imagination" and does not have the "sharp edge" of things that have been experienced. At the point where we now stand it is therefore important to be very cautious about extrapolating from qualitative arguments. It is quite possible – indeed likely – that, even when the transition is over and a new steady state is reached, a large part of manufacturing will still be highly standardized in the old pattern. This is especially likely at the level of the elementary building blocks of both material goods and software. It is difficult to conceive a viable industrial system without modularity. In a sense industry is distinguished from craft by modularity.

We must proceed along two methodological paths in parallel if we hope to understand the economic and social implications of CIM. One path starts from the present and extrapolates *quantitatively* from current trends. The other path consists of trying to sketch, from incomplete but persuasive evidence, the decisive *qualitative* features of the coming industrial system. From such a sketch we can try to work our way back to an understanding of the adjustment process that will take us from here to there.

Each approach is subject to certain limits. The present–forward method is always deceptive in that it necessarily evaluates the effects of the "new" by standards derived from the "old" system. The projection–backward approach is, on the other hand, inherently speculative and undisciplined in nature.[2] The combination of the two approaches in a sensible way requires constant and careful attention to their interrelationships. Thus, with each step forward in the technology, the short-run cost–benefit argument will also lead implicitly to long-run arguments.

A brief discussion of the indirect implications of increasing flexibility will elucidate the analytical procedure that we intend to follow. Let us start with the standard cost–benefit evaluation problem. Cost reductions to users are, of course, essential preconditions to any significant technology diffusion. Without some current factor saving, no matter how optimistic the technological expectations may be, few entrepreneurs facing the discipline of the marketplace would opt to buy new machines. However with the new type of flexible machines, entrepreneurs acquire a new capability (see Cohendet and Llerena, 1989). To take full advantage of their added flexibility (and offer a wider variety of products), they will simultaneously have to do (or commission) more market research and perhaps more product R&D than before. Market search is also becoming increasingly important since one has to create or adjust to or both micro markets of very specific needs. User-producer interactions are important in such a setting to both R&D and marketing activities (Lundvall, 1988). In general, the increased flexibility of the firm requires and stimulates an information-intensive production regime (Willinger and Zuscovitch, 1988). Jobs in the firm will have to be redefined to make use of the potential creativity and innovativeness of employees, not only to perform existing tasks better but to invent new ones.

Job redefinition, in turn, will require a different wage-incentive scheme that stimulates information exchanges among employees and groups. This will call for further redefinitions at the level of the organization as a whole as to achieve synergies, cross-fertilization, etc. Matrix organization is more efficient for this purpose than the usual hierarchical forms. Management complexity will then increase because the firm will have to adopt hybrid organizational forms to satisfy multiple objectives including both greater creativity in the development of new products and improved short-run production efficiency (see Shachar and Zuscovitch, 1990). Multicultural firms are not easy to manage, because social tensions very quickly build up when direct productivity measures are applied for one group and indirect team-related compensation schemes are adopted for another. This is not just speculation. Big high-tech firms have already faced this problem for some time.

Intrafirm adjustments are only the beginning. The firm's external relations change too. Along with the usual competitive relations, cooperative arrangements are likely to be intensified. As complexity grows and informational requirements (and costs) increase with it, interfirm links are multiplied. Under different forms – value-added partnership arrangements, joint ventures, common R&D programs, joint use of marketing networks, etc. – such links determine a higher and higher proportion of the firm's strategic behavior. Even the nature of relatively standard types of interfirm relations, such as subcontracting, is rapidly changing. In the past most subcontracting relations were based upon the use of cheap labor to produce nonstrategic components for bigger firms. Now, more technological specialization is likely to be required; the subcontractor has to be equipped with the latest automation and communication technologies to be an effective partner to the parent firm. Networks are created through these multiple forms of partnership, and again we encounter hybrid organizational forms with respect to markets and industries.[3]

Economic theory does not yet offer many conceptualizations of such competitive–cooperative structures, let alone rules for determining optimal arrangements (see Kodama, 1990). What are the simple behavioral rules in such a context? How should contracts be written in a network context? How does technological learning occur in such complex systems? Does the role of industrial policymakers change in such a setting? We have some ideas about this evolution because of previous experience with technological networks such as the European space program. However, our understanding of the matter remains fragmentary; most of the issues already raised by industrial economics have to be reassessed in the new context.

We know already from standard input–output (I–O) approaches and their extensions that the industrial system is an interactive one. Changes that occur at the firm level or at the industry level will directly affect the industries that use the former products or processes. Subsequently, price changes or efficiency gains will indirectly affect the next circle of suppliers and customers through the interindustry system. But I–O models are vastly oversimplified. Most of the important technologies have a fundamentally "transverse" nature in the sense that they apply to many industries. This is certainly the case for flexible manufacturing and CIM.

Not only do these technologies themselves apply in many industrial sectors, but the need for flexibility is transmitted upstream and downstream. As the chemical industry must supply a greater variety of materials, it must offer more flexibility itself. Although the production process is very different from what we observe in the mechanical industries, chemical manufacturers increasingly look forward to the use of multipurpose equipment and small-batch production of custom products in comparison with the dedicated capital goods and commodity products characteristic of much of the industry today. Flexible capital goods are more electronics intensive than dedicated equipment. Moreover, the business-service sector seems to thrive whenever industry faces new challenges.

Taking into consideration changes in input–output relations is not the end of the story. Production is only a part of the economic system. Take R&D, for instance. Much research activity is still done in public institutes and universities. A more intensive use of R&D information will call for a better integration of research and commercial activities and will create the need for new forms of cooperation between public institutions and private firms. This underscores the need for new kinds of partnerships and networks previously mentioned. Progress in some of the sciences also seems to call for more flexible modes of production. In both new materials applications and biotechnology, the trend toward mastering the microscopic properties of matter so as to match material properties exactly with applications is unmistakable. This permits a much larger menu of tailor-made industrial applications and further strengthens the case for information-intensive paradigm, the heart of which is flexible manufacturing.

If we look upon this new production system as a whole, it is not at all clear that it can function satisfactorily in sustaining an economic growth regime. Standardization was a key issue because it was a condition of access to increasing returns to scale in manufacturing. These scale economies were and still are the main engine for economic growth. In a diversified and complex production regime, it is not at all clear that a comparable principle can be set into motion. Certainly economies of scope that apply the same production factor to many uses are possible and effective, but as such they do not offer the potential for continuously rising income. In a way, they rely on the wealth accumulated in the previous regime; they do not clearly offer a new mechanism for wealth accumulation. We suggest that informational returns to scale are potentially available, but the effectiveness of these is uncertain for the moment (Ayres and Zuscovitch, 1989). We can now go beyond the frontier of the economic system.

Education, both general and professional, will also have to change in such a context since the requirements of the firms change. The workers themselves need to be more flexible, which means greater emphasis on basic but versatile education and on continuously revised learning tools rather than on old-style training (which is often not even transferable from one machine to another).

At more sophisticated levels, too, the boundaries between narrow professional specialties is breaking down. At most engineering schools, materials engineering in the past meant metallurgy. However, in the future this field must also encompass plastics, composite materials, ceramics, and adhesives. Similarly, manufacturing engineering – which once meant, in effect, machine-shop theory and practice – must now explicitly address a wide range of topics from "design for manufacturability" to scheduling and even programming. This means more emphasis on general principles common to many fields and less emphasis on specifics. It is indeed possible that a large part of the latter kind of knowledge will become firm-specific, and relationships between university and industry will change.

These changes will, of course, influence wages, incentive schemes, and work organization. It is clear that a new macroeconomic "paradigm" does imply all that. In fact it implies even more. A productive system that allows variety and diversity in the satisfaction of human needs implicitly gives greater weight to the individual. Individuality of tastes was a nuisance to manufacturers during the mass-production era. Individuality of worker interests and abilities was equally disregarded. But it will become increasingly important in both domains: as an attribute of workers and as an aspect of consumer demand.

This kind of change cannot easily be reflected in a standard macroeconomic model. But let us not be misled: macroeconomics never has been derivable from any microeconomic theory of the firm, still less one based on realistic engineering principles (e.g., in the production function) or realistic human behavior. We have always approached macroeconomics by starting from certain "stylized facts," such as the downward rigidity of wages and the instability of investment behavior, and we will probably continue to do this. In a way our macroeconomic representation was consistent with the basic rules of the production system. Recall that Keynes' effective demand, which was the keystone for his macroanalysis, was actually formed by the *entrepreneurs' expectations*. In a standardized production setting, the primacy of supply conditions was clearly established. In the context of new information-intensive production, we should aim at the same goal at least. The features of the macro system are still uncertain. We can speculate that new relevant aggregates may emerge. The national accounting systems will have to take into account the intangible assets in a significant manner. Higher levels of R&D activities will certainly be required to feed growing and diversified needs. Different composites of labor and leisure will lead to different patterns of consumption. The main variables' variance will certainly affect macro behavior and not only average or representative behavior. Cyclical crisis of overinvestment may be less violent as efficiency is less scale dependent. Prophecy is a difficult profession. It is clear that stylized facts or observed regularities are basically a result, an *ex post* truth, and we will have to wait. What is both needed and feasible, however, is an explicit attempt to maintain consistency between our heuristic scenario of evolution and our microanalysis.

Any description of the theoretical and practical difficulties of a satisfactory understanding of the full range of economic implications of CIM, from micro cost reduction to the macroeconomic paradigmatic change, can lead to deep frustration. We are only beginning the analytical process, and the number of "black holes" is still very impressive. All we can do for the moment is to apply these two complementary approaches.

To summarize: we shall begin with the foreseeable implications at the firm level and thence the industry level, at least insofar as each can be apprehended quantitatively in current production and productivity terms. From this perspective the changes will be manifest as increased factor efficiency. Since this level of analysis is the most accessible, it will be the focus of Part I (Overview). We shall then move to interindustry analysis (input–output), especially the international aspects, in Part II (Micro Effects). Part III (Meso Effects: Industrial Structures) will consider the labor and skills implications of flexible manufacturing at the interindustry level of detail. The last part of the book (Macro and International Effects) will attempt to address the more dynamic aspects of the problem. In the final chapters we will offer a simulation model and a theoretical discussion of the prospects for a new techno-economic system. In this way we hope to give a reasonable picture of where we stand without ignoring the more interesting issue of where we go from here.

Notes

[1] See discussion of demand-pull-induced innovation, e.g., Gilfillan (1937) and Schmookler (1966).
[2] These two approaches have been characterized as *alpha* and *omega* forecasting in *Uncertain Futures* (Ayres, 1979).
[3] For a very stimulating analysis of the network behavior in information technology in Japan see Imai and Baba (1989).

References

Ayres, R.U., 1979, *Uncertain Futures: Challenges to Decision-Makers*, John Wiley and Sons, Inc., New York, NY.

Ayres, R.U., and E., Zuscovitch, 1989, "Information, Technology and Economic Growth: Is There a Viable Accumulation Mechanism in the New Paradigm?" *Technovation*.

Cohendet, P., and P. Llerena, 1989, Flexibilité, Information et Decision, *Economica*.

Gilfillan, S.C., 1937, "The Prediction of Inventions," in W. Ogburn *et al.*, eds., *Technological Trends & National Policy, Including the Social Implications of New Inventions*, National Research Council/National Academy of Sciences National Resources Committee, Washington, DC.

Imai, K.-I., and Y. Baba, 1989, "Systemic Innovation and Cross-Border Networks," in *International Conference on Science, Technology, and Economic Growth*, OECD, Paris, France.

Kodama, F., 1990, "Rivals Participating in Collective Research: Its Economic and Technological Rationale," Presented at the International Conference on Science and Technology Policy Research, Shimoda, Japan.

Kuznets, S., 1977, "Two Centuries of Economic Growth: Reflection on US Experience," in S. Kuznets, *Growth, Population and Income Distribution: Selected Essays*, W.W. Norton, New York and London.

Lundvall, B.-A., 1988, "Innovation as an Interaction Process: From User-Producer Interaction to the National System of Innovation," in G. Dosi *et al.*, eds., *Technical Change and Economic Theory*, Frances Pinter, London, UK.

Schmookler, J, 1966, *Invention and Economic Growth*, Harvard University Press, Cambridge, MA.

Schumpeter, J.A., 1934, *Theory of Economic Development*, Oxford University Press, Oxford, UK. (Originally German, 1911.)

Shachar, J., and E. Zuscovitch, 1990, *Technological Learning and Efficient Organization Structure in High-Tech Environments*, Monaster Center for Economic Research, Discussion Paper, July, Beersheba, Israel.

Rosenberg, N., 1976, *Perspectives on Technology*, Cambridge University Press, Cambridge, MA.

Willinger, M., and E. Zuscovitch, 1988, "Towards the Economics of Information-Intensive Production Systems," in G. Dosi *et al.*, eds., *Technical Change and Economic Theory*, Frances Pinter, London, UK.

Chapter 2

CIM, Economy, and Society: Clues for Empirical Analysis

Kimio Uno

2.1 High Technology and the Economic System

The global community is changing its contour rapidly as it approaches the twenty-first century. IIASA's Computer Integrated Manufacturing (CIM) Project has tried to assess the impact of technological change already under way and which is expected to become widely diffused during the twenty-first century in our global community. The nations of the world have become more interdependent today than at any time in history, and economic analysis cannot be concluded by simply looking at a single national example. Especially important is the case of computer-communication-control (C-C-C) technology, which has found its way into new products, new production processes, and new social organizations. Indeed, C-C-C has become the basis of our society today.

The world is beleaguered by environmental disruption and resource exhaustion. Continued economic expansion appears incompatible with the limits imposed by the environment. Here too, technological change centering on C-C-C and the reorganization of our economies incorporating such change seems to be the only feasible solution to safeguard sustainability of our global community.

In industrialized countries, the structure of the economy is changing, radically reflecting the increased introduction of electronic devices leading to robotization and automation. Information as capital, input, and output, although intangible, is changing the social structure to a horizontal network rather than a hierarchical command system. The occupational profile is also undergoing rapid change in both the industrial sectors and the occupational categories. This may create mismatch and adjustment problems. The role played by gender and by age may also be affected. Among nations differentials in the diffusion of high technology result in trade friction, investment friction, and technology friction, and point to the danger of collision of interest.

In industrializing countries, where population pressure and income differentials are pressing, industrialization and export efforts that take advantage of cheap labor can no longer be among development options. The electronics revolution in industrialized countries with its capacity to replace human labor with machinery precludes such a strategy. For resource-exporting countries, resource conservation measures in industrialized countries result in stagnant exports. What is more, environmental concern on a global scale constrains the drive for industrialization.

Industrialized countries and industrializing countries, market economies and planned economies are all connected by international trade, direct foreign investment, technology flows, population migration, and a global environment. Is the high-tech growth path a blessing for us all? This seems to be the question we ultimately have to address. To this end, we start by collecting empirical evidence of the socioeconomic impact of C-C-C technology.

Research and development (R&D), technological progress, capital formation, production, exports, direct foreign investment, employment, environmental impact, and so on are interwoven. For example, if new technology emerges as a result of an R&D effort, the existing production facility will become obsolete and new capital formation will be carried out.

In an economy where demand is expanding, the introduction of new technology will be facilitated. As a consequence, production will become more efficient, due, for example, to labor saving or resource saving or both, resulting in lower prices. At the same time, the quality of the products may well facilitate export promotion. Increased production and the resultant increase in employment will eventually make the labor market tighter, inviting wage increases. Price changes signal the direction of desirable technological change. Labor is a scarce factor of production in industrialized countries, so is the environment in a wide sense including not only raw materials but also clean air and water. Technological progress will be induced in the direction of conserving these scarce factors.

The complex interrelationship among these diverse aspects can only be captured by an econometric model that incorporates relevant variables. However, the construction of an econometric model to suit such purposes is a difficult task. First, disaggregated analysis is needed to focus on the industrial sectors related to C-C-C technology, but the construction of a multisector model is always an extremely demanding job. In a newly emerging field such as the one discussed in this chapter, the problem is often compounded by the lack of data.

Second, international linkages are needed to assess worldwide repercussions of the diffusion of high technology. Ideally, this must be accomplished by constructing multisectoral national models and linking them to form a world model. Sectoral classification and other specifications have to be consistent among countries. Only then can they be linked by trade matrices.

In the following section, some of the key factors are discussed which link technology with the functioning of an economy using a multisector industrial model. An overview of the methodology and results provided by some contributors to the volume is then presented.

2.2   Clues for Empirical Analysis

Research and development: R&D is an important economic activity in today's economy. Technological knowledge is undoubtedly a key factor for an economy to gain production efficiency while achieving resource conservation. However, economic theory and statistical accounts have not dealt with R&D in a systematic manner (Uno, 1989b). The benefit of R&D activity does not remain totally within the sector which undertakes it. Rather, the benefit spills over to the entire economy through improved quality and lower prices of intermediate inputs and capital equipment.

A practical method was proposed by Terlecskyj (1980, 1982). Starting from the R&D expenditures undertaken in individual industrial sectors, he tried to distribute the benefit to the purchasing sectors by using information obtained from input–output tables.

Based on Terlecskyj's methodology and empirical data from Japan, which include annual R&D data, annual input–output tables, and fixed capital formation matrices from 1970, 1975, 1980, and 1985, an estimation was made on the annual flow of R&D benefit accruing to industry and final demand (Statistical Data Bank Project, 1990). Industry is divided into 36 standardized sectors, and final demand is broken down into private consumption, government consumption, private investment, government investment, exports, and so on. A fixed capital formation matrix is employed to distinguish the sector where capital formation is carried out. This process is needed because fixed capital formation among final demand items in the input–output tables only shows the amount of output of a particular sector going into investment.

In addition, due consideration must be given to technology imports. Both domestic R&D and technology imports are appropriately lagged to take into account the time required for R&D to create new products and new production processes.[1]

Table 2.1 is a summary of R&D activity relevant to the machinery sector. Comparisons of R&D activities and benefit received reveal that the precision instruments, electrical machinery, and general machinery sectors are typical examples of net providers of R&D benefits to other industrial sectors.

The diffusion of computer-communication-control technology: Electronics technology is already having a tremendous impact on our society, and yet the full impact is still to unfold. As an indicator of such development, one can look at production statistics of semiconductors, the progress of digital communication, or the development of data base technology. All these represent an electronics revolution. In this section, we examine the diffusion process of mainframe computers, NC metalworking machines, industrial robots, and automatic vending machines as representing C-C-C technology.

In the COMPASS model, diffusion rates in respective sectors are explained by the relative price of labor and capital ($WHRi$) and the real capital formation (IPi),

$$\begin{aligned} &RKCOMi \\ &RKNCi \quad = f(WHRi, IPi) \ , \\ &RKROBOi \\ &RKVENDi \end{aligned}$$

where $RKCOMi$, $RKNCi$, $RKROBOi$, and $RKVENDi$ represent the diffusion of computers, NC metalworking machines, industrial robots, and automatic vending machines, respectively.

The reasoning behind the specification is that the diffusion of new technology would be accelerated as the relative price of capital became lower, due to either lower wage levels or reduced capital equipment prices. Capital

Table 2.1. R&D activity conducted and R&D benefit received in machinery sectors, in billions of yen, 1970 prices.

Year	General machinery	Electrical machinery	Motor vehicles	Precision instrument
R&D activities conducted				
1975	201.4	637.1	250.2	40.4
1976	230.8	653.4	285.2	39.9
1977	230.2	625.1	274.8	36.6
1978	205.2	597.5	254.8	39.0
1979	238.5	611.6	275.5	53.2
1980	244.8	632.3	330.6	64.3
1981	241.5	707.1	394.3	66.7
1982	260.8	777.4	409.4	73.9
1983	284.4	838.6	428.9	82.7
1984	296.7	989.3	506.5	85.6
1985	336.4	1146.7	550.1	100.3
1986	378.8	1289.1	604.8	117.4
1987	418.3	1486.7	656.1	130.7
R&D benefit received				
1975	115.0	165.0	181.7	10.6
1976	106.7	179.5	185.4	11.0
1977	121.7	176.8	196.1	11.2
1978	113.8	166.6	187.3	13.0
1979	123.8	169.6	192.7	14.6
1980	128.8	182.9	216.5	16.6
1981	128.2	200.6	229.5	16.1
1982	145.8	231.0	238.2	18.3
1983	157.2	263.5	254.8	20.2
1984	171.6	331.1	289.1	20.4
1985	189.6	389.4	320.3	23.7
1986	219.5	474.8	357.1	28.0
1987	244.7	540.6	391.7	31.1

Source: Statistical Data Bank Project, 1990. For details of methodology, see Uno, 1989b.

formation exhibits fluctuations reflecting the changes in the demand–supply gap or interest rates or both. It can be assumed that the introduction of new technology would be easier when investment is being carried out to cope with rising demand. Empirical examination of the function was fairly good.

Tables 2.2 and *2.3* represent the diffusion of NC metalworking machines, industrial robots, and FMS in Japanese industry based on a MITI survey

Table 2.2. Number of NC machines in use.

Sector		1967 Survey[a] Number[c]	1973 Survey[a] Number[d]	1981 Survey[b] Number[d]	1987 Survey[b] Number	1987 Survey[b] Value (in millions of yen)
	All industries	769	4,861	19,549	70,465	670,388
31	Iron and steel	47	NA	NA	1,572[e]	15,092[e]
32	Nonferrous metals	11	NA	NA		
33	Fabricated metals	74	NA	NA	2,537	24,344
34	General machinery	264	2,783	9,378	28,195	284,984
35	Electrical machinery	85	832	3,185	11,932	101,343
36	Transport equipment	257	948	5,404	21,981	213,330
36M	Transport equipment, motor vehicles		604	4,293	20,123	189,853
37	Precision instruments	71	298	1,582	3,520	22,850
	Diffusion rate (= B/A)					
	All industries	0.1%	0.9%	3.7%	10.9%	33.5%
31	Iron and steel	0.0%	NA	NA	9.0%[e]	33.7%[e]
32	Nonferrous metals	0.1%	NA	NA		
33	Fabricated metals	0.1%	NA	NA	7.4%	34.1%
34	General machinery	0.1%	1.2%	4.8%	12.1%	40.5%
35	Electrical machinery	0.1%	0.9%	3.9%	13.2%	35.9%
36	Transport equipment	0.1%	0.5%	2.6%	9.7%	26.0%
36M	Transport equipment, motor vehicles		0.4%	2.3%	9.6%	24.7%
37	Precision instrument	0.1%	0.5%	3.3%	8.9%	40.1%

[a] Establishments with more than 100 employees.
[b] Establishments with more than 50 employees.
[c] Total includes ordinance and accessories sector.
[d] Sectors 34, 35, 36, and 37 only.
[e] Casting and forging only.
Source: Compiled from the Ministry of International Trade and Industry, *Survey on Machine Tools Installation*, various issues.

Table 2.3. Number of industrial robots and FMS in use in number of units.

Sector		1967 Survey[a] Number	1973 Survey[a] Number[c]	1981 Survey[b] Number[d]	1987 Survey[b] Number
	Industrial robots				
	All industries	NA	3,058	14,158	47,308
31	Iron and steel		53	NA	750[e]
32	Nonferrous metals		150	NA	
33	Fabricated metals		120	NA	1,372
34	General machinery		205	1,253	5,380
35	Electrical machinery		423	3,859	16,475
36	Transport equipment		1,610	8,383	20,901
36M	Transport equipment, motor vehicles		1,592	8,315	20,229
37	Precision instruments		495	663	2,094
	Flexible manufacturing system				
	All industries	NA	NA	NA	259
31	Iron and steel				
32	Nonferrous metals				
33	Fabricated metals				
34	General machinery				171
35	Electrical machinery				44
36	Transport equipment				40
36M	Transport equipment, motor vehicles				21
37	Precision instrument				2

[a]Establishments with more than 100 employees.
[b]Establishments with more than 50 employees.
[c]Total includes ordinance and accessories sector.
[d]Sectors 34, 35, 36, and 37 only.
[e]Casting and forging only.
Source: Compiled from the Ministry of International Trade and Industry, *Survey on Machine Tools Installation*, various issues.

in 1987.[2] The number of NC machines in 1987 stood at 70,465, of which 28,195 belonged to the general machinery sector; 11,932 to the electrical machinery sector; 20,123 to the motor vehicles sector; and 3,520 to the precision instrument sector.

The diffusion rate of the number of machines was 10.9% in 1987. Among the industrial sectors, general machinery and electrical machinery sectors had the highest rates, recording 12.1% and 13.2%, respectively. However, a more relevant figure seems to be the diffusion rate in value terms because NC machines are more efficient and more expensive than conventional metalworking machines. Data on this sector are available only for 1987, and the diffusion rate in value terms is recorded at 33.5% for all industries.

In the case of industrial robots, the 1987 survey reports that 47,308 units were operating in Japanese industry. The 1973 figure was 3,058 (of which 2,735 were in the machinery sectors), and the 1981 figure was 14,158. The same source reveals that Japanese industry possessed 259 FMSs as of 1987.

The survey also reveals vintages of metalworking machinery. The percentages of units installed within the last three years prior to the survey stood as follows: NC metal-cutting machines, 37.4%; NC secondary metalworking machinery, 35.9%; industrial robots, 46.6%; flexible manufacturing system, 47.1%. The figures indicate how rapid the diffusion of new technology has been. At the same time, such trends seem to justify the use of average age of capital stock as a proxy for quality improvement of production equipment, a practice which was followed in the COMPASS model.

Structural change in capital formation: New technology is embodied in new capital equipment and thus the diffusion of C-C-C has its impact on an economy through fixed capital formation. Here we turn to private fixed capital formation, and examine the changes in sectoral composition of the supplying sector for this particular demand category. The basis of our analysis is annual input–output tables of the Japanese economy in a recent 10-year period (Uno, 1990, pp. 56–66 and pp. 195–205). The sectoral ratio of the supplying sector for final demand is sometimes called converters. *Table 2.4* shows converters for private fixed capital formation, focusing on machinery sectors which include general machinery, electrical machinery, motor vehicles, other transport equipment, and precision instrument. The construction sector, which traditionally comprises the bulk of capital formation, is listed for comparison. The table shows the share of each sector. The total is defined by unity.

The table reveals that the share of the machinery sector has steadily increased during the 1980–1985 period though the figures for 1975 and 1980 are not particularly different, and a closer examination of annual figures confirms this. It is clear that the structural change of private fixed capital formation started circa 1980 in Japan; this new trend is characterized by the rising share of the machinery sectors which include electrical machinery and general machinery. The share of precision instrument also has risen

Table 2.4. Supplying sectors for private fixed capital formation (unit: ratio of total).

Code	Sector names	1975	1980	1981	1982	1983	1984	1985
34	General machinery	0.1394	0.1230	0.1328	0.1297	0.1321	0.1378	0.1665
35	Electrical machinery	0.0806	0.0872	0.1217	0.1070	0.1252	0.1417	0.1317
36M	Motor vehicles	0.0518	0.0535	0.0540	0.0576	0.0581	0.0581	0.0487
36S	Other transport equipment	0.0116	0.0177	0.0112	0.0187	0.0177	0.0166	0.0237
37	Precision instrument	0.0067	0.0094	0.0090	0.0098	0.0102	0.0110	0.0132
	Machinery sectors total	0.2899	0.2872	0.3286	0.3223	0.3429	0.3650	0.3835
15	Construction	0.6054	0.6097	0.5562	0.5655	0.5532	0.5295	0.4947
	Total[a]	1.0000	1.0000	1.0000	1.0000	1.0000	1.0000	1.0000

[a]Total is the sum of sectors 34, 35, 36M, 36S, and 37 (machinery) + 15 (construction) + other sectors (such as furniture and fixture, fabricated metals, and wholesale and retail trade).
Source: Uno, 1990.

steadily. Thus, during 1980 and 1985, the share of the machinery sectors total has risen almost 10 percentage points, from 28.7% to 38.3%. Over the same period, electrical machinery has gained most in terms of growth rate, expanding its share from 8.7% to 13.2%. The general machinery sector increased its share from 12.3% to 16.7%. The share of precision instruments is inherently smaller than these other sectors, but it also has increased shares from 0.9% to 1.3%. In contrast, the construction sector share dropped from around 60% in 1975 and 1980 to less than 50% in 1985. This is evidence of the nature of technological change that is taking place in our economies.

Table 2.4 is based on current prices. It is important to note that the price of C-C-C-related products has declined during this period. Current price figures provide a picture which is a mixture of rapid diffusion of C-C-C technology and declining prices. What is lacking is an appropriate measure of quality improvement, though there are sporadic attempts to estimate quality-adjusted price indices in this field. According to such results the price of mainframe computers has dropped from a base level of 100 in 1970 to 21 in 1980 and 7 in 1987; in the case of industrial robots, it has dropped from 100 in 1970 to about 40 in 1985. In contrast, the deflator for electric machinery on an SNA (system of national accounts) basis showed a decline of about 7%, whereas that for general machinery has risen by about 60% during the 1970–1985 period.[3] The importance of C-C-C-related sectors such as electric machinery and precision instrument must have risen at a much faster rate than is suggested by the current price figures.

High technology and exports: Various surveys show that the introduction of high technology is usually motivated by labor-saving incentives. Other factors include quality improvement of products, increases in production capacity, shortening of lead times, and greater flexibility in design and production. Some of these factors contribute to a lowering of prices, but some are not reflected in product prices. A typical non-price factor is an improvement in quality. However, capacity increase (presumably without much increase in marginal cost, because automatic operation is possible without increases in labor input), shorter lead time, and added flexibility all work toward quick response to changes in the market.

Japanese exports continued to expand despite a rapid appreciation of the yen exchange rate after the G5 (Group of Five) agreement in 1985. One suspects that non-price factors such as quality and flexibility, made possible by the C-C-C technology, are the main contributing incentives.

To examine this hypothesis in an empirical context, we introduced variables representing the diffusion of C-C-C technology in export functions of the multisector industry model. The model explains an industry's exports (in real terms ERi) by the world industrial production ($YWORLD$), relative price which is the ratio of export prices adjusted for the exchange rate fluctuation ($REX \times PEi$) and the world price ($PUNi$), and proxy variables describing the diffusion of mainframe computers, NC machines, and industrial robots ($RKCOMi + RKNCi + RKROBOi$). The exchange rate in this specification is represented by dollar per yen:

$$ERi \ = \ f(YWORLD, REX \times PEi/PUNi, RKCOMi \\ + RKNCi + RKROBOi) \ .$$

The function was tested for general machinery, electrical machinery, motor vehicles, and precision instruments sectors, and the results were promising. It should be noted that the movement of prices already captures the effects of high-tech diffusion. That is, export prices in the COMPASS model are obtained as functions of the corresponding producing-sector prices. This in turn is explained as a function of unit labor costs and unit material cost (Uno, 1987, pp. 233–256). Material cost is then specified reflecting the input structure of respective industrial sectors. When C-C-C-related prices go down, benefit accrues to the sectors that use these products in their production processes and intermediate inputs. Lower prices of electronics-related goods thus spread to other sectors of the economy.

In addition, as the significance of C-C-C technology becomes widely recognized internationally, countries not only import C-C-C-related goods but also try to initiate their production. This means that Japan's position as a supplier of these goods and technology is heightened even more. This is reflected in the high elasticity of Japanese exports *vis-à-vis* the growth of the world economy. The causal chain is closed because R&D in Japan is quite active, especially in C-C-C-related civilian sectors. It is shown that R&D expenditures by the electric machinery sector has expanded quite rapidly and exceeds one-quarter of total R&D spending, and that a large portion is also in the electronics field (Uno, 1989b, pp. 220–224).

Table 2.5 shows the export structure of Japan in the 1980s together with those for other countries and regions. The largest component of Japan's exports is machinery (SITC 7). The proportion of machinery shares as a percentage of the total was 58% in 1980, 68% in 1985, and 74.4% in 1988. This compares favorably with Europe where the percentage was approximately 33% (1985) and with the United States where it was about 46%. The ratio in Asian countries (excluding the Middle East), where the wave of industrialization is accelerating, was 14% in 1980, but jumped to 22% by 1985.

Table 2.6 lists the share of respective countries and area in total world exports. In machinery exports, Japan's share has reached about 20% in recent years, a figure nearly five percentage points higher than in 1980. This figure is already higher than the figure for the United States where it is 16%. The share of Asia (other than Middle East countries) was recorded at 4% in 1980 and 7% in 1985.

Table 2.5. Commodity composition of exports of selected countries and areas, in percent.

| Origin/Destination → SITC ↓ Class / commodity | Year | World | Market economies | | | | Centrally planned economies | Europe total | USA | Japan |
| | | | Developed | Developing | | | | | | |
				Total	OPEC					
0–9	Total									
	All commodities	1980	100.0	100.0	100.0	100.0	100.0	100.0	100.0	100.0
		1982	100.0	100.0	100.0	100.0	100.0	100.0	100.0	100.0
		1983	100.0	100.0	100.0	100.0	100.0	100.0	100.0	100.0
		1984	100.0	100.0	100.0	100.0	100.0	100.0	100.0	100.0
		1985	100.0	100.0	100.0	100.0	100.0	100.0	100.0	100.0
0&1	Food, live	1980	10.0	10.2	10.1	1.2	8.1	9.6	14.0	1.2
	animals, and	1982	10.2	10.2	11.4	1.3	8.9	10.0	13.0	1.0
	beverages	1983	10.2	9.9	12.3	1.6	6.6	9.7	13.7	0.9
	tobacco	1984	9.8	9.4	12.2	1.9	6.1	9.5	12.8	0.8
		1985	9.2	8.7	11.8	2.1	6.4	9.2	10.7	0.7
2&4	Crude	1980	6.9	6.8	7.1	1.7	7.4	4.2	11.9	1.2
	materials	1982	6.1	6.0	6.3	1.2	6.1	3.7	10.0	1.0
	oils and fats	1983	6.2	6.0	6.8	1.4	5.9	3.9	10.2	1.0
	(fuels	1984	6.5	6.3	7.4	1.6	6.1	4.3	10.5	0.9
	excluded)	1985	6.2	5.7	7.6	1.7	6.2	4.0	8.8	0.8

	Year								
3 Mineral fuels, lubricants, and related material	1980	24.1	7.0	61.4	94.8	25.8	8.3	3.7	0.4
	1982	23.2	8.4	55.8	94.1	30.0	9.8	6.2	0.3
	1983	21.0	8.2	49.3	92.5	31.0	9.8	4.9	0.3
	1984	19.7	7.9	45.1	91.7	31.3	9.7	4.4	0.3
	1985	18.7	8.0	43.3	91.1	29.9	9.4	4.8	0.3
5 Chemicals	1980	7.0	9.7	1.7	0.5	5.0	11.1	9.6	5.1
	1982	7.2	9.5	2.2	0.7	5.5	11.2	9.6	4.5
	1983	7.6	10.0	2.5	1.0	5.5	11.8	10.1	4.6
	1984	7.7	10.0	2.6	0.7	5.6	12.1	10.6	4.4
	1985	7.9	10.0	2.8	0.9	6.2	12.2	10.3	4.3
7 Machinery and equipment transport	1980	25.6	34.7	5.3	0.5	26.5	32.0	39.0	58.4
	1982	27.7	36.8	7.3	0.8	25.0	32.8	42.0	61.3
	1983	28.7	37.0	9.3	0.9	25.3	32.0	42.1	63.8
	1984	29.7	37.8	11.0	1.1	24.7	31.5	42.4	66.6
	1985	31.1	39.1	11.9	1.2	25.1	32.5	45.7	67.8
6&8 Other manufactured goods	1980	24.0	29.4	13.5	1.2	19.1	32.9	17.9	32.4
	1982	23.1	27.0	16.1	1.6	17.6	30.6	15.9	30.8
	1983	23.6	26.8	18.7	2.3	17.1	30.7	15.5	28.3
	1984	23.9	26.3	20.6	2.7	17.4	31.0	14.7	26.0
	1985	24.2	26.4	21.4	2.8	17.0	31.1	14.4	25.0

Source: United Nations, *Yearbook of International Trade Statistics*, various years.

Table 2.6. Origin of world exports, in percent.

Origin/Destination → SITC Class commodity	Year	World	Market economies				Centrally planned economies	Europe total	USA	Japan
			Developed	Developing						
				Total	OPEC					
0–9 Total All commodities	1980	100.0	62.9	28.3	15.3	8.8	40.1	10.8	6.5
	1982	100.0	63.2	26.7	12.1	10.1	38.5	11.2	7.5
	1983	100.0	63.6	25.4	10.0	11.0	38.4	10.8	8.1
	1984	100.0	64.3	25.1	8.9	10.6	37.4	11.1	8.9
	1985	100.0	65.5	23.9	8.0	10.5	39.0	10.7	9.1
7 Machinery and equipment transport	1980	100.0	85.0	5.9	0.3	9.1	50.0	16.5	14.8
	1982	100.0	83.9	7.0	0.4	9.1	45.6	17.0	16.6
	1983	100.0	82.1	8.2	0.3	9.7	42.9	15.9	18.0
	1984	100.0	81.8	9.3	0.3	8.9	39.8	15.9	20.0
	1985	100.0	82.3	9.2	0.3	8.5	40.7	15.7	19.8

Source: United Nations, *Yearbook of International Trade Statistics*, various years.

Employment aspect: Probably the most important social aspect of C-C-C technology is its impact on employment. Changing skill levels and occupational profiles, with its implications for changes in wages, employment opportunities, education and training, etc., will then become evident. In this regard, the most comprehensive picture can be obtained from an industry-occupation matrix that shows the number of workers in each industry classified by type of job. Employing time series data which supplement structural information provided by an industry-occupation matrix, one can fit an employment function that includes variables representing C-C-C technology. Denoting the number of workers in sector (i) by NLi, real value of production by $VQREALi$, and relative price of labor and capital in the respective sector by $WHRi$, and including proxy variables describing the diffusion of mainframe computers, NC machines, and industrial robots as before, the employment function can be written as

$$NLi \;=\; f[VQREALi, WHRi, RKCOMi + RKNCi$$
$$+\, RKROBOi, NLi(-1)] \;.$$

Empirical examination of the function reveals a labor-saving effect of C-C-C technology for general machinery, electrical machinery, motor vehicles, other transport equipment, and precision instruments (Uno, 1989b, pp. 145–164). Among services, employment in wholesale and retail is being affected by computerization and the increased use of automatic vending machines. The business- and personal-service sectors as well as public administration are influenced by labor-saving factors through computerization.

Professional and technical workers, clerks, sales representatives, and production process workers are affected by C-C-C technology of one kind or another. Occupational categories not affected include managers and officials and transport and communication workers. One item which deserves particular attention is the fact that women appear to be more vulnerable to the labor-saving effects of new technology than men, presumably reflecting the relatively unskilled nature of their jobs.

However, to examine the total effect of C-C-C on employment, we need a comprehensive model because investment may be activated or exports may be promoted by such technological change, resulting in larger outputs of a particular industry, which counters the labor-saving effect. In fact, this is the reason why we need to employ an economywide model with appropriate sectoral disaggregation.

2.3   An Overview of the CIM Analyses

The discussion thus far has described the possibility of incorporating C-C-C-related variables in a framework of a multisectoral industrial model. From this discussion we may conclude that such an attempt is indeed feasible. One can construct an econometric model which is suitable for analyzing CIM diffusion and its economic and social impacts. This allows us to analyze CIM diffusion in a dynamic context, simultaneously and consistently taking into account other important variables such as wage levels, prices, investment trends, interest rates, input structures, exports, exchange rates, and the global economy. One can even envisage such models to link individual countries through trade matrices to form a global model. Before such an endeavor is undertaken, however, one must construct building blocks relevant to various aspects of CIM technology pertaining to its economic and social impacts. The papers in this volume actually represent such an effort. This section provides an overview of some of the work in the volume.

Macroeconomic models incorporating CIM: Three papers attempt to deal with CIM diffusion in the framework of economic models. They analyze the impact of CIM diffusion neither at the micro (individual firm) level nor at the macro (national aggregate) level, but at the disaggregated industry level. This is the level where one can establish linkages between technology, on the one hand, and the economy and society, on the other, based on empirical data.

Dobrinsky's contribution (Chapter 14), entitled "Macroanalysis of the Economic Impact of CIM Technologies: An International Comparison Based on a Global Econometric Model," draws upon the experience of the Bonn–IIASA Research Project (Krelle, 1989). In essence, the model consists of 23 macroecocomic models linked by a system of international trade. He chooses 11 countries that are relevant for CIM-related research. He then focuses on five subsectors within fabricated metal products, i.e., final metal products, nonelectrical machinery, electrical machinery, transport equipment, and professional and scientific equipment. He assumes that there are four levels of technology: level 0, nonautomated, stand-alone machines; level 1, stand-alone NC machine, industrial robots; level 2, FMS and FMC; level 3, CIM. Based on the judgments of experts, his supposition is that, compared with level 0, the productivity of level 1 is twice as high, level 2 is five time as high, and level 3 is 30 times as high in the base year (1984). This ratio is altered toward the year 2000, reflecting technological improvements. Diffusion scenarios are drawn for individual countries, and standard macroeconomic figures are calculated.

The second contribution in this category is entitled "An Econometric Analysis of Future Technological Change: Japan, USA, and FRG" by Yamada (Chapter 10). Constructed along the Kinoshita–Yamada (1988) model, he employs a framework that incorporates information obtained from input–output tables in a multisectoral industry model (the so-called Leontief–Keynesian type). He tries to modify the inherently static nature of input–output tables by explaining the deviation of actual output from calculated ones by the changes in relative industry prices. However, the model remains basically demand oriented, and the production side is not treated explicitly. Thus, among various channels through which the impacts of the introduction of CIM can be traced, he was able to focus on three factors, including the acceleration of capital formation, labor saving, and changes in input coefficients. To carry out simulation experiments based on identical assumptions, Yamada then adopts the relative productivity and diffusion scenarios from Dobrinsky's analysis. Kinoshita and Yamada earlier adopted the results on labor-saving effects from Saito (1988). The approach adopted here has the advantage of standardizing the assumptions among models. The cost of this approach was that the labor-saving effect could not be endogenously explained. As for the third factor, he assumes secular changes in input–output coefficients.

The third contribution, entitled "Impact of CIM on the Economy: Simulations Based on Macromodels," by Tomaszewicz *et al.* (Chapter 11), employs a framework similar to Yamada's. Their framework is also a Leontief–Keynesian-type multisectoral model. Unlike Yamada, however, they do not attempt to adjust the input–output relationship over time. Thus, the main causal chain runs through CIM diffusion to capital enhancement and productivity gains, resulting in labor saving. The cases in question are the United States and Japan. Tomaszewicz *et al.* also adopt diffusion scenarios from Dobrinsky, and the basic country data were adopted from Yamada, providing an opportunity to compare the structure and the performance of the three models.

The diffusion of industrial robots: The missing link between the CIM technology and the economy is provided in a most convincing manner in Chapter

4 by Mori, entitled "The Enhancement of Labor: Robots in Japan." He discusses the industrial robot diffusion in Japan. He begins by specifying a production function as

$$Y = Y[K, F(L, R), x] \ ,$$

where Y is real output; K is conventional capital stock; L is labor; R is stock of industrial robots; and x is other unspecified inputs. He is essentially saying that labor and industrial robots are alternatives. Thus, he was able to reduce the problem into one of standard cost minimization. The key variables then become the prices of labor and industrial robots. For the former, he uses annual labor cost per worker. For the latter, he proceeds to estimate a quality-adjusted price index based on the different capacities and unit prices of robot types. His major findings include the fact that labor substitutability of robots has decreased from more than 4.0 in earlier years to 1.1 and 1.3 recently. He says that this finding is supported by opinions obtained through micro-level surveys.

Dealing with industry-level technological change: Economic models, discussed earlier, suffered from an inherently static nature in input–output framework. Furukawa (Chapter 9) studies this point and predicts the changes in input coefficients and capital formation for the year 2000. In his contribution, entitled "Input–Output Structural Changes and CIM," he examines Japan, the USA, and the Federal Republic of Germany. His prediction of input structures is based on input–output tables from three points in time for each country. He employs a framework widely known as the RAS method which was initiated by Brown (Stone and Brown, 1962). This is essentially a mathematical method of arriving at a consistent input–output structure in some future time based on commodity technology changes (R) and the industry technology changes (S).

Furukawa observes that in the initial period (circa 1985), the percentage shares of the machinery sectors in capital formation in the three countries are almost identical, from 36% to 39%, although the share of the industrial machinery sector is larger in Japan and the FRG than in the United States where the share of the automobile sector is the largest.

As for the impact of CIM diffusion in the future, assuming that past trends will be strengthened in both input coefficients and capital formation, more investment is expected to be spent on electrical machinery such as microcomputers, communication networks, and industrial robots. More importantly, he points to the rising importance of the business-service sector and the educational-service sector, which is induced by software development and R&D activities.

Industry-occupation profile: Structural change in the economy invites changes in occupational structures in different sectors. Workers are replaced by machinery on the shop floor, but more personnel are required for software development resulting in less input of production workers and more input of technical workers. Brautzsch attempts to compile industry-occupation matrices for major industrial countries, during different time periods, in Chapter 12, "Occupation-by-Sector Matrices: Methodological Problems and Results." [His data set does not cover Japan, one of the focal points of CIM diffusion analysis, but Japan is considered in other recent work (Uno, 1989b).] This is actually an enormous task which requires a lot of labor in adjusting country practice to fit into a consistent classification scheme in both occupation classification and industry classification. Only then, can one start analyzing the causes of occupation shift within individual sectors. Any change in final demand or intermediate inputs can be gauged by the input–output framework, which can now be linked to the industry-occupation matrix to arrive at employment impacts of technological change. Although much still remains to be done in this field, this is an important development in linking the technology and economy, on the one hand, and the social impact, on the other.

Overcoming the rigid framework: Empirical studies based on econometric model building and input–output tables are based on past statistical data. Although established methods have carried out simulation experiments for the future, or extrapolated input–output structures at some point in time in the future, these methods are rather rigid in their formulation. This rigidity precludes arbitrary assumptions as to the exogenous variables or endogenous relations among variables. When dealing with entirely new phenomena, such as the diffusion of CIM technologies, this rigid framework becomes powerless because of the lack of statistical data and lack of theory which appropriately explains the new phenomenon. Fleissner and Polt attempt to fill this methodological gap by employing system dynamics in Chapter 16, entitled "Switching from a Fordist to a CIM Accumulation Regime: A System Dynamics Model." This method allows us to introduce feedback loops as they are judged fit from an analytical point of view, not necessarily supported by the past observations but deemed likely, and important, by experts.

Their model consists of two sectors, a Fordist sector and a CIM sector. They enumerate stylized facts about the differences in the two regimes (economies of scale and economies of scope); decreasing inventories; lower price elasticity of demand due to, for example, differentiated products; lower capital-output ratios reflecting the ability to respond to change in demand without massive new investment.

They then assume the share of CIM-related investment in total as reaching 60%, following a logistic curve. The CIM sector reaches its saturation level after 20 years, having reached the point of inflection after 9.5 years. The share of CIM at the end of the simulation period is about 20% of total capital stock. They also provide simulation results for standard macroeconomic variables, but their contribution lies in providing a tool for analysis to tackle this very complex process. Needless to say, this methodology can incorporate empirically established relations among variables based on an econometric method, case studies, and so on; such a development is indeed needed.

Organizational aspects related to CIM diffusion: Not all aspects of CIM introduction are amenable to quantitative analysis. Ebel's contribution (Chapter 5), entitled "Management, Jobs, and Employment," looks at some of the prerequisites for successful introduction of CIM and the institutional changes that are likely to follow. Topics covered include a new management style, skill requirements, new types of work organization, hazards in the working environment, the need for new types of work force for CIM, and the impact on industrial relations. He makes the point that the technocentric approach results in the neglect or underrating of the human factor in production.

He contends that as a prerequisite of CIM introduction redesigning of the information and restructuring of data flow are required. Production processes and organizational boundaries will have to be altered. Taylorist views of the workers, where they are told what to do and how to do it, have to be replaced by a new approach where workers' skills and motivation are constantly improved by investing in the people operating the system. This requires a management style that allows for greater worker autonomy. This also implies bottom-up decision making and initiative. Workers will need different skills; people will be required to handle a great deal of technical information and respond on the spot.

Thus, CIM has to be accompanied by new types of work organization characterized by broader work content, better information flows, and more decision making at the bottom. However, in reality, there is a shortage of qualified personnel, and training institutions often lag behind the newly emerging social needs. The solution has to be found in further training of the work force.

The transition to CIM will, however, not be accomplished without creating tension and conflict because of the fear of pay reductions, reduced promotion prospects, machine-paced work, replacement of skills by data bases and expert systems, etc. Adversarial industrial relations can easily derail the CIM introduction. The solution offered by Ebel is genuine consultation at all levels, in an atmosphere of social dialogue and goodwill at the enterprise level.

CIM in the north–south context: Vuorinen's paper, entitled "Flexible Automation and LDCs" (Chapter 18), puts the CIM technology in the context of developing economies. As a result of the introduction of industrial robots and NC metalworking machines in the industrialized countries, the full-employment ceiling – especially that for skilled workers – has virtually disappeared. By replacing workers with machinery, industrialized countries are now able to continue production of hitherto labor-intensive products. A shift in technological paradigm means investment opportunities. Thus, CIM is interpreted as a blessing in the industrialized north. However, the differentials among industrialized countries in the CIM diffusion can be a cause of trade friction, investment friction, and technological friction.

The situation may be more serious in the developing south. In many instances, social infrastructures needed for CIM introduction, such as training facilities, opportunities for learning by doing, a free flow of information, technical and managerial advisers, and financial mechanisms to facilitate investment, are lacking. CIM incorporates computer-communication-control technology which is discontinuous from traditional production technology. Due to the emergence of the global market, it is extremely difficult for LDCs to participate in the world market unless they themselves produce goods of globally acceptable quality. Cheap labor can no longer be a leverage through which economic development can be promoted. On the other hand, interest rates in LDCs tend to stay higher than in the rest of the world; this is due to a shortage of savings and risk premiums, hampering the diffusion of capital-intensive technology. Vuorinen's chapter is an attempt to look into some of these issues.

2.4   Further Remarks

This chapter has tried to examine achievements of IIASA's CIM Project, putting emphasis on methodology.

Economics as a discipline has tended to treat technological change as exogenous. Changes in productive efficiency are either completely neglected or treated as a function of time. In a world where R&D plays an important part in business activities and product and process innovations abound, this practice must be changed in a fundamental way.

From this perspective, the CIM Project is an attempt to establish a firm linkage between technology and the economy. In so doing, most authors chose to follow established theoretical frameworks in economics and tried to widen the horizon of its application. This was a wise choice.

Not all questions have been answered. The diffusion study and econometric model building will have to be merged so that the (quantifiable) factors working behind the process can be produced from within. The ratio of CIM diffusion may be underestimated if it is calculated by the number of machines because automated units can be operated for longer periods of time, can produce high-quality products, need less setup time, but are usually more costly. Treatment of (often not quantifiable) intangibles such as software, data bases, and other knowledge-related elements needs to be improved. The same could be said of workers. Rather than counting their numbers, it is now essential to evaluate their ability to form a team and communicate among themselves to cope with constant dialogue with the production system. A production system is not something given to the workers, but it is something to be created through daily improvements in know-how and hardware. This may imply a breakdown of the man–machine dichotomy.

The relationship among business firms is also changing since the new technology needs more service inputs. A large R&D element, increased computerization and automation encompassing not only production but also logistics, and an emphasis on workers as team players seem to point to the advantage of long-term commitments as opposed to short-term profit maximizing.

The new technology also requires horizontal network organization rather than a vertical command-driven type. In a borderless economy, the roles of business firms operating across national boundaries seem to be important in understanding technology transfer. Formal accounting of linkages of economies will have to be gauged more precisely, for example, by internationally linked input–output tables.

CIM technology seems to be conducive to resource conservation if this segment of industry comes to dominate the production sphere of our economies. CIM is widely applied in assembly process, but it can find its way in disassembling various wastes, such as used cars and consumer durables, making industry more amenable to the environment. Perhaps a new project is needed to respond to these questions in a consistent manner.

Notes

[1] According to a survey by the Science and Technology Agency of the Japanese government, the gestation period is 3.54 years on the average, with considerable differentials among industrial sectors (see Uno, 1989b, pp. 232–234).

[2] MITI, *Survey on Machine Tools Installation*.

[3] Quality-adjusted deflator for mainframe computers is from Cartwright (1986), and that for industrial robots from Mori (1987). Sectoral deflators are from the Economic Planning Agency.

References

Cartwright, D.W., 1986, "Improved Deflation of Purchases of Computers," *Survey of Current Business*, March:7–10.

Cole, R., Y.C. Chen, J.A. Barquin-Stolleman, E. Dulberger, N. Helvacian, and J.H. Hodge, 1986, "Quality-Adjusted Price Indexes for Computer Processors and Selected Peripheral Equipment," *Survey of Current Business*, January:41–50.

Coombs, R., P. Saviotti, and V. Walsh, 1987, *Economics and Technological Change*, Macmillan Publishers, London, UK.

Economic Planning Agency, various issues, *Kokumin Keizai Keisan Nenpo* [Annual Report on National Accounts], Ministry of Finance Printing Bureau, Tokyo, Japan.

Freeman, C., and L. Soete, eds., 1987, *Technical Change and Full Employment*, Basil Blackwell Inc., Oxford and New York.

Kinoshita, S., and M. Yamada, 1988, "The Impact of Robotization on International Economy" in Y. Kaya, ed., *The Impact of Robots and Related Technology on Industry and Economy and the Policy Options*, National Institute for Research Advancement, Tokyo, Japan (in Japanese).

Krelle, W., ed., 1989, *The Future of the World Economy: Economic Growth and Structural Change*, Springer-Verlag, Berlin, Heidelberg, New York.

Mori, S., 1987, *Social Benefits of CIM: Labor and Capital Augmentation by Industrial Robots and NC Machine Tools in the Japanese Manufacturing Industry*, WP-87-40, IIASA, Laxenburg, Austria.

Saito, M., 1988, "The Impact of Robotization on the Japanese Economy" in Y. Kaya, ed., *The Impact of Robotics and Related Technology on Industry and*

Economy and the Policy Options, National Institute for Research Advancement, Tokyo, Japan (in Japanese).

Statistical Data Bank Project, 1990, *Fixed Capital Formation and R&D Benefit*, Statistical Data Bank Project Report No. 65, University of Tsukuba, Tsukuba, Japan.

Stone, R., and A. Brown, 1962, *A Computable Model of Economic Growth*, A Program for Growth 1, Chapman and Hall, London, UK.

Terlecskyj, N.E., 1980, "Direct and Indirect Effects of Industrial Research and Development of the Productivity Growth of Industries," in J.W. Kendrick and B.N. Vaccara, eds., *New Development in Productivity Measurement and Analysis*, University of Chicago Press, Chicago and London.

Terlecskyj, N.E., 1982, "R&D and US Industrial Productivity in the 1970s," in D. Sahal, ed., *The Transfer and Utilization of Technical Knowledge*, Lexington Books, Lexington, MA.

Uno, K., 1987, *Japanese Industrial Performance*, Elsevier, Amsterdam, Netherlands.

Uno, K., 1988, "Gijutsu Henkato Paradaimuno Henka" [Technological Change and Paradigm Change], in M. Shinohara, ed., *Kokusai Tsuka, Gijutsu Kakushin, Choki Hado* [International Currency, Technological Innovation, and Long Wave], Tokyo Keizai Shinposha, Tokyo, Japan.

Uno, K., 1989a, "Economic Growth and Environmental Change in Japan – Net National Welfare and Beyond," in F. Archibugi and P. Nijkamp, eds., *Economy and Ecology: Towards Sustainable Development*, Kluwer Academic Publishers, Dordrecht, Boston, London.

Uno, K., 1989b, *Measurement of Services in an Input–Output Framework*, Elsevier, Amsterdam, Netherlands.

Uno, K., 1990, *Annual Input–Output Tables in Japan, 1975–1985*, Statistical Data Bank Project Report No. 63, Institute of Socio-Economic Planning, University of Tsukuba, Tsukuba, Japan.

Part II

Micro Effects

Chapter 3

Labor Productivity and CIM

Robert U. Ayres and Luigi Bodda

3.1 Sources of Past Gains in Manufacturing Productivity

The direction and pace of change in any technology can only be forecast on the basis of a solid grasp of the historical background. If the changes now apparent in the field of manufacturing technology are truly portents of a second (or third) industrial revolution, as argued in this book, then it is not inappropriate to look back, at least briefly, at the changes that have taken place since the first industrial revolution, in the late eighteenth century.

The major innovation of the first industrial revolution (c. 1770–1830) was the substitution of steam power for water power and muscle power. This was of great importance in the UK, where good sites for water power were scarce to begin with and were essentially exhausted by the end of the eighteenth century. Horses, too, were expensive to maintain because of the high price of feed. However, in the USA, where animal feed was plentiful and water power was more readily available, steam power was introduced initially only for river and then for rail transport.

Table 3.1. Productivity increases: 1836 to 1897.

Item	Period	Increased output per man-hour (multiplier)
Metal Products		
Pitchforks (steel)	1836–1896	15.60
Plows, iron, and wood	1836–1896	3.15
Rakes, steel	1858–1896	5.96
Axle nuts (2″)	1850–1895	148.00
Carriage axles	1856–1896	6.23
Carriage axles (4″ steel)	1862–1896	6.23
Tire bolts (1 3/4″ x 3/16″)	1856–1896	46.90
Carriage wheels (3′6″)	1860–1895	8.41
Clocks, 8–day brass	1850–1896	8.30
Watch movements, brass	1850–1896	35.50
Shears, 8″	1854–1895	5.51
Saw files, 4″ tapered	1872–1895	5.51
Rifle barrels, 34 1/2″	1856–1896	26.20
Welded iron pipe, 4″	1835–1895	17.60
Nails, horseshoe, no. 7	1864–1896	23.80
Sewing machine needles	1844–1895	6.70
Other Products		
Bookbinding, cloth (320 pp.)	1862–1895	3.80
Men's shoes, cheap	1859–1895	932.00
Women's shoes, cheap	1858–1895	12.80
Hat boxes, paperboard	1860–1896	3.22
Wood boxes (18″ x 16″ x 9″)	1860–1896	9.73
Paving bricks	1830–1896	3.89
Buttons, bone	1842–1895	4.04
Carpet, Brussels	1850–1895	7.95
Men's overalls	1870–1895	10.10
Rope, hemp	1870–1895	9.74
Sheet, cotton	1860–1896	106.00
Electrotype plates	1865–1895	2.91
Chairs, maple	1845–1897	6.43

Source: Ayres, 1984, data from US Department of Labor.

The immediate economic benefits of steam power (versus water power), even in the UK, were quite modest – of the order of 0.25% per annum (p.a.) added to the annual growth of GNP – at least up to the 1830s when railroad building began in earnest (von Tunzelmann, 1978). Mechanization,

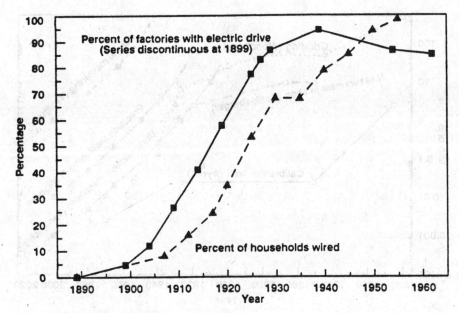

Figure 3.1. Electrification in the USA.

the application of mechanical power (from water or steam) to drive textile machinery and woodworking or metalworking machines, seems to have been more significant in the long run. Mechanization and economies of scale made possible enormous increases in manufacturing productivity throughout the nineteenth century (*Table 3.1*). However, the direct application of massive amounts of steam power to a single factory drive shaft peaked in the early 1900s, as electric drive began to be widely adopted (*Figure 3.1*).

Nevertheless, total installed horsepower per unit of output continued to grow at an average rate of 1.1% p.a. from 1899 until about 1920 (Schurr, 1984). It declined thereafter until 1953, and has increased slightly since then. Factory electrification (i.e., machine tools driven by electric motors) was highly beneficial in terms of flexibility of operations and plant layout. In fact, the adoption of electrified unit drive appears to be a major factor in the rapid improvement in US productivity growth that occurred after World War II (Schurr, 1984).

Yet, there were other major contributions to productivity gains since 1800. The most important milestone in the history of manufacturing, by many accounts, was the ability to produce truly *interchangeable parts* (e.g., Hounshell, 1984). This had been an explicit goal of military technology

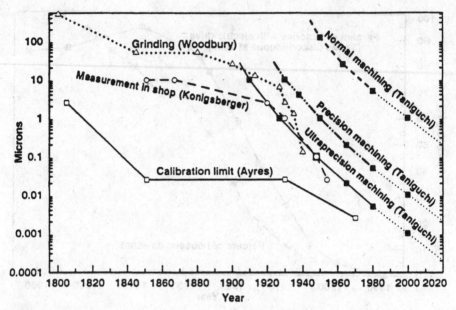

Figure 3.2. Measures of precision in metalworking.

since 1717 (France). Interchangeability was achieved only by degrees. A very crude form was claimed, for instance, by Eli Whitney (c. 1805), but it was not a practical reality even for weapons manufacturing until the late 1840s. Samuel Colt's famous exhibit at the Crystal Palace in London (1851) created a media sensation and undoubtedly marked a significant step in mechanization. It resulted in contracts for Colt to build munitions factories of his design for the British government.

Underlying the achievement of interchangeability was a series of innovations in precision, metalworking, and measurements by Wilkinson, Stowell, North, Whitney, Whitworth and Fitch, and others. The trend toward increased precision in measurement and in cutting has continued to the present, and even accelerated since World War II (*Figure 3.2*).

On the other hand, there is little or no evidence of major improvements in machine tool design since 1900. Modern production machine tools tend to be much bigger and more powerful than earlier counterparts, but they are scarcely more precise. Yet machine output per labor-hour input has increased enormously over the same time. For example, a 36-inch vertical boring mill in 1950 operated by one worker could produce the same output in one day that would have required 50 such machines (and 39 operators) in

1890. Similarly, a 20-inch engine lathe with one operator in 1950 produced the same output as 30 machines (and 50 operators) in 1890 (Tangerman, 1949). Similarly, the centennial issue of *American Machinist* (1977) cited a theoretical turned part – a steel axle – that would have required 105 minutes to machine in 1900, as compared with less than a minute in 1975.

Yet, based on machine tool attributes listed in catalogs, machine tool productivity – with characteristics held constant – should have *declined* more or less continuously at about 2% per year since the 1890s (Alexander and Mitchell, 1985). The most likely explanation of the Alexander–Mitchell paradox is that harder metals introduced since 1900 permit higher cutting speeds and less frequent tool changing. Prior to the mid-nineteenth century, the hardest available metal for cutting was carbon steel made by the crucible process (c. 1740) and "case-hardened" by heat treatment. A major step forward was the introduction in 1868–1882 of manganese-wolframite-based "self-hardening" alloys by Robert Mushet (Tylecote, 1976). These were the predecessors of "high-speed" tungsten steels developed especially by Frederick W. Taylor and Mansell White (c. 1900). This last development resulted in an approximately 70% increase in the maximum cutting rate from 1900 to 1915. The introduction of cemented tungsten carbide cutting tools resulted in cutting-speed increases of the same magnitude between 1915 and 1925.

Another major innovation was tungsten-titanium carbide, introduced by McKenna in 1938. Somewhat surprisingly, although few new cutting tool alloys have been introduced since then, tool fabrication (e.g., hardcoating) techniques have resulted in surprising further gains. Maximum cutting rates increased by no less than a factor of 10 from 1925 to 1975 (*Figure 3.3*). Interestingly, rapid improvements in cutting technology are still continuing, but the most recent gains are primarily due to advances in gas-bearing technology that will permit cutting speeds, in principle, at least 10 times greater than the 3,000 sfpm (speed, feet per minute) achieved by off-the-shelf machine tools in 1977 (*American Machinist*, 1977). Machine tools have, once again, become a dynamic technology.

Continuing gains in cutting speed have not been matched by comparable improvements in other areas of manufacturing, unfortunately. In the early nineteenth century, manufacturing labor was predominantly concerned with wood or metal cutting and forming, but by 1900 progress in metalworking together with increased product complexity had changed the nature of the problem. The assembly of a complex product such as a clock, sewing machine, or bicycle – supposedly made from standardized interchangeable parts

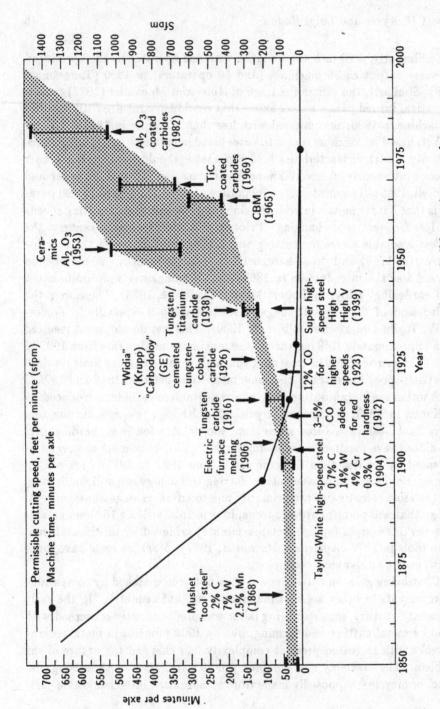

Figure 3.3. Machining speeds for steel axle: machining time and permissible cutting speed.

– typically constituted a labor-intensive activity requiring highly skilled "fitters." This was particularly true in Europe, where the greater availability of skilled labor resulted in a greater emphasis on high-quality (better-finished) manufactured products as compared with the USA, where there was a greater emphasis on large-scale production at minimum cost.

By some accounts Henry Ford's historic contribution to *mass production* was achieved primarily by enforcing rigid quality control in parts manufacturing – utilizing the scientific management methods of Frederick W. Taylor (1911) – thus, finally eliminating the need for fitting. Ford himself stressed the combined principles of "power, accuracy, economy, system, continuity, and speed." Ford engineers certainly looked everywhere for opportunities not only to subdivide the manufacturing process into many individual tasks, and to increase the efficiency of tasks by application of Taylor's methods, but also to substitute machines wherever possible for human workers. "Bringing the work to the man" was one of the ways to increase efficiency. Conveyor belts and gravity feeders began to be used extensively in the Highland Park plant by 1913. The moving assembly line (c. 1916) was the logical outcome of this rationalization.

Ford's assembly-line methods did, in fact, sharply reduce the cost of assembly as compared with parts manufacturing in the 1920s. However, in a fundamental sense, the assembly line is nothing more than a scheme to permit a more effective division of labor. The technology of assembly itself has changed very little until the last decade or so, except to the extent that assembly-line workers have gradually acquired power-assisted tools (such as wrenches).

In summary, while the mechanization of parts manufacturing has not yet reached any physical limits, its contributions to gains in manufacturing productivity were becoming negligible by the 1970s. In fact, logistics, assembly, and quality control now account for most of the direct costs of manufacturing – quite apart from *indirect* costs of R&D, engineering, finance, marketing, personnel management, and the like. To reduce costs significantly – and remain competitive – a completely new technology of production seems to be needed. This imperative will become increasingly manifest over the next several decades.

The historical factors resulting in productivity growth in manufacturing, from the first industrial revolution to the 1970s, can be summarized as follows:

- Division and specialization of labor.
- Application of mechanical power (from water or steam engines).
- Development of better engineering materials (iron and steel).
- New tools suitable for mechanization (turning, milling, and grinding).
- Methods of precision measurement (calipers, gauges, and comparators).
- Interchangeability of parts (elimination of "fitting").
- Electrification of machines to increase efficiency and flexibility.
- Harder materials for faster cutting tools (alloys and ceramics).
- "Scientific management" and vertical integration (Taylorism–Fordism).
- Mechanical integration (automatic transfer machines).
- Statistical quality control (SQC) and total quality control (TQC).

None of these factors involved computers or information technology in any fundamental way. Most had already reached a stage of maturity (or saturation) such that further gains would be expected to be very slow and costly, at best. The major exceptions – areas of rapid progress – were (and are) the following: (1) the development of new engineering materials and corresponding new methods of fabrication (e.g., ceramics and composites), (2) greater precision, and (3) faster cutting speeds.

3.2 Manufacturing Productivity Trends

At the aggregate level, advances in technology are reflected in manufacturing productivity. Of course, manufacturing productivity (defined as output per unit factor input) has been increasing more or less continuously since the beginning of the industrial revolution.

It is possible to observe the long-term trend of labor productivity for different countries divided into two effects: increases in labor productivity and declines in hours worked for each worker (Maddison, 1987, p. 651). The reductions of the hours worked in the last century for selected countries are reported in *Table 3.2*.

The yearly hours worked per person have decreased in the past 114 years by 0.6% per year in France and the UK, 0.5% per year in the USA and the FRG, and "only" 0.3% in Japan. Japanese workers work 24% more hours per year than American workers and 30% more than British workers. The time spent at work (in contracted or paid time) is now about one-half of what it was in 1870. The average of lifetime hours of paid employment for the male work force has decreased from 166,000 in 1870 to approximately

Table 3.2. Yearly hours worked per person (1870–1984).

Year	France	FRG	Japan	Netherlands	UK	USA
1870	2945	2941	2945	2964	2984	2964
1890	2770	2765	2770	2789	2807	2789
1913	2588	2584	2588	2605	2624	2605
1929	2297	2284	2364	2260	2286	2342
1938	1848	2316	2391	2244	2267	2062
1950	1989	2316	2289	2208	1958	1867
1960	1983	2081	2450	2177	1913	1795
1973	1785	1805	2213	1825	1688	1710
1984	1554	1676	2149	1640	1518	1632

Source: Maddison, 1987, p. 686.

Table 3.3. Projection of male working time.

Year	Hours per week	Weeks per year	Years per life	Hours per year	Lifetime hours
1961	46.3	48.5	50.4	2246	113,200
1976	42.7	46.9	47.4	2003	95,000
1981	41.7	45.9	46.0	1914	88,100
1991	39.4	45.5	44.6	1786	79,600
2001	37.3	44.3	42.8	1655	70,800

Source: Armstrong, 1984, p. 40.

88,000 at present. The average number of hours for paid female employees has decreased from 55,000 to the current level of 40,000, despite the fact that more women are now employed. Lifetime working hours are expected to decline at a yearly rate of 1.1% to the year 2000 and beyond (*Table 3.3*).

The percentage of awake time spent working has decreased significantly since 1880: in 1880, 42% of awake time was spent working; in 1945, 28%; and in 1985, 18%. Working time as a ratio of awake lifetime is expected to continue to decrease, reaching the 14% to 16% range in the year 2000 (Clutterback, 1989, p. 1).

In the meantime we have had a dramatic increase in production per worker (*Table 3.4*). The GDP yearly growth rate for the whole period has been 2.8% in Japan, 2.1% in France, 2.0% in the FRG, 1.6% in the USA, 1.5% in the Netherlands, and 1.2% in the UK. The sum of the two effects (reduced hours worked and increased output per worker) gives us the increase of the labor productivity per hour worked. Japan shows the highest growth with 3.1% for the 114 years (6.2% in the postwar period 1950–1984), followed by

Table 3.4. GDP per worker, in 1984 US dollars.

Year	France	FRG[a]	Japan	Netherlands	UK	USA
1870	3,090	3,320	1,090	6,020	6,360	5,690
1890	3,930	4,510	1,470	NA	7,920	8,260
1913	5,720	6,460	2,100	8,730	9,420	11,700
1929	7,370	7,100	3,360	11,210	10,330	15,460
1938	7,650	8,920	4,230	11,020	11,130	14,450
1950	9,030	8,500	3,480	13,620	12,550	20,410
1960	13,780	14,840	6,620	18,770	15,380	25,090
1973	25,540	25,160	18,570	30,610	22,260	32,760
1984	32,290	32,310	25,470	33,980	26,060	34,780

[a]Data prior to 1945 refer to Germany, including the eastern region that was subsequently administered separately until 1990 as the German Democratic Republic.
Source: Calculation on Maddison data, 1987.

Table 3.5. Gross fixed investment as percent of GDP.

Year	USA	Japan	FRG	France	UK	Italy	Canada
1965	16.4	18.0	27.0	23.0	20.0	28.0	17.0
1970	15.5	27.0	25.7	23.0	21.0	30.0	16.0
1975	14.7	23.0	21.6	23.0	20.0	25.0	17.0
1980	16.2	22.0	22.6	22.0	18.0	24.0	20.0
1985	17.4	23.0	20.0	20.0	18.0	22.0	18.0

Source: Webster and Dunning, 1990, p. 18.

France with 2.7%; the FRG, 2.5%; the USA and the Netherlands, 2.1%; and the UK, 1.8%. However, while Japan was nearly at the European level in GDP per worker by 1984, this output was achieved by working more hours.

To keep the Japanese case in perspective we must recall that Japan started from the lowest hourly productivity of the countries considered. In 1984 it was "only" at $11.85 (1982 US dollars) compared with a range in the other industrialized countries between $17.20 for the UK and $21.30 for the USA. Of course, since 1984 the dollar/yen exchange rate has declined by more than a third. Adjusting for this change, Japanese productivity is now comparable to the European level. On the other hand, the price level (cost of living) in Japan is now considerably higher than in Europe and the USA.

During the period 1965–1985 Japan has increased its share of gross fixed investment as a percentage of GDP from 18% to 23%, reaching the highest share among the advanced industrialized countries (*Table 3.5*).

The average growth rate of the US labor productivity for the whole period has been 1.63% per year. During the period 1981–1987, there has been an

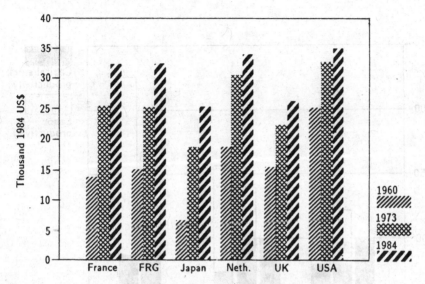

Figure 3.4. Labor productivity level (GDP per worker).

average productivity growth of 0.93% per year reaching the current value of $34,200 (1982 US dollars) GDP per worker.

Recent trends in labor productivity for the major Western countries at the level of economy as a whole are shown in *Figure 3.4*. Comparable data for manufacturing productivity broken down by a factor (labor and capital) are only available for a few countries. *Figure 3.5* compares labor and capital stock productivity for 1966, 1976, and 1986 for the manufacturing sectors of the USA, the UK, and the Federal Republic of Germany. It is noteworthy that, while labor productivity has consistently increased, capital productivity has not. (In fact, it has generally decreased.) The productivity data for the USA are available in more detail in *Figure 3.6* and *Figure 3.7* and show the same pattern.

There is a trade-off between labor and capital productivity. Normally labor productivity increases while capital productivity decreases. The manufacturing sector shows, after the year 1965, higher growth in labor productivity than the total business sector. *Table 3.6* shows the growth rate of the US labor, capital, and multifactor productivity for the business and manufacturing sectors.

Productivity growth permits increased wages. In the last century real wages increased at an average yearly growth rate of 1.81%. If we divide this time into three periods, we observe a growth rate of 1.23% during the

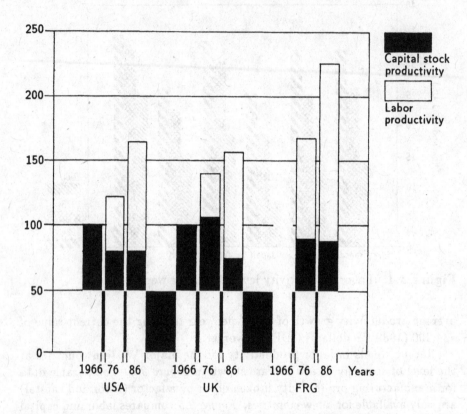

Figure 3.5. Labor and capital productivity: the USA, the UK, and the FRG.

period 1880–1920, 2.94% during the period 1920–1960, and 0.73% during the period 1960–1980 (Cyert and Mowery, 1988, pp. 332–333). In the other industrialized countries the trend has been quite similar. *Table 3.7* shows the average yearly growth rate of GDP, per capita productivity, and real wages in some European countries.

Technological progress in conventional production technologies, such as those noted at the beginning of this chapter, has not ceased by any means. Potential economic gains from increments in machine size, speed, or accuracy seem to be less and less significant. By contrast, the economic gains from flexible programmable automation and CIM are just beginning to be felt.

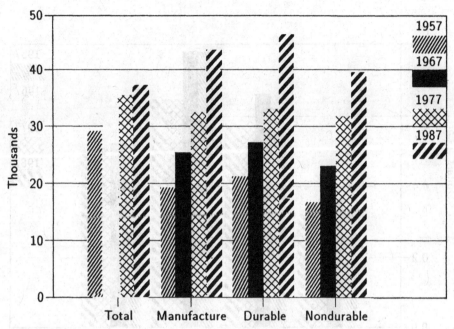

Figure 3.6. Labor productivity in the USA (1982 US dollars).

Table 3.6. Productivity growth in the USA.

Period	Business sector			Manufacturing sector		
	Labor	Capital	Multifactor	Labor	Capital	Multifactor
1948–1965	3.25	0.81	2.39	2.92	0.78	2.26
1965–1973	2.14	−0.69	1.12	2.48	−0.91	1.46
1973–1979	0.61	−0.88	0.10	1.37	−1.85	0.52
1979–1986	1.40	−1.05	0.52	3.42	−0.06	2.53

Source: Bayly and Chakrabarti, 1988, p. 3.

Several factors seem to be jointly responsible for the new emphasis on flexibility. In Western society, broadly speaking, basic material needs have been fulfilled during the 1960s and 1970s. A higher income per capita stimulates a more diversified pattern of consumption. "Customization" is the new watchword of final goods manufacturers.

On the supply side, the two oil crises of the 1970s set in motion long-term shifts in the technological content of industrial structure. The importance of traditional materials is declining, for instance. The important sectors are

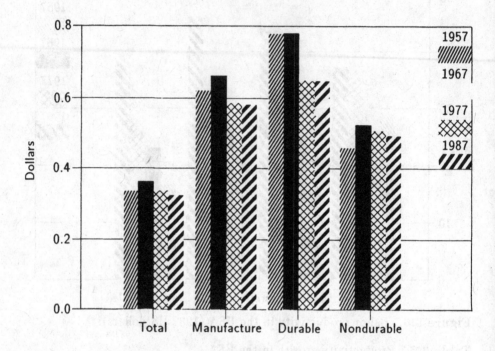

Figure 3.7. Capital stock productivity in the USA (1982 US dollars).

Table 3.7. GDP, productivity, and real wages growth rate, in percent.

Country	GDP			Per capita productivity			Real wages		
	1960/ 1973	1973/ 1979	1979/ 1985	1960/ 1973	1973/ 1979	1979/ 1985	1960/ 1973	1973/ 1979	1979/ 1985
France	5.6	3.1	1.1	4.9	2.9	1.6	5.0	4.0	1.5
UK	3.1	1.4	1.0	2.9	1.1	1.8	3.3	1.3	1.8
Ireland	4.4	4.1	2.1	4.3	2.7	2.3	5.4	3.3	−0.1
Belgium	4.9	2.4	1.0	4.2	2.3	1.8	5.0	3.9	0.2
Spain	6.6	2.5	1.5	5.4	4.2	3.3	6.4	2.9	1.2
Italy	5.3	2.6	1.3	5.6	1.6	0.9	6.5	2.4	1.2
FRG	4.5	2.4	1.1	4.2	2.9	1.6	5.3	2.6	0.2

Source: Boyer, 1988, pp. 20 and 210.

microelectronics, telecommunications, advanced materials, and biotechnology. Such a deep restructuring generates a high degree of uncertainty for current investment decisions. The flexibility that is derived from new computerized tools of production enables firms to respond more effectively to these pressures from both the supply side and the demand side.

Two other factors drive this evolution. One is the increased competition in US and world markets resulting from the rise of the Japanese and other East Asian export-oriented economies. Among other effects, this has destroyed the postwar hegemony of General Motors in the (world) auto industry. It has also made obsolete GM's formerly dominant strategy of gradual "managed" innovation (the annual model change), with its emphasis primarily on exterior appearance. As mass-production methods are easily transferable to newly industrialized countries the new competitive edge must be derived from higher added value. This implies differential products with a significant content of R&D and market information, produced in smaller batches. In current market conditions the relative importance of responsiveness to the marketplace has increased sharply, requiring correspondingly greater emphasis on variety and customization. The second, and related, trend is toward increased product complexity, not only in the auto industry, but throughout the manufacturing industry.

The consequences of the two trends are also twofold. In the first place, the rate of product design change has accelerated, in the auto industry and elsewhere. Whereas in the 1950s and 1960s the auto industry phased in design changes rather slowly so as to permit mass-production facilities a 20-year useful life before major renovation and retooling, today this is no longer possible. But, on the other hand, increased complexity has made the design process increasingly expensive and risky. This led to the so-called "productivity dilemma": an apparent contradiction between the need to cut manufacturing costs by maximizing standardization and specialization and the need to introduce new and improved products (Abernathy, 1978).

The possibility of a way out of this dilemma through increasing the *flexibility* (economies of scope) of manufacturing technology was only clearly recognized by industry in the 1980s, although the facilitating technologies – computers and numerically controlled machines – began filtering into the manufacturing world as early as 30 years ago.

In summary, it must be emphasized that labor productivity gains have been achieved in the past largely by increasing the magnitude of capital investment (*Figure 3.8*). This is the primary reason for static or declining

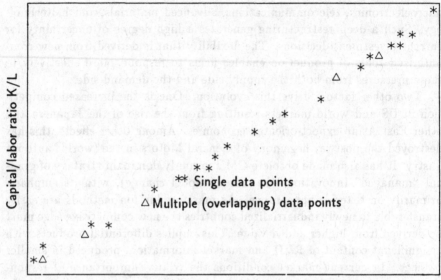

Figure 3.8. Productivity as a function of the capital/labor ratio: USA (1950–1982). Source: Abel and Szekely, 1989; data from Ross, 1985.

levels of capital productivity. However, as the capital *intensity* of the economic system as a whole increases, capital costs account for an increasing fraction of total costs. Indeed, if savings rates remain constant and capital-output ratios continue to decline (or even remain constant), a time must inevitably come when the replacement of depreciated capital consumes all of the surplus available for investment in the economy. At that point, economic growth would come to a halt.

Indeed, the worldwide productivity slowdown that has occurred since the 1970s strongly suggests that this might be happening, especially in the USA. Are we doomed to a no-growth economy for the indefinite future?

It is to be hoped that the adoption of CIM offers the way out of this dilemma by eliminating some of the constraints on capital productivity. The specific mechanisms that can be identified in this connection are (1) sharply reduced inventories of goods and work in progress and (2) sharply increased output per machine, via increased operating speeds and increased utilization rates.

3.3   Direct Benefits of CIM

Evidently improved product quality and increased flexibility in the use of
capital are beneficial to users of CIM. However, the argument thus far is only
qualitative. To carry it a step further, one must define quality and flexibility
more precisely and formulate them in conventional economic variables and
models.

To organize the discussion, it is helpful to consider five possible kinds of
economic benefit.

(1) *Labor saving.* Some CIM technologies (most notably robots) can be re-
    garded as direct substitutes for semiskilled human labor. This means,
    for example, that robots (sometimes called "steel-collar workers") can
    also be regarded as additions to the labor force, although their "wages"
    are partly operating costs and partly capital costs. Similarly, CAD sys-
    tems can be regarded (in part, at least) as direct substitutes for drafting
    engineers.

(2) *Capacity augmenting/capital saving.* Some CIM technologies, such as
    scheduling systems and programmable controllers (PCs) with sensory
    feedback, can be regarded as capital savers or capacity augmenters. This
    is the case to the extent that they increase the effective utilization of
    existing machine tools and other capital equipment (e.g., by permitting
    unmanned operation at night) or permit shorter delivery times, faster
    turnarounds, and reductions in the inventory of work in progress. The
    productivity of capital is thus increased.

(3) *Capital sharing/saving.* The major benefit of flexibility is that it permits
    faster response to changing market conditions or superior ability to differ-
    entiate products. The major reason for slow response is the widespread
    use of dedicated, specialized ("Detroit") automation in mass production.
    Here, the lowest possible marginal unit cost is achieved at the expense of
    very high fixed capital investment and large write-offs in case the prod-
    uct becomes obsolete and cannot be sold. Flexibility in this context is
    the ability to adapt (or switch) capital equipment from one generation
    of a product to the next. The term flexibility is also widely used in a
    rather different context – to describe a futuristic concept analogous to an
    automated job shop, capable of producing "parts on demand." In either
    case, capital is shared among several products rather than dedicated to
    a single one. Evidently capital sharing is practically indistinguishable

Table 3.8. Major source of pay-back.

	CAD	CAM	Robots
Pay-back time (years)	3.18	4.94	2.05
Labor costs	85.70	46.20	100.00
Other costs	0.00	0.00	0.00
Quality	0.00	0.00	0.00
Market factors	14.30	53.80	0.00

Source: McAlinden, 1989, p. 84.

from capacity augmentation. However, it is perhaps slightly preferable to model it as an extension of the lifetime of existing capital or (in some cases) as credit for capital recovery.

(4) *Product quality improvement.* The term "quality" is not very precise, since it comprises at least two aspects: product *reliability* (defect reduction) and product *performance*. The latter can be disregarded, here, as being an aspect of product change. It is postulated that several CIM technologies, especially those that use "smart sensors" in conjunction with programmable controllers, will eventually reduce the in-process error/defect rate. Moreover, these technologies will also permit more complete and more accurate testing and inspection of workpieces and final products.

(5) *Acceleration of product performance improvement ("market share").* As noted above, improved product performance can be distinguished, in principle, from improved product reliability (quality) resulting from reduced error/defect rates. The latter is a function of the manufacturing process only, whereas the former requires changes in the actual design of the product. It was pointed out that one benefit of flexibility is that it reduces the cost of each product change. A further benefit is that, as a result, product redesigns are likely to be more frequent and firms with CIM have the opportunity to respond more quickly to market conditions and to improve market share. The problem for an economist is to find empirical evidence of a relationship between the cost of product redesign and retooling and the non-price factors contributing to gains in market position. This appears to be a relatively unplowed field of research, to date.

Recent US experience shows that noncapital savings resulted in an overall average pay-back time of 2.8 years for CIM-type investments. The range is

Table 3.9. Labor-cost change per unit.

	Overall	CNC/CAM	MH robot	Welding robot
Mean	−46.3	−43.9	−61.9	−54.8
Median	−63.4	−63.1	−98.4	−71.6

Source: McAlinden, 1989, p. 68.

Table 3.10. Capacity increase, in percent.

Rate	Overall	CAM	MH robot	Welding robot
Simple throughput	14.2	60.0	0.0	0.0
Full throughput	20.0	100.6	0.0	0.0

Source: McAlinden, 1989, p. 53.

from two years for robots (1.7 years for material-handling robots for the fabricated metals sector) and up to five years for CAM applications (McAlinden, 1989, p. 73). Labor-cost reduction was the main source of pay-back overall for robots and CAD applications. In fact, in many instances the new technologies require setup and maintenance procedures not required with the old manual technologies. Robots in stand-alone applications often replace workers performing exactly the same operations at the same cycle times with no change at the system level in comparison with previous practices.

Table 3.8 shows the pay-back times for selected application and the main sources of pay-back. The category "other costs" in *Table 3.8* refers to energy and raw-material savings. In the iron and steel industries, energy and scrap savings are indicated as major benefits. This is less important for the engineering sectors. The CAM technologies are a major factor in market profit increases, providing a more flexible automation technology. Labor costs show a sharp material-handling decline for all applications, with a greater effect on materials-handling (MH) and welding robots. *Table 3.9* shows the average and median of the unit labor-cost reduction for various applications.

However, capacity augmentation is more important in the case of CAM applications. *Table 3.10* shows the percentage capacity increase for two (simple and full) throughput rate hypotheses, the throughput rate being the reciprocal of the cycle time. Capital productivity is discussed in Section 3.4.

The direct labor impact of industrial robots (IR) has been variously estimated (*Table 3.11*). An overall average of two jobs lost per robot installed has been suggested by one study (Edquist and Jacobsson, 1988, p. 115). Japanese experience is the most thoroughly documented. According to the

Table 3.11. Labor substitution potential of industrial robots (IR).

Robot type	1 shift (B)	(S)	2–3 shifts (B)	(S)
Industrial robots (average)	1.5	1.40	4.0	3.0
Assembly robots	2.1	1.60	6.2	3.3
Tool manipulators	0.9	0.43	1.7	0.8

B = Battelle Frankfurt; S = Sociological Institute Göttingen.
Source: Haustein and Maier, 1981.

Japanese experience the employment reduction for robot use is 0.9–1.0 workers per shift. With a multishift use of IR, the average total reduction has been estimated as 1.75 workers per robot (Tchijov, 1988, p. 263).

In general, the labor replaced is semiskilled or unskilled, while some skilled jobs are created. There is a sharp reduction in the number of jobs devoted to the productive cycle (from 94% to 53% of the total) while setup labor increases from 5% to 27% and maintenance labor from 1% to 20% (McAlinden, 1989, p. 65).

As CNC machines, robots, and CAM are brought together in FMC and FMS, the benefits are multiplied. Several specific examples for which before-and-after comparisons can be made explicitly are illustrated in *Tables 3.12(a–h)*. In some cases there are significant savings in capital (reduced floor space, lower inventory). In some cases greater product variety was the goal; in others it was reduced lead time. The impact of FMS on direct labor is dramatic. For small systems the labor saved is roughly proportional to the reduction in stand-alone machine tools. Several examples are given in *Tables 3.12(a–h)*. The IIASA FMS data base (*Table 3.13*) suggests an overall personnel reduction of 74%; there is a replacement of unskilled with skilled labor (Edquist and Jacobsson, 1988, p. 117).

3.4   Impacts of CIM on Capital Productivity

The magnitude of potential savings in inventory costs varies dramatically, depending on how the problem is viewed. Within the manufacturing sector, *per se*, the ratio of value-added to inventory ranges from 3.5 for Japan and 3.0 for the USA to 1 for some of the Eastern Europe countries. Most West European countries seem to range between 2 and 2.3 (*Figure 3.9*).

This ratio may be regarded as a measure of the efficiency of the goods transport system (USA) or the efficiency of manufacturing organization

Table 3.12(a). Benefits of FMS (Case No. 1).

	Before	After
Types of parts per month	543	543
Number of pieces per month	11,120	11,120
Floor space (m^2)	16,500	6,600
CNC machine tools	66	38
Non-NC machine tools	24	5
Total machine tools	90	43
Operators	170	36
Distribution and production control workers	25	3
Machining time per part[a] (days)	35	3
Unit assembly	14	7
Final assembly	42	20
Total time	91	30

[a]Including queuing (work-in-progress).
Source: Jaikumar, 1989.

Table 3.12(b). Benefits of FMS (Case No. 2).

Auto engines	Before	After
Number of engines per month	5,000	5,000
Floor space (m^2)	6,000	10,000
Machine availability (%)	?	95
Capital cost (%)	100	60
Work force	?	50
Lead time		no change
Work-in-progress		no change

Source: Merchant, 1989.

Table 3.12(c). Benefits of FMS (Case No. 3).

Sewing machines	Before	After
Number of products per month	8,000	8,000
Number of different models	30	30
Number of different parts	60	60
Work force per shift	28	3
Minimum lot size	150–200	1
Changeover time	3 hours	1 minute
Finished parts inventory	6 months	2 days

Source: Merchant, 1989.

Table 3.12(d). Benefits of FMS (Case No. 4).

Machine tools manufacturing	Before	After
Floor space (m^2)	6,500	3,000
Number of machines	68	18
Staff, total	215	12
Inventory (million dollars)	5.0	0.218
System cost (million dollars)	14	18
Throughput time (days)	90	3

Source: Fleissner, 1987, p. 101.

Table 3.12(e). Benefits of FMS (Case No. 5).

	Before	After
Number of machine tools	8	4
Number of processes	3	1
Number of workers	10	2
Machine utilization (%)	50	75
Factory availability (hours per day)	16	24
Lead time (days)	6	1

Source: Stokes (1988), p. 383.

Table 3.12(f). Benefits of FMS (Case No. 6).

	Before	After
Machine tools	90	43
Operators	195	39
factory	170	36
production control	25	3
Floor space (m^2)	16,500	6,600
Machining time (days)	35	3
Unit assembly time (days)	14	7
Overall assembly time (days)	42	20
Total process time (days)	91	30

Source: Warnecke, personal communication.

(Japan). Either way, of course, it is a measure of unavailable capital. In the USA, this amounts to roughly 33% of annual value-added or 14% of annual output. For the year 1989, the monetary value of capital tied up in manufacturing inventory was more than $360 billion (*Statistical Abstracts of the US*, 1989).

Table 3.12(g). Benefits of FMS (Case No. 7).

Index	Before	After
Machine tools	50	6
Operators	70	16
Floor space (m^2)	1500	350
Throughput time (days)	18.6	4.2
Output (%)	95	98
Operations per part	15	8
Machine tool utilization (%)	20	73
Operates 24 hours/day, 11 hours unmanned		

Source: Warnecke, personal communication.

Table 3.12(h). Benefits of CIM (Case No. 8).

Index	Before NC	With CIM
Manufacturing time	100	47
Throughput-time	100	75
Personnel	100	47
Machine tools	100	47
Floor space	100	58
Investment	100	90
Tooling	100	70
Total annual costs	100	76

Source: Warnecke, personal communication.

Worldwide the CIM "market" for 1988 was roughly $20 billion (exclusive of computers) of which the largest share was NC and CNC machine tools: $820 million was for CAD/CAE, $338 million was for electronic inspection systems, $274 million was for NC systems, $414 million was for robot systems, and $180 million was for vision systems. The data-processing market ($81 billion) cannot be readily broken down into end-use segments, but the fraction that can reasonably be attributed to CIM is surely not as high as 10%. The total worldwide investment in CIM (for 1988), therefore, was less than $30 billion, of which the USA accounted for some one-third or $10 billion. Even assuming 20% annual growth for the next decade, annual CIM outlays would only reach $60 billion by the year 2000.

Each 1% reduction in manufacturers' inventories would "create" $3.75 billion in capital in 1990 (1988 dollars) and as much as $5 billion by 2000. If inventory levels could be cut by just 25% in the next decade, the total capital made available for other purposes would be in excess of $100 billion –

Table 3.13. Average FMS benefits by area of application.

Indicators	Metal cutting	Metal forming	Welding assembly	Overall average
Pay-back time (years)	3.80	3.10	3.60	3.80
Lead time (% decrease)	−80.39	−89.47	−90.10	−81.48
In-process time (% decrease)	−86.30	−75.00	−75.61	−83.05
Inventory (% decrease)	−74.36	a	a	−76.19
Work-in-progress (% decrease)	−72.97	a	a	−75.00
Personnel (% decrease)	−76.19	−44.44	−75.61	−74.36
Number of machines (% decrease)	−75.00	a	a	−75.61
Floor space (% decrease)	−67.74	−41.18	−54.55	−64.29
Capacity utilization (% increase)	80.00	a	a	80.00
Unit cost (% decrease)	−41.18	−62.96	−58.33	−44.44

aNumber of observations is not enough for averaging.
Source: IIASA.

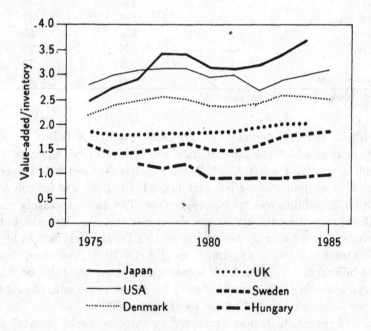

Figure 3.9. Value-added/inventory in manufacturing industry. Source:
Dimitrov and Wandel, 1988.

Table 3.14. Average expected benefits for CIM.

Improvement	Expected (%)
Productivity increase	100
Costs decrease	30
Product-quality increase	100
Design lead-time decrease	65
Shipment lead-time decrease	50
Capital utilization increase	200
Inventory work-in-progress decrease	70
Average weekly hours decrease	20

Source: CIRP Delphi Survey, 1985.

enough to finance the first six years of CIM investment (assuming $10 billion for 1990 and rising at 20% p.a. to $25 billion in 1995).

Although it is unlikely that gains in capital productivity from manufacturing inventory reductions, *per se*, would suffice to pay for the entire investment in CIM, there are further potential gains to be exploited. Indeed, for every $1 of manufacturers' inventories in the USA there was another $1.13 (1987) in wholesale and retail trade inventories. In principle, information technology could also reduce both wholesale and retail inventories by a large factor. Indeed, the recent rise in mail orders and retail merchandising over cable TV points strongly in this direction. Given an efficient computerized order-processing and delivery system, coupled with a manufacturing system geared to rapid turnaround and customized products "on demand," there is no fundamental reason for most shops to hold large stocks of merchandise on the premises, or even in regional warehouses. The economic implications are staggering, not only in terms of increased capital productivity but in terms of the elimination of a significant part of the distributional chain – and the associated costs.

In 1985, the Scientific Technical Committee on Optimization (STCO) of the College Internationale pour l'Etude Scientifique des Techniques de Production Mechanique (CIRP) conducted a Delphi-type survey of its members, to study the *ultimate* performance improvement that could be expected from CIM technology as compared with current (1985) levels. Eighty questionnaires were sent out, eliciting 40 responses to the first round and 46 to the second. The results are displayed in *Table 3.14*.

A recent study on the impact on labor in Midwest manufacturing automatization has been issued by the US Commerce Department's Economic Development Administration for some selected sectors: iron and steel (SICs

Table 3.15. Impact of automatization on jobs.

Industrial sector	Initial displacement	Sales gains	Net jobs displacement
Iron and steel	54,000	13,900	40,100
Machining and tooling	106,200	13,300	92,900
Fabrication metals	437,500	25,700	411,800
Auto	25,600	5,600	20,000
Total	623,300	58,500	564,800

Source: McAlinden, 1989.

331-332), metalworking machinery and tooling (SIC 354), metals (SIC 34), and auto industries (SIC 371) to verify the indirect impact of automatization (McAlinden, 1989). The main results of the study were the following:

- The unit labor time falls from 98% to 25% in the automated operation with a reduction of the total factory labor force up to 42% and a total cost reduction between 0.5% in the auto and 32.2% in the machine and tooling industries.
- If all the cost savings were transformed into price reductions, utilizing the US Commerce and Labor Department's long-run price elasticity of demand, there could be a sectoral sales increase between 1.1% (autos) and 26.2% (machines and tooling).
- The net number of jobs displaced for the areas and the sectors considered will be quite negative: of 623,300 job displaced by technology only 58,500 will be gained by the increased sales with a net job displacement of 564,800. *Table 3.15* shows the impact on jobs in the sectors considered (McAlinden, 1989).

References

Abel, I., and I. Szekely, 1989, *Technical Progress and New Logistics Techniques*, WP-89-21, IIASA, Laxenburg, Austria.

Abernathy, W.J., 1978, *The Productivity Dilemma*, Johns Hopkins University Press, Baltimore, MD.

Alexander, A.J., and B.M. Mitchel, 1985, "Measuring Technological Change of Heterogeneous Products," *Journal of Technological Forecasting & Social Change* **27**(1–2) May.

American Machinist, 1977, "Metalworking: Yesterday and Tomorrow" [100th Anniversary Issue] McGraw Hill, New York, NY.

Armstrong, P.J., 1984, "Work, Rest or Play? Changes in Time Spent at Work," in *New Technology and Future of Work and Skills*, Frances Pinter, London, UK.

Ayres, R.U., 1984, *The Next Industrial Revolution: Reviving Industry Through Innovation*, Ballinger Publishing Company, Cambridge, MA.

Bayly, M., and A. Chakrabarti, 1988, *Innovation and the Productivity Crisis*, The Brookings Institution, Washington, DC.

Boyer, R., 1988, *The Search for Labor Market Flexibility: The European Economies in Transition*, Clarendon Press, Oxford, UK.

CIRP, 1985, Delphi Study, Paris, France.

Clutterbuck, D., 1989, *Information 2000: Insights Into the Coming Decades in Information Technology*, Pitman Publishing, London, UK.

Cyert, R., and D. Mowery, 1988, *The Impact of Technological Change on Employment and Economic Growth*, Ballinger Publishing Company, Cambridge, MA.

Dimitrov, P., and S. Wandel, 1988, *An International Analysis of Differences in Logistics Performance*, WP-88-31, IIASA, Laxenburg, Austria.

Edquist, C., and S. Jacobsson, 1988, *Flexible Automation, the Global Diffusion of New Technology in the Engineering Industry*, Basil Blackwell, New York, NY.

Fleissner, P., 1987, Zur wissenschaftlich-technischen Revolution der Gegenwart, in P. Fleissner, ed., *Technologie und Arbeitswelt in Österreich: Trends bis zur Jahrtausendwende*, Springer-Verlag, Vienna, Austria.

Haustein, H.-D., and H. Maier, 1981, *The Discussion of Flexible Automation and Robotics*, WP-81-152, IIASA, Laxenburg, Austria.

Hounshell, D., 1984, *From the American System to Mass Production, 1800–1933*, Johns Hopkins University Press, Baltimore and London.

Houthakker, H.S., and L.D.S. Taylor, 1970, *Consumer Demand in the United States: Analysis and Projections, 1929–1970*, Harvard University Press, Cambridge, MA.

Jaikumar, R., 1989, *From Filing and Fitting to Flexible Manufacturing: A Study in the Evolution of Process Control*, WP-89-1, IIASA, Laxenburg, Austria.

Maddison, A., 1987, "Growth and Slowdown in Advanced Capitalist Economies: Techniques of Quantitative Assessment," *Journal of Economic Literature* **25** (June):649–698.

Marchetti, C., 1989, La lenta marcia dei colletti bianchi, *Dimensione Energia*, n. 34/35.

McAlinden, S.P., 1989, Programmable Automation, Labor Productivity, and the Competitiveness of Midwestern Manufacturing, June, US Commerce Department's Economic Development Administration, Washington, DC.

Merchant, M.E., 1989, "CIM – Its Evolution, Precepts, Status and Trends," Report, ECE/IIASA, Paper presented at the ECE/IIASA Seminar on CIM, Botevgrad, USSR.

Ross, H. 1985, *Bonn-IIASA World Model, Data Base*, University of Bonn, Discussion Paper No. B-28, B-29, Bonn, Germany.

Statistical Abstracts of the US, 1989, Manufacturing and Trade, Sales and Inventories, Table B-55.

Schurr, S.H., 1984, "Energy Use, Technological Change and Productive Efficiency," in Hollander and Brooks, eds., *Annual Review of Energy*, Series: Annual Reviews of Energy 9, Annual Reviews, Inc., Palo Alto, CA.

Stokes, B., 1988, "The 21st-Century Factory," *National Journal*, February 13.

Tangerman, E.J., 1949, "Do Machine Tools Cost Too Much?" *American Machinist*, September 8.

Tchijov, I., 1988, "CIM Introduction: Some Socioeconomic Aspects," *Technological Forecasting and Social Change*, **35**.

Tylecote, R.F., 1976, *A History of Metallurgy*, The Metals Society, London, UK.

von Tunzelmann, G., 1978, *Steam Power and British Industrialization to 1860*, Clarendon Press, Oxford, UK.

Warnecke, H.-J., Personal communication.

Webster, A., and J. Dunning, 1990, *Structural Change in the World Economy*, Routledge, London, UK.

Chapter 4

The Enhancement of Labor: Robots in Japan

Shunsuke Mori

In Chapter 3 the benefits of CIM technologies were classified into five categories: (1) labor saving; (2) capacity augmenting and capital saving; (3) capital sharing and saving; (4) product quality improvement; and (5) acceleration of product performance improvement. It is clear that in the short run, the first three benefits contribute immediately to the profitability of firms. However, structural problems associated with industry reallocation and employment might occur during the period of penetration of the new systems. In the long run, these benefits – as well as product quality improvement and the acceleration of product performance improvements – will be passed on to consumers through price reductions and increased consumer surplus value (Ayres and Funk, 1987). This discussion has been extended to the international economy in UNECE (1985) and in Kaya (1986).

The approaches taken by previous econometric studies are mainly of two kinds. One deals with labor substitutability and interaction among industries at the national level based on a macroeconomic model. The I–O model, in particular, has been used to evaluate the impacts of computer aided systems (Leontief and Duchin, 1985; MITI, 1985). The plausibility of these I–O studies depends on how the labor and capital coefficients are determined. In these studies, however, they are given as "appropriate" values because of lack of basic statistics. It is also difficult to include engineering and managerial issues.

The other approach, at the microeconomic level, is based on factory-level surveys. Although the coverage of such surveys is restricted, detailed engineering information and qualitative opinion of the managers can be obtained, as well as economic effects (Ayres and Miller, 1983; Jaikumar, 1984; JIRA, 1984, 1985; Flexibility in Manufacturing Systems, 1986). Based on such data, one can subsequently discuss the detailed effects including potential labor displacement. However, because survey studies do not provide historical trends, another method is needed to evaluate the penetration behavior and market growth. It is also difficult to guarantee consistency between enterprises in the sample and the total national economy.

The purpose of this chapter is to evaluate the macroeconomic benefits of industrial robots based on national statistics. By comparing the empirical results with the factory-level survey data, one can verify the compatibility of the macro-level model with the micro-level survey results. This may permit the application of other detailed results from the micro level to the macro level. This possibility is also discussed in the chapter.

4.1   The Production-Function Approach

The formulation of our benefits-evaluation model begins with a production function that involves four heterogeneous production factors, namely, $Y(K, L, R, x)$. Here Y, K, and L represent output in real terms, conventional capital stock, and labor, respectively; R denotes the stock of industrial robots; and x represents other inputs. It is postulated that L and R are essentially alternative "kinds" of labor, and that they are separable from K and x, *viz.*,

$$Y = Y[K, F(L, R), x] . \tag{4.1}$$

In the remainder of this section, K and x are ignored for convenience. (They are reintroduced later.) Evidently, $F(L, R)$ can be interpreted as an augmented equivalent labor force. It may therefore be plausible to impose the following conditions:

$$F(L, 0) = L \tag{4.2}$$

$$dF/dL > 0 \tag{4.3}$$

$$dF/dR > 0 \ . \tag{4.4}$$

Linear homogeneity and the second-order differentiability of $F(L,R)$ are also postulated. One of the simplest functional forms that satisfies these conditions is

$$F(L,R) = (L^a + A \times R^a)^{(1/a)} \ , \tag{4.5}$$

where A must be positive to meet the condition (4.2). Equation (4.5) is a special form of the well-known CES production function (Allen, 1967). It should be noted that, because of condition (4.3), other familiar production functions, such as the Cobb-Douglas and the trans-log types, cannot be adopted for our purposes.

The optimal strategy for economic management, in view of equation (4.5), is formulated as follows:

$$\min P_L L + P_R R \tag{4.6}$$

subject to

$$F(L,R) = M \ , \tag{4.7}$$

where M, P, and P_R denote total demand for labor input, annual labor cost per worker (wage), and annual cost per industrial robot, respectively. The equilibrium condition for (4.6) yields a well-known equation (Allen, 1967):

$$A(R/L)^{a-1} = (P_R/P_L) \ . \tag{4.8}$$

One can now estimate the parameters A and α by employing a least squares method. Based on these parameters, the impacts of industrial robots also are evaluated based on the following equations. Let L_R, E_R, B_R, and R_R denote labor force augmentation, equivalent workers per industrial robot, net profit yielded by one industrial robot, and net benefit rate of industrial robots, respectively. These are defined as follows:

$$L_R = F(L,R) - L \tag{4.9}$$

$$E_R = \{F(L,R) - L\}/R \ , \tag{4.10}$$

where R denotes the industrial robot population. Then,

$$B_R = P_L \times F(L,R) - (P_L L + P_R R) \ . \tag{4.11}$$

The first term on the right-hand side represents the labor cost to achieve the same labor force without using industrial robots. Finally, dividing equation (4.11) by R,

$$R_R = B_R/R \ . \tag{4.12}$$

Table 4.1. Data availability on industrial robot (IR) shipment statistics.

Item	1970–1973	1974–1977	1978–1983	1984–1985
Total IR production	A	A	A	A
IR production by type	NA	A	A	A
IR shp. by type and industry (in value)	NA	A	A	A
IR shp. by type and industry (in unit)	NA	NA	A	A

A = available; NA = not available.

4.2 Data Sources and Availability

Fortunately, the statistics on the shipments of industrial robots in Japan are available from JIRA (1984, 1985). Data availability on industrial robot penetration in Japan is summarized in *Table 4.1*. These data enable us to estimate the benefits from the macroeconomic point of view. It should be noted that no import statistics on them are available for Japan since most items are not yet distinguishable in the trade statistics code (SITC). Only export statistics on industrial robots are available.

The next step is to develop a price index for industrial robots. Since the capability and unit price are quite different among robot types, a divisia price index P (Jorgenson and Griliches, 1967) is appropriate, namely,

$$\frac{\dot{P}}{P} = \sum_{i=1}^{N} S_i \frac{\dot{P}_i}{P_i} \ , \tag{4.13}$$

where N, P_i, and S_i denote the number of different types, the price of the i-th type of robot, and the investment share of the i-th type of robot in the total robot investment. Unfortunately, the divisia index is not applicable before 1973 since industrial robot production data by robot type became available only from 1974 onward, as shown in *Table 4.1*. The average unit price for industrial robots was employed to calculate the index for pre-1974 data. Combining the two one can obtain a price index for industrial robots. This is exhibited in *Table 4.2* with estimated industrial robot population, capital stock of industrial robots, labor cost, and number of human workers.

The next problem is to estimate additional system costs consisting of peripheral equipment, operation training cost, engineering cost, and so on. These additional system costs depend on the type of industrial robot, and quite often exceed its original price (see Miller, 1983). JIRA (1984) reported

Table 4.2. Price index (1980=1), capital stock (in 1980 billion yen), and population of industrial robots (units); average labor cost per worker (in million yen) and number of workers (in thousands).

	Industrial robots			Human workers	
Year	Price index	Capital stock	Popu- lation	Labor cost per worker	Number of workers
1970	1.77857	6.61205	1,300	0.95474	10,933.00
1971	1.19354	15.25860	3,000	1.10263	10,950.70
1972	1.69316	23.90520	4,700	1.25622	11,039.30
1973	1.75534	36.62060	7,200	1.45976	12,020.50
1974	1.15822	63.82040	11,400	1.95296	11,283.00
1975	1.31297	88.70830	15,800	2.34680	10,153.10
1976	0.88558	128.55100	23,000	2.57803	10,212.20
1977	0.865553	180.71400	30,300	2.88791	10,017.40
1978	0.982675	241.29300	37,465	3.14260	9,704.05
1979	0.945306	349.88600	49,293	3.31669	9,889.37
1980	0.000000	531.90600	65,032	3.54410	10,054.20
1981	0.944806	755.90300	81,332	3.74214	10,303.60
1982	0.838704	1,056.52000	98,474	3.95995	10,218.40
1983	0.768663	1,421.91000	117,222	4.06842	10,542.70
1984	0.704833	1,960.29000	141,467	4.26936	10,719.80

Table 4.3. Ratio of initial system cost to the price of industrial robots.

Robot type	Price of industrial robot	Cost of peripheral equipment	Other cost (training, engineering)	Total
Manual manipulator	1.0	1.38	0.32	2.7
Fixed-sequence robot	1.0	2.29	0.31	3.6
Variable-sequence robot	1.0	0.94	0.06	2.0
Playback robot	1.0	0.81	0.19	2.0
NC robot	1.0	1.00	0.50	2.5
Intelligent robot	1.0	0.54	0.16	1.7
Average	1.0	1.13	0.27	2.4

Source: JIRA, 1984.

the average ratio of total initial investment for the robotization to the industrial robot price, *per se*, to be approximately 2.4 based on several hundred interviews. This is displayed in *Table 4.3*.

In practice, training and engineering costs tend to decrease in proportion to the penetration level, because of the learning effect among multiple users, as well as increased user friendliness. According to Miller (1983), the development cost for each successive application decreases by 10% for similar applications. In principle, when the above effect is taken into account in macroeconomic investigations, one must define the penetration level of industrial robots in one user by robot type and process type. For simplicity, the ratios in *Table 4.3* are assumed to be constant over time.

According to JIRA (1985), the average lifetime of industrial robots is about seven years. Based on the above data and assumptions, the capital stock of industrial robots can be estimated in monetary value (in 1980 billion yen). These data are exhibited in *Table 4.2* together with labor cost per worker and number of workers in the whole manufacturing industry extracted from *Yearbook on Labor Statistics* (Ministry of Labor, 1970–1984).

To estimate the parameters α and A through equation (4.5), we need an annual expense r for industrial robots. This is derived from the equation

$$P_R \frac{(d + \delta + \Theta)}{(1 - T_X)} = r P_{IR} \; , \tag{4.14}$$

where P_R, denotes the robot price, d is the depreciation rate, δ is the operation and maintenance expense rate l, Θ is the interest rate, and T_X is the real property tax rate.

Since the lifetime of an industrial robot is assumed to be seven years, the effective depreciation rate (d) is 14.3% (straight line). Net interest rates after 1970 ranged between 5% and 10%.

The real property tax rate on industrial robots *per se* is unknown. However, according to Noguchi (1985), the rate of local tax involving real property to the total *effective* corporate tax is 12.3% and the total effective tax rate was 51.6% in 1983, *where the effective tax rate involves both corporate tax on the profits (about 40% in 1983) and others on the real estate, capital stock, etc.* Since in 1983 gross output was 83,832 billion yen and capital stock of the whole manufacturing industry was 155,980 billion yen, the rate of local real estate tax to the capital stock was 3.4% in 1983.

JIRA (1984) has published maintenance costs per total initial investment for industrial robots – by robot type – based on interviews. Weighting in terms of 1984 stock value by robot type, the mean maintenance cost rate is calculated to be 4.5% as shown in *Table 4.4*. Other capital costs are not available in the JIRA reports. However, Unimation Inc. estimated "other operating costs" to be $0.80 per hour for a typical $50,000 industrial robot in 1980 (Ayres and Miller, 1983). Assuming a rather optimistic total of 6,000 operating hours per year, as claimed by Unimation, yields an estimate

Table 4.4. Annual maintenance cost as a percentage of total initial industrial robot investment, by robot type.

Robot type	Age of industrial robot (year)							Mean
	0	1	2	3	4	5	6	
Manual manipulators			NA					NA
Fixed-sequence robot	7.5	3.5	2.5	3.5	4.5	5.5	5.5	4.6
Variable-sequence robot	6.0	3.3	2.7	3.3	4.7	4.7	4.7	4.2
Playback robot	5.9	4.3	4.4	6.7	5.8	6.5	7.8	5.9
NC robot	4.2	3.5	2.5	2.3	3.5	4.0	4.3	3.5
Intelligent robot	5.0	4.0	2.0	2.0	3.0	4.0	4.0	3.4
Average, 1984 basis	5.4	3.7	3.2	4.2	4.5	5.3	5.6	4.5

Source: JIRA, 1984.

of 4% based on the total initial investment for an industrial robot. Summarizing, the total expense rate for industrial robots (r) appears to range between 29.0% and 34.1%. In our model, the benefits of industrial robots are evaluated for two cases: 30% (low case) and 35% (high case).

4.3 Evaluation of the Benefits of Industrial Robots

Based on the model and data described in Section 4.2, we can estimate the parameters of equation (4.7). If r is assumed to be constant, then P_{IR} can be substituted for the estimation P_R using equations (4.8) and (4.14), *viz.*,

$$(A/r) \times R/L)^{a-1} = (P_{IR}/P_L) \tag{4.15}$$

using the Cochran–Orcutt method for statistical estimation (Johnston, 1963).
The result is as follows:

$$\log(P_{IR}/P_L) = -0.411 \times \log(R/L) -2.543 \tag{4.16}$$
$$(8.18) \qquad (13.2)$$

$$R^2 = 0.837; \; R'^2 = 0.825; \; D.W. = 1.87; \; \rho = 0.399 \; .$$

Hence one obtains the numerical values

$$a = 0.589 \tag{4.17}$$

Table 4.5. Equivalent workers per unit and benefit rate of industrial robot (total manufacturing industry).

Year	Low case		High case	
	Equivalent workers per unit	Benefit rate (%)	Equivalent workers per unit	Benefit rate (%)
1970	4.305	27.6	5.017	31.9
1971	3.051	30.3	3.561	35.4
1972	2.542	12.0	2.968	14.1
1973	2.210	10.8	2.579	12.6
1974	1.887	31.1	2.201	36.2
1975	1.582	26.8	1.846	31.2
1976	1.355	35.9	1.582	42.0
1977	1.248	34.4	1.455	40.2
1978	1.181	28.1	1.378	32.8
1979	1.126	24.2	1.314	28.3
1980	1.100	17.7	1.283	20.6
1981	1.093	15.7	1.275	18.3
1982	1.096	15.3	1.279	17.9
1983	1.111	14.2	1.297	16.6
1984	1.120	13.4	1.308	15.6
1985	1.107	11.4	1.292	13.4

and

$$A = 0.079 \ . \tag{4.18}$$

It follows that

$$F(L, R) = (0.079 \times r \times R^{0.589} + L^{0.589})^{1.698} \ . \tag{4.19}$$

Using equation (4.19), one can now re-evaluate the equivalent labor force, $F(L, R)$, and labor force augmentation, LR. The results are shown in *Table 4.5*. The levels of equivalent workers per industrial robot E_R for r equal to 30% (low case) and r equal to 35% (high case) are shown in *Figure 4.1*. Discounting the gross benefits of industrial robots B_R by GDP deflator, the real gross benefits of industrial robots in 1980 prices are then calculated.

JIRA (1984) surveyed average labor reductions per shift for each industrial robot for 277 companies. JIRA (1985) also published average working hours of industrial robots by process type and the distribution of process type for each industry sector. Therefore, one can calculate the average shift number of industrial robots by industrial sector. The average labor reduction

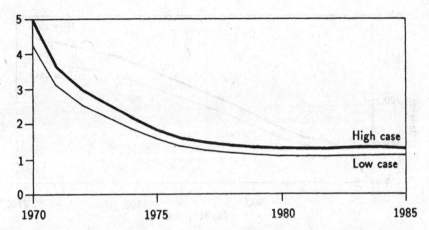

Figure 4.1. Equivalent workers per industrial robot unit: high case ($r=35\%$); low case ($r=30\%$).

Table 4.6. Comparison of equivalent workers per industrial robot unit between the high-low case method and JIRA's factory-level survey in 1984.

| Industry | | Estimated | | JIRA survey | |
		Low case	High case	Per unit and shift	Per unit
A	Total manufacturing	1.120	1.310	1.1	1.51
L	Fabricated metal	1.470	1.720	0.9	1.21
M	General machinery	0.784	0.914	0.9	1.20
N	Electric machinery	1.540	1.800	1.3	1.75
O	Transportation machinery	1.250	1.460	1.1	1.50
P	Precision machinery	0.714	0.834	1.0	1.46

per industrial robot is obtained by multiplying these two values. The results are shown in *Table 4.6*. Some interesting implications can be derived from a comparison of *Table 4.5* and *Table 4.6*. Although the equivalent number of workers per industrial robot depends on the annual expense rate r, the values corresponding to $r=30\%$ and $r=35\%$ in *Table 4.5* are consistent with the average labor reduction per robot after 1977.

It is often pointed out that in practice the capability of one industrial robot is basically equivalent to that of one worker per shift, although the robot can work longer hours and can therefore replace several workers in a multishift operation. This observation is supported independently by the

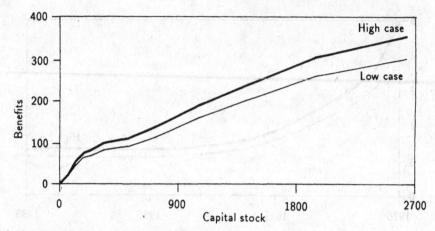

Figure 4.2. Trend of industrial robot benefits versus capital stock of industrial robots: high case (r=35%); low case (r=30%).

macroeconomic analysis presented here. The imputed capability of industrial robots in the early 1970s is probably exaggerated. The reasons may be the following:

- The population of industrial robots, especially the more basic types, may have been underestimated in the early 1970s.
- In the first stages of penetration, robots were substituted for workers in tasks where workers were least effective for various reasons. This point is clear from both the practical and engineering points of view.
- Since the procedure used to estimate the production function is not based on actual performance of industrial robots but on managerial assessments, it may well be concluded that these results partly reflect the "robot boom" atmosphere in Japanese industry in the early 1970s.

A historical relation between the capital stock of industrial robots and their net benefits is shown in *Figure 4.2*. One observes that, after 1979, the benefits of industrial robots increase almost linearly to the growth of their capital stock, i.e., the marginal effect of industrial robot investments is still diminishing in recent years, although the share of high-level robots has been increasing (see *Figure 4.3*). This observation suggests that the present generation of industrial robots has already penetrated its most favorable markets. This conclusion is consistent with independent results on robot diffusion by Tani (1991).

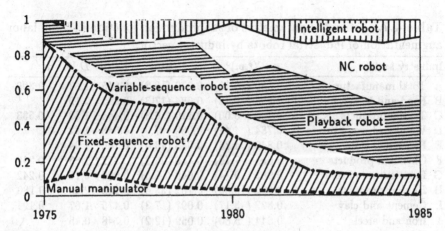

Figure 4.3. Historical shipment share of industrial robots by type.

4.4 Benefits of Industrial Robots in 16 Manufacturing Industries

In this section, the method described in Section 4.3 is applied to 16 manufacturing industry sectors, separately. The expense rate and lifetime of industrial robots are assumed to be the same as those indicated earlier. Unfortunately, statistics on industrial robot shipments by sector before 1974 are not available, therefore it is possible that the stock of industrial robots in the mid-1970s may have been underestimated. Shipment data in units are available only after 1978. Therefore a reasonably accurate estimate of the Japanese population of industrial robots by sector only became possible after 1983.

The estimated parameters of the labor augmentation subproduction function (4.5) are summarized in *Table 4.7*. In the case of sector G (the petroleum and coal industries), the estimators are not statistically significant. The results on equivalent workers per industrial robot unit in 1985 are exhibited in *Table 4.6*, compared with those of the factory-level survey given by JIRA (1984). One can observe that these values are roughly comparable. The equivalent workers per industrial robot unit and the average benefit rates to the capital stock of industrial robots in 1985 are summarized in *Table 4.8*. It may be noteworthy that the results on equivalent workers per industrial robot unit among industries show a different pattern from those of the net benefit rate.

Table 4.7. Estimated parameters of subproduction function (4.5) on labor augmentation of industrial robots by industry sector.

Industry sector	a	$(t.v.)^a$	A	$(t.V.)$	R^2	D.W.	ρ
A Total manufacturing	0.589	(8.18)	0.079	(13.2)	0.825	1.87	0.399
B Food and tobacco	0.805	(9.72)	0.084	(17.3)	0.895	2.10	b
C Textile products	0.760	(3.00)	0.110	(6.3)	0.444	1.32	0.533
D Wood and wood products	0.784	(4.41)	0.082	(7.9)	0.648	1.82	b
E Paper and pulp	0.849	(4.76)	0.083	(13.4)	0.706	1.94	b
F Chemical products	0.693	(7.79)	0.109	(24.9)	0.857	1.63	0.243
G Petroleum and coal	0.983	(0.14)	0.048	(6.3)	0.000	2.16	0.242
H Rubber products	0.852	(13.80)	0.101	(33.6)	0.955	2.40	−0.180
I Cement and clay	0.822	(3.17)	0.097	(7.3)	0.475	1.67	0.191
J Iron and steel	0.644	(5.80)	0.059	(12.2)	0.748	0.45	b
K Nonferrous metals	0.788	(4.00)	0.143	(15.1)	0.600	1.59	0.350
L Fabricated metal	0.739	(5.30)	0.113	(13.3)	0.731	1.50	0.407
M General machinery	0.717	(8.55)	0.086	(20.0)	0.878	1.78	0.252
N Electric machinery	0.779	(7.66)	0.156	(22.8)	0.852	1.76	0.209
O Transportation machinery	0.691	(5.69)	0.119	(17.6)	0.758	1.62	0.376
P Precision machinery	0.702	(5.12)	0.109	(12.0)	0.716	1.67	0.220
Q Other manufacturing	0.695	(16.40)	0.050	(26.1)	0.960	1.70	b

^{a}t-statistics.

^{b}Since D.W. statistics of OLS $yt = axt + b + \epsilon t$ indicated strong bias from auto correlation, the Cochran-Orcutt method

$$(yt - \rho yt - 1) = a(xt - \rho xt - 1) + b(1 - \rho) + (\epsilon t - \rho \epsilon t - 1)$$

is employed except for industry sector B, D, E, and Q. In case of J (iron and steel industries), D.W. statistics do not converge around 2.0.

One can observe that the effect of industrial robots on the primary metal industry is relatively higher than the effect on other industries. The reason may be partly that cheap low-level industrial robots (i.e., manual manipulators) have effectively substituted for workers in the casting and die-casting processes where labor costs and the share of jobs which have two to three shifts (70.4%) are high (JIRA, 1985).

In the case of sector F (the chemical products industry), the share of low-level industrial robots (fixed- and variable-sequence robots) is also high (about 90%) according to JIRA (1985). Here, the number of equivalent workers per industrial robot is low, yet the benefit rate is about average. According to JIRA (1985), industrial robots in the plastic-forming industry are mainly utilized as product extractors; these robots are inexpensive. For

Table 4.8. Equivalent workers per unit and benefit rate of industrial robot in 1985, by industry sector.

Industry sector		Equivalent workers per robot unit		Benefit rate of robot stock (%)	
		Low case	High case	Low case	High case
A	Total manufacturing	1.11	1.29	11.40	13.40
B	Food and tobacco	1.80	2.10	4.36	5.09
C	Textile products	2.44	2.84	12.20	14.20
D	Wood and wood products	2.88	3.36	5.66	6.60
E	Paper and pulp	1.11	1.30	2.67	3.11
F	Chemical products	0.30	0.36	11.40	13.30
G	Petroleum and coal[a]	—	—	—	—
H	Rubber products	1.18	1.37	5.12	5.98
I	Cement and clay	1.15	1.37	6.94	8.11
J	Iron and steel	2.08	2.43	22.70	26.50
K	Nonferrous Metals	1.72	2.01	10.70	12.50
L	Fabricated metal	1.37	1.60	9.90	11.60
M	General machinery	0.82	0.96	9.14	10.70
N	Electric machinery	1.65	1.93	6.63	7.76
O	Transportation machinery	1.19	1.39	12.30	14.30
P	Precision machinery	0.70	0.82	9.99	11.70
Q	Other manufacturing	1.26	1.47	9.81	11.40

[a]The parameters of labor augmentation function are not statistically significant.

instance, in 1985, 5,852 fixed-sequence robots were installed in this industry at a mean price of 1.27 million yen, each. By contrast, the average price of q fixed-sequence robot is 2.27 million yen. Therefore, it might be concluded that industrial robots in this sector are used mainly as a part of the process line rather than for direct labor substitution.

The number of equivalent workers per industrial robot in sector N (electrical machinery) is relatively higher than the number in other machine industry sectors, while its benefit rates are slightly lower than those in the other machinery sectors. This point may be compatible with the fact that relatively expensive high-level industrial robots (NC robots and "intelligent robots") are mainly used in this sector. The electrical machinery sector (N) presents quite a contrast to the case of iron and steel industries (J) and the chemical products industry (F).

One also can observe that the number of workers per industrial robot is relatively high in the light industry sectors. The share of high-level industrial robots in these industries is also high. For instance, more than 50% of total industrial robots implemented in 1984 in the food and tobacco industries (B) were playback robots.

4.5   Simulation Model for Evaluating the Impact of Labor Enhancement

Returning to the production function (4.1), a simplified formulation of (4.1) is employed as a preliminary study. Namely,

$$
\begin{aligned}
Y(K,L,R,x) &= K^{\alpha}F^{(1-\alpha)}\beta e^{ct} \\
&= K^{\alpha}(L^a + AR^a)^{(1-\alpha)/a}\beta e^{ct} ,
\end{aligned}
\tag{4.20}
$$

where c denotes the technological progress rate. It is assumed that c involves the effects of x, that is, inputs other than K, L, and R. Hereafter, the optimal strategy on K, L, and R is discussed under the scenarios of the total production of the industry, denoted by V, and the input prices P_K, P_L, and P_R.

A serious issue regarding new technology penetration levels might be the unemployment problem. Although the two studies cited in Sections 4.3 and 4.4 conclude that the total unemployment created is not serious, the problem with regard to labor market structural change may still be important. Therefore to see the robotization effects clearly, two employment policies are imposed in the simulations and then compared:

POLICY-1: All the inputs $(K, L$, and $R)$ are generated endogenously.
POLICY-2: L is exogenously given; K and R are generated endogenously.

It should be noted that the optimal input strategies are different between these policies. For the POLICY-1 simulation, optimal inputs on K, L, and R are determined through the following procedure:

$$
\min P_R R + P_K K + P_L L
\tag{4.21}
$$

subject to

$$
Y(K,L,R) = V .
\tag{4.22}
$$

The equilibrium conditions are

$$
(P_R/Pi_L) = A(R/L)^{(1-a)}
\tag{4.23}
$$

$$K = \{\alpha/(1-\alpha)\}(P_L L + P_R R)/P_K \ . \tag{4.24}$$

Hence one obtains optimal input ratios $(R/L)^*$ and $(K/L)^*$. Since the production function (4.1) is linear and homogeneous, one can calculate the optimum input K^*, L^*, and R^*.

The optimum input strategy in the case of POLICY-2 is derived as follows:

$$\min P_K K + P_R R \tag{4.25}$$

subject to

$$Y(K, L, R) = V(L : \text{given}) \ . \tag{4.26}$$

It should be noted that the production function (4.1) is not a linear homogeneous function of K and R when L is exogenously given. The equilibrium conditions of these equations yield the following equation of R:

$$P_R = A \times P_K (V/\beta)^{(1/\alpha)} e^{-ct/\alpha} \{L^\alpha + A \times R^\alpha\}^{-1-\{((1-\alpha)/\alpha\}/\alpha} \times 1-\alpha)/\alpha \ . \tag{4.27}$$

Employing numerical methods, one can obtain the optimum input of industrial robots R^{**} under the POLICY-2 simulation. Inserting it into (4.1), K^{**} is also calculated.

The price of the output products P_Y is defined by

$$P_Y = (P_R R + P_K K + P_L L)/Y(K, L, R) \ , \tag{4.28}$$

which provides basic information for the evaluation of the robotization policies. The next step is to determine the parameters α, β, and c. The simulations on A (total manufacturing industry), N (electric machinery industry), and O (transportation machinery industry) are described in Section 4.6.

The price of the capital services P_K (Ayres and Miller, 1983) is defined by

$$P_K = \frac{d+\Theta}{1-TAX} \times P_I \ . \tag{4.29}$$

Effective tax rate, TAX, and depreciation rate, d, are extracted from Japan Economic Planning Agency, National Accounts (1970–1985) as well as total capital stock $K0$. K is defined by

$$K = K_\Theta - R \ . \tag{4.30}$$

To simplify, Θ is fixed at 5% (which is the low case described in Section 4.2) in the simulations in Section 4.6.

Table 4.9. The estimated parameters α, β, and c of the production functions by industry.

Industry sector	α	(S.D.)	c	(t.v.)	β	(t.v.)	R^2	D.W.
A Total manufacturing	0.270	0.0409	0.0396	10.40	0.840	22.80	0.884	1.77
N Electrical machinery	0.241	0.0420	0.1740	17.20	0.599	5.75	0.952	0.88[a]
O Transportation machinery	0.260	0.0338	0.0296	5.41	2.660	20.30	0.653	0.85[a]

[a]Since the Cochran-Orcutt method did not give revised result, OLS estimators are adopted for the simulations.

One can now estimate α from equation (4.5) and the historical statistics on K, L, R, P_L, P_R, and P_K for the period 1970 to 1985; β and c are estimated from

$$\beta\, e^{ct} = (V/u)/\{K^{\alpha}(L^a + AR^a)^{(1-\alpha)/a}\}^n \; , \qquad (4.31)$$

where u denotes capacity utilization rate. The variables V and u are extracted from Japan Economic Planning Agency, National Accounts (1970–1985) and MITI Statistics (1970–1985), respectively. The estimated parameters are summarized in *Table 4.9*.

4.6   Simulation Results for Three Manufacturing Industries

The next step is to determine the exogenous scenarios. Since the price of the capital stock P_K was almost constant after 1980, it is assumed to be constant throughout the period. Although the price of industrial robots declines about 7% per year after 1980 (*Table 4.2*), JIRA (1985) points out that this trend will not continue in the future; therefore, our model assumes that the price of industrial robots P_R is constant after 1985.

The average growth rates of production in real terms and of wages in nominal terms are calculated from 1980 to 1985 and are shown in *Table 4.10*. The growth rate of wages is assumed to be constant in the future. It may be too optimistic to assume that the economic and technological growth rates that occurred in the first half of 1980s will continue in the future. As a result of these considerations, the two cases are simulated and compared for each employment policy.

Table 4.10. Average growth rate of production in real terms and average growth rate of wages in nominal terms from 1980 to 1985.

Industry sector		Economic growth rate (real)	Growth rate of wages (nominal)
A	Total manufacturing	5.85%	4.7%
N	Electrical machinery	21.20%	3.7%
O	Transportation machinery	5.93%	5.3%

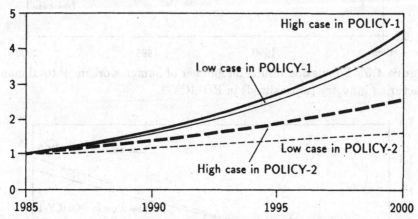

Figure 4.4. Simulation results on industrial robot capital stock in total manufacturing industry (normalized).

- High-level scenario: The economic and technological growth rates in the first half of the 1980s last throughout the simulation period.
- Low-level scenario: The economic and technological growth rates in the first half of 1980s fall by 50% after 1986.

For the POLICY-2 simulation, labor input (L) is fixed at the 1985 value. The simulation results are shown in *Figures 4.4* through *4.12*. In the case of POLICY-1 one can observe that the industrial robot stock of sector A (total manufacturing industry) ranges between 2.56 (low case) and 2.68 (high case) in 1995 (*Figure 4.4*). This is quite compatible with the forecast by JIRA (1985), i.e., between 407,000 units and 537,000 units. The number of human workers under POLICY-1 declines to between 90.1% (high case) and 84.3% (low case) by the year 2000 (*Figure 4.5*). This may not be an exaggerated value.

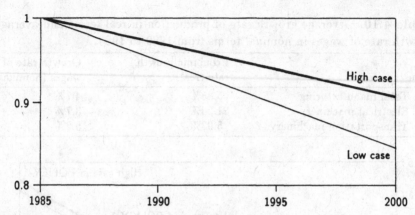

Figure 4.5. Simulation results on number of human workers in total manufacturing industry (normalized) in POLICY-1.

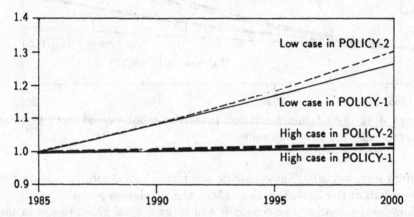

Figure 4.6. Simulation results on output price index in total manufacturing industry.

In the case of sector N (electrical machinery industry) in POLICY-1, the capital stock of industrial robots ranges between 4.61 (low case) and 4.83 (high case) in 1995, while the JIRA (1985) estimates it to be 3.01 (*Figure 4.7*). The reason for this difference might be as follows: since JIRA's forecast is an extrapolation of managers' assessments on the potential substitutability of industrial robots, the effects of economic growth might be underestimated.

In the case of sector O (transportation machinery industry) in POLICY-1, the capital stock of industrial robots ranges between 5.11 (low case) and 5.86 (high case), which may be rather exaggerated compared with that

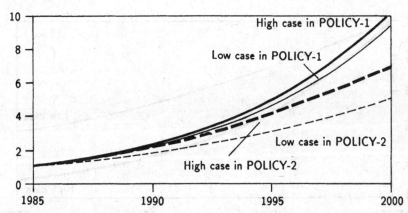

Figure 4.7. Simulation results on industrial robot capital stock in electric machinery industry (normalized).

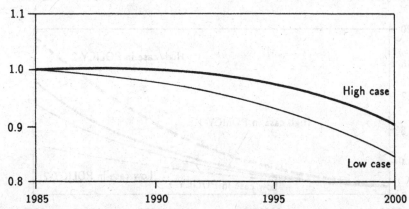

Figure 4.8. Simulation results on number of human workers in electric machinery industry (normalized) in POLICY-1.

of JIRA's forecast of 2.19 in 1995 (*Figure 4.10*). Since the technological progress rate c is relatively low while the growth rate of wages is relatively high, it is necessary to conclude that labor-saving strategies may be undertaken.

The output price paths of POLICY-2 simulations are almost the same as those of POLICY-1 simulations for all cases. For instance, the output price for total manufacturing industry (*Figure 4.6*) of POLICY-2 in the year 2000 is at most 3% (0.2% per year) higher than that of POLICY-1. In other cases, the difference is negligible. Therefore it could be concluded

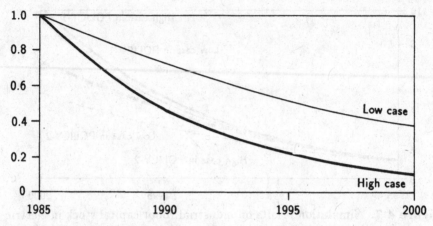

Figure 4.9. Simulation results on output price index in electric machinery industry. The results in POLICY-1 and POLICY-2 are almost identical.

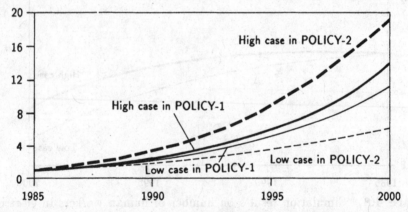

Figure 4.10. Simulation results on industrial robot capital stock in transportation machinery industry (normalized).

that, although the optimum robotization strategy may affect about 10% of human workers, unemployment problems may not be serious because the economic inefficiency of full employment policies is relatively small.

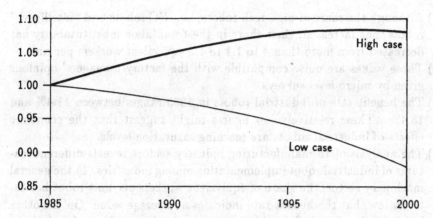

Figure 4.11. Simulation results on number of human workers in transportation machinery industry (normalized) in POLICY-1.

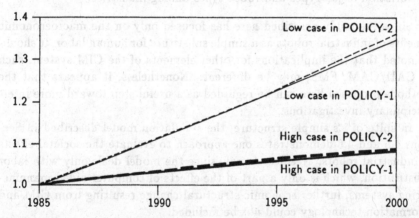

Figure 4.12. Simulation results on output price index in transportation machinery industry.

4.7   Conclusions

In this chapter, I proposed a production-function approach to evaluate the macroeconomic effects of industrial robots in the Japanese manufacturing sector. Although the formulation of the model is quite simple, it has provided results that are compatible with factory-level survey data. The main findings of Sections 4.1–4.4 are the following:

(1) Although the share of high-tech robots, i.e., NC robots and "intelligent" robots, has increased, their share in the total labor substitutability has decreased from more than 4 to 1.1 to 1.3 equivalent workers per unit.

(2) These values are quite compatible with the factory managers' opinions given by micro-level surveys.

(3) The benefit rate of industrial robots in 1985 ranges between 11.4% and 13.4%. These relatively low figures might suggest that the economic effects of industrial robots are reaching saturation levels.

(4) The analysis of 16 manufacturing industry sectors reveals different features of industrial robot implementation among industries. In the general machinery sector, the share of equivalent workers per unit robot is relatively low, but the benefit rate indicates an average value. On the other hand, in the electrical machinery sector, the former factor is relatively high, but the economic benefit rate is low. These differences reflect the variation of job types and robot types among industries.

Since the study described here has focused only on the macroeconomic benefits of industrial robots as a simple substitute for human labor, it should be noted that the implications for other elements of the CIM system, such as CAD/CAM, FMS, may be different. Nonetheless, it appears that the methods proposed here can be regarded as a useful step toward more inter-disciplinary investigations.

In spite of its simple structure, the simulation model described in Sections 4.5 and 4.6 demonstrates one approach to evaluate the societal effects of industrial robots. Needless to say, since the model deals only with labor substitution, which is only a part of the effects of computer aided manufacturing systems, further economic structural change resulting from CIM and information technology could not be included.

Although it is recognized that the economic and societal impacts of new computer aided manufacturing systems may be quite large and important, their quantitative effects have not been evaluated empirically. This is because not only the basic statistics are not yet well established, but also their effects are often qualitative and sometimes quite ambiguous. In other words, the incentive for the penetration of these technologies are not for economic reasons alone. Therefore, it should be emphasized that when one wants to discuss social and economic impacts not only studies from an engineering point of view but also investigations on micro-level surveys, macro-level statistics, and wider sociological aspects are needed.

References

Allen, R.G.D., 1967, *Macro-Economic Theory*, Macmillan, London, UK.

Ayres, R.U., and J. Funk, 1987, *The Economic Benefits of Computer-Integrated Manufacturing*, WP-87-39, IIASA, Laxenburg, Austria.

Ayres, R.U., and S.M. Miller, 1983, *Robotics: Applications and Social Implications*, Ballinger Publishing Company, Cambridge, MA.

"Flexibility in Manufacturing Systems," 1986, *OMEGA International Journal of Management Science* 14(6):465–473.

Jaikumar, R., 1984, *Flexible Manufacturing Systems: A: Managerial Perspective*, Harvard Business School Working Paper 1-784-078, Cambridge, MA.

Japan Economic Planning Agency, National Accounts, 1970–1985.

JIRA (Japan Industrial Robot Association), 1984, Research Report on the Economic Effects Analysis of Industrial Robots Implementation, June, Tokyo, Japan.

JIRA, 1985, Long Range Forecasting of Demand for Industrial Robots in Manufacturing Sector, June, Tokyo, Japan.

Johnston, J., 1963, *Econometric Methods*, McGraw-Hill, New York, NY.

Jorgenson, D.W., and Z. Griliches, 1967, "The Explanation of Productivity Change," *Review of Economic Studies* 34(99):249–283.

Kaya, Y., 1986, *Economic Impacts of High Technology*, CP-86-8, IIASA, Laxenburg, Austria.

Leontief, W., and F. Duchin, 1985, *The Future Impact of Automation on Workers*, Oxford University Press, Oxford, UK.

Miller, S.M., 1983, *Potential Impacts of Robotics on Manufacturing Costs within the Metalworking Industries*, PhD Thesis, Carnegie-Mellon University, Pittsburgh, PA.

Ministry of Labor, 1970–1984, *Yearbook of Labor Statistics*, Tokyo, Japan.

MITI (Ministry of International Trade and Industry), 1970–1985, *Yearbook of International Trade and Industry Statistics*, Tokyo, Japan.

MITI, 1985, *The Recent Trends of Productivity Improvement Technologies*, Tokyo, Japan.

Noguchi, Y., 1985, "Tax Charge for Japanese Companies," *Contemporary Economics* 61:48–64.

Tani, A., 1991, "The Diffusion of Robots," in R. Ayres, W. Haywood, and I. Tchijov, eds., *Computer Integrated Manufacturing*, Volume III: *Models, Case Studies, and Forecasts of Diffusion*, Chapman and Hall, London, UK.

UNECE, 1985, *The Diffusion of Electronics Technology in the Capital Goods Sector in the Industrialized Countries*, United Nations, Geneva, Switzerland.

Chapter 5

Management, Jobs, and Employment

*Karl-H. Ebel**

5.1 Management Approaches to CIM

Management attitudes on how to cope with CIM vary. In part they mirror the national idiosyncrasies and different industrial backgrounds. Thus the technocentric approach results in management strategies that neglect or underrate the human factor in production. This tendency is frequently aggravated by short-term profit considerations, which are major stumbling blocks to sound technology planning and management. The introduction of CIM requires long-term strategic thinking. From the managerial point of view it is essentially an organizational quandary: how to maintain order out of potential chaos. Equipment needs to be carefully selected and compatibility ensured. However, the fundamental problem is how to reshape existing production processes, alter organizational boundaries, and make them permeable. This requires redesigning the information and data flow to foster decentralized decision making. The difficulties of actually accomplishing this

---

*The views expressed are the author's and do not necessarily reflect those of the International Labour Office.

Table 5.1. Characteristics of Taylorist and non-Taylorist systems.

Taylorist systems	Non-Taylorist systems
Centralization of control and decision making	Decentralization of control and decision making
Operator told how work is to be carried out	Operator told how work is carried out
Standardization of working methods	Operator develops own working methods
One best way of doing work	Many good ways of doing work
Functional specialization	Multiskilled operator
Simplification of work	Variety of complex tasks
Imposing methods and solutions on people	Consultation and participation

Source: Kidd and Corbett, 1988.

in existing organizations, to make them more effective and efficient, should not be underestimated.

It must not be forgotten that the relatively slow productivity growth in manufacturing observed in recent years has much to do with haphazard computerization and system incompatibilities within organizations. A down-to-earth comment by Bessant expresses the problem well: "When you put a computer into a chaotic factory the only thing you get is computerized chaos" (Bessant and Rush, 1988).

At the same time, the introduction of CIM may well act as an antidote to poor management practices. This is where the human factor comes in. Industrial case studies and experience accumulated so far clearly indicate that a pragmatic management approach – which advances step by step, builds up the skill, responsibility, and motivation of the work force, invests in people operating the systems, and relies on the human factor to make it flexible – has consistently paid off.

This implies going beyond Taylor's vision of "a system in which the workman is told in minute detail just what he is to do and how he is to do it, with any improvements which he makes upon the orders given to him being seen as fatal to success" (Taylor, 1907). The principal characteristics of non-Taylorist as opposed to Taylorist systems are given in *Table 5.1*.

The conviction that this is really the best course seems to be lacking in many management circles; otherwise a more systematic and consistent effort would be made to enhance the human factor. At any rate, it has been found that, in general, CIM is not introduced primarily to "humanize"

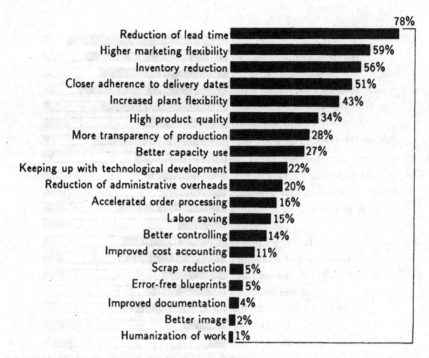

Figure 5.1. Frequency of basic enterprise objectives, in percent (each expert specified five objectives). Source: Köhl *et al.*, 1989.

work. The motives and expectations of management mostly have to do with stock reduction, greater transparency of the organization, reduction of lead time, closer adherence to deadlines, reduction of personnel, greater marketing flexibility, increased capacity use, higher-product quality, or the need to keep up with technological developments – all leading to cost reduction and higher productivity. An expert survey conducted in the Federal Republic of Germany in 1987 is unequivocal on this point, as shown in *Figures 5.1* and *5.2* (Köhl *et al.*, 1989).

Improved working conditions thus tend to be an incidental by-product of CIM, and to assume a very low priority. In fact, working conditions may even be neglected or grow worse, particularly where automated machinery is used to accelerate the pace of work, where only residual tasks are entrusted to workers, or where the new tasks created by computerization involve a high degree of stress. In this context it is significant that in the United States there is evidence of a considerable increase in repetitive strain injury

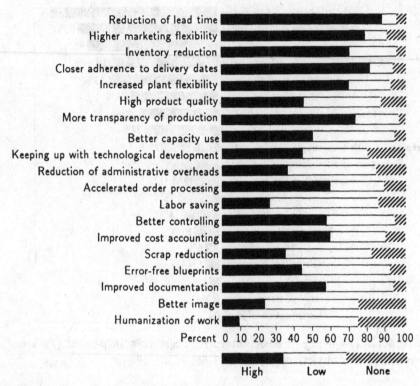

Figure 5.2. Influence of CIM technology on reaching objectives, in percent. Source: Köhl *et al.*, 1989, p. 123.

attributed to poor job design. It is, for instance, alleged that about half of the members of the United Autoworkers Union, (some 500,000 workers) suffer from such health disorders. Keyboard operators tend to be affected more than those in other occupations (*AFL-CIO News*, 1989b). This may well indicate that, instead of alleviating physical strain, automation has increased the work pace of residual tasks by machines. If confirmed by further research, this would certainly constitute a rather perverse result of technological advancement.

5.2   A New Management Style

A management style that allows production personnel greater autonomy and restores initiative to the shop floor may well mean a break with entrenched principles and thus be seen as a threat to vested interests and the power structure in an organization. Although it is not surprising that as a rule everything is done to avoid such clashes, it is, in fact, possible to switch to new technology without making fundamental organizational changes, and to keep established divisions and top-heavy hierarchies in place. Information technologies may also be used to institutionalize and even reinforce ineffective and counterproductive management practices such as excessive centralization of decision making or abusive monitoring of individuals. This is, of course, costly and leads to mediocre results while prolonging the life of organizational dinosaurs. It defeats the primary purpose of CIM, namely, the integration of all functions, which requires "vertical and horizontal synchronization of departments, people, machinery, and processes in the flow of information and material" (Warnecke, 1987). In such a system the necessary flexibility can be achieved through decentralization of information and responsibility within a given framework to achieve small and fast control loops. This enables the production system to respond rapidly to market demand, particularly in the case of multiple product options.

The effective introduction of CIM requires a clear strategy that has the backing of both senior management and the rank and file. Obviously nothing much can be done without the consistent support of top management; however, the stumbling block may prove to be middle managers, who often dread the destabilizing effect of advanced technology more than skilled workers and technicians since they stand to lose part of their influence when established hierarchies are dismantled and all needed information is available directly "on-line" to all participants in the production process. This emphasizes the need for top-level technology management, a function frequently neglected because legal, financial, and marketing aspects tend to dominate decision making at the top. It is not enough to let middle management acquire new technology and then resort to crisis management at the top when bottlenecks occur in the organization or when the middle managers lack the necessary skills for handling the technology. It is, of course, not enough for top management to announce a CIM strategy; this can be implemented only at the lower levels of the hierarchy with the commitment of everybody involved. Such commitment must be assured through consultation.

The implementation of such a strategy is by no means easy and much depends on the specific national, cultural, and industrial relations and the labor market in which enterprises operate. It will succeed only if the work force is adequately prepared and willing to cooperate. Little will be gained from thrusting an unwanted autonomy and a participative structure ordained by top management on a reluctant work force unable to perceive the advantages of new forms of organization and to recognize its own interest in adopting a new approach. Indeed, many workers may find it convenient to rely on detailed instructions without assuming much responsibility and may refuse to accept "multiskilling." Taylorism relies on this attitude, which has no place in CIM systems. However, as it is widespread many efforts to introduce autonomous group work may be doomed to failure.

Some commentators have traced existing problems to the lack of managerial competence (see Hodson and Hagan). Managers' knowledge of advanced manufacturing systems is frequently limited, even when they have received technical education or are engineers. Owing to the advances in manufacturing technology, especially information technology, the professional knowledge and experience that managers have acquired rapidly become obsolete unless they are continually exposed to shop-floor experience and the latest techniques. Consequently, potential users of automation equipment, often fearing that they will not be able to muster the updated know-how needed for operating the equipment, rely on outside consultants and equipment suppliers. They naturally tread carefully in unknown territory and avoid incalculable risks.

Managers are under pressure to justify the high expenditure required for implementing CIM. By the standards of a short-term return-on-investment (ROI) approach, the financial feasibility of most CIM projects is doubtful despite the hypothetical long-term economic advantages outlined above. In fact, there are no generally agreed upon methods for making reliable cost-benefit analyses of CIM, and the cost of full-scale CIM implementation is often considered to be prohibitive, especially when the cost of tailoring the system to the enterprise's specific uses is added to the cost of the equipment. It is feared that CIM will be inefficient to use and expensive to maintain because technical change constantly requires the replacement of parts of the system, a particularly difficult task in an integrated system. In present flexible manufacturing systems fixed costs constitute about 70% of the total outlay. This is one indication of the high risks that management takes when installing CIM.

It also must be considered that the implementation of major organizational changes is not without costs. They are, in fact, substantial and are said to exceed the costs of building the system, i.e., the acquisition of hardware and development of the corresponding software, by two to three times; they also have a tendency to grow (Roberts and Hickling, 1989).

There is, of course, the other side of the coin: risks are balanced by opportunities if the expected economic benefits of CIM materialize. Moreover, in the future the capital outlay is bound to decrease as cheaper systems come on the market. This will make CIM technologies more attractive to small- and medium-sized enterprises. It also has been estimated that CIM plants could break even at 30% to 35% of capacity utilization as against 65% to 70% in the case of conventional plants. In addition, the planning of CIM and at least partial implementation could help management to improve the organization of the production process and the flow of communication. A leaner management will be able to cope better with essential planning and control tasks and to devote more time to functional tasks such as product development and delivery and customer service. This will enhance productivity and speed up reaction to market demand.

The most promising prospect of all is probably that CIM technologies will enable management to dissociate factory operation time from actual working time through a wide range of flexible working arrangements. This will considerably enhance the utilization of fixed capital and increase the return on high investments, thus diminishing risks and making enterprises more competitive. It also is likely to ease recruitment problems, particularly among parents who will be able to choose work schedules to fit in with their family responsibilities.

To implement CIM, management's task is to overcome hierarchical rigidities and organizational resistance to change, from the shop floor up through all layers of the organization. To streamline an organization and make it fit for CIM may be a considerable challenge but may be well worth the effort. The findings of several surveys concur on this point; manufacturers who have introduced advanced manufacturing systems attribute between 40% and 70% of the total improvement achieved to better logistics and organizational changes. In other words, the main benefit does not necessarily stem from the sophisticated and integrated technology itself but from the reform of management and production practices and from a more transparent and efficient organization (Haywood and Bessant, 1987).

5.3   Skill Requirements for CIM

The foregoing observations imply that there should be a rise in the level
of skills of shop-floor workers or that different skills will be needed, despite
the fact that some of their present skills will become obsolete. In particular, trends toward the division of labor will be reversed. CIM requires
versatile artisans and technicians, hardware and software experts, mechanical and communications engineers, and, in general, people who understand
production methods and systems and are capable of handling a great deal of
technical information and making decisions on the spot. These requirements
go far beyond simple machine-minding, since only qualified people can ensure
maximum utilization of the costly equipment. There is little room for unskilled workers such as assemblers, laborers, machine loaders, and transport
workers. CIM also renders redundant clerical workers engaged in ordering
parts and materials and scheduling the work load of machines.

Specific skill requirements and training needs depend to a large extent on
the organizational changes triggered by CIM and the particular CIM configuration and network strategy chosen by individual enterprises. There will be
new working procedures, task structures, and work content, as well as new
relationships among different technical functions. It is possible that in some
instances skill requirements will be reduced, when equipment components
embodying information technology simplify or facilitate tasks (for example,
maintenance of electronics equipment by replacing components such as electronic cards instead of difficult repairs on site). On the whole, however,
troubleshooting and maintenance work tend to become more complex because of the integration of electronic, electrical, mechanical, pneumatic, and
hydraulic functions.

Computer aided design (CAD) is an important component of CIM. It
speeds up the design and production of drawings and enables the performance of operations that would be impossible without a computer such as
the rapid production of design variations, direct inclusion of calculations,
simulations of functions, creation of very complex designs (e.g., computer
chips), and direct transmission of machining data. It also expedites routine
work such as detail drawing, information searches, calculations (finite element analysis), and the establishment of lists of workpieces and control work.
As regards the new skill requirements for designers and drafting engineers,
a comparative ILO study concluded:

> The main new skill requirements for staff working with CAD systems appear
> to be "computer literacy" and higher mathematical and analytical skills,

especially a good understanding of the principles of analytical geometry and the application of coordinate systems, together with an open mind and a high degree of accuracy and attention to detail. Traditional draftsman skills nevertheless appear to retain their validity and importance – at least for the time being – since the main tasks remain unchanged. Because they are relieved of much routine work draftsmen and designers should have more time for creative work, but in practice existing software and rules and macros stored in data banks may seriously limit their options. (Ebel and Ulrich, 1987)

In the Federal Republic of Germany a detailed study of a large FMS points out the extent to which maintenance skills (especially the newer ones such as systems analysis and diagnostics) contribute to the utilization of advanced manufacturing systems. An analysis of more than 6,000 hours of operation revealed that more than half of the system downtime was due to unscheduled stoppages or breakdowns. Of the time taken to repair and bring the system back into operation, about half was taken up in diagnosis. The conclusion was that:

> The more complex and automated the systems were, the higher the skills level of the maintenance specialists had to be to achieve reasonable failure rates and implement facility improvements ... and ... the lower the personnel levels were (producing with automated facilities), the broader the educational background of these workers (operators and maintenance) had to be. (Handke, 1982)

Middle-level managerial jobs also are bound to decrease or diminish in CIM systems because of the general dissemination and free flow of information. There tend to be fewer hierarchical levels and demarcation lines and fewer coordinating tasks. The emphasis is on planning, creativity, anticipating problems, communications, teamwork, and interaction, and much less on giving instructions. Excessive monitoring of workers (which is technically possible) is best avoided because it can antagonize the very people needed to run the systems. The new role of team leader requires from managers a subtle combination of human, conceptual, and technical skills (Crocker and Guelker, 1988).

The implementation of CIM requires continuous learning and alertness at all levels. Since CIM is essentially based on microelectronics and information technology, it therefore demands much theoretical understanding, abstract thinking, and comprehension of work methods. The development of cognitive skills, as well as analytical, problem-solving, and logical capabilities, has a high priority; such qualifications are in scarce supply. In addition,

organizational, economic, social, and communication skills are called for as
an increasing proportion of resource management is carried out on the shop
floor.

Higher skill requirements and greater versatility – and finally more in-
tellectually demanding and interesting jobs – will be the rule in CIM installations, particularly for the core workers who keep the systems operating; it
is also true that the skill content of jobs depends largely on the particular
category of enterprise, the production process, and the work organization
chosen. For some workers it might appear to be quite a different proposi-
tion. Certain optimistic views and generalizations about multiskilling, job
enlargement, and job rotation need to be taken with a pinch of salt. When
certain "skills" can be acquired in just a few days, multiskilling may turn
out to be a euphemism for work intensification. To quote an automobile
worker: "The jobs are just the same as before, you just do more of them"
(Turnbull, 1988).

5.4   New Types of Work Organization

Computer integrated manufacturing enables management to integrate functions and tasks and to create new types of jobs. It makes job enlargement and
enrichment a realistic possibility through broadened work content, better information, and more decision-making power. However, there is a caveat. The
combination of functions must make sense and must not overburden people
operating the system if acceptance problems are not to hinder the realization
of the full potential of CIM technologies (Esser and Kemmner, 1989).

As work processes are progressively integrated and many specialized jobs
are abolished because of the reduced division of labor, work organization
tends toward the group pattern. Relatively autonomous groups are organized
and their members execute complementary tasks; they must be versatile
enough to handle a variety of jobs to keep the system running smoothly, and
they must have the ability to cooperate and communicate beyond narrow
technical boundaries. They also are given a certain amount of autonomy
in the choice of tasks and in planning their work. In this way existing
qualifications can be used more efficiently and mutual coaching takes place.
Such teamwork, if properly organized, results in greater job satisfaction.

There is a further argument for increasing the autonomy of workers at
all levels: making CIM work requires initiative, alertness, and considerable

creativity from all concerned. This can flourish only when people have sufficient autonomy in their work to apply their skills, knowledge, and talents to challenging tasks and projects.

However, workers also need time to get used to increased responsibility and autonomy, which some may perceive as a mixed blessing. Unclear or incomplete instructions can be frustrating and may make people insecure. It should be obvious that increased responsibility cannot be sprung on workers without adequate preparation.

There clearly is room for innovative work organization; but, in the real world of manufacturing, including enterprises introducing CIM technologies, little progress in new forms of work organization is evident. The inertia of existing organizational patterns appears to make the restructuring of organizations the most difficult part of CIM implementation. Divesting technical offices, for instance, of control functions and giving them to the shop floor usually goes against the grain of the established power structure (Köhler *et al.*, 1989).

Participative work organization is, therefore, by no means an automatic outcome of introducing CIM. Management must consciously seek to overcome outdated, unmotivating, and unsuitable hierarchical forms of organization, and this means shedding old power relationships – often a painful process fraught with pitfalls. Those who have a vested interest in maintaining the status quo are often in a strong position. In this context the results of a recent sample survey carried out in the Federal Republic of Germany deserve attention. Most enterprises continued to try to remove planning and control functions from the shop floor through automation, concentrating them instead in the technical offices and at the supervisory level. Only a minority of enterprises attempted new organizational forms aimed at making systematic use of empirical shop-floor knowledge and competence by adapting the level of automation and reducing the division of labor (Schultz-Wild *et al.*, 1989).

However, it should be some comfort to managers deciding to base CIM on functional integration using human skills that this type of production is usually less capital intensive than full-scale CIM since it is less computerized and requires less expensive software; the fact that many decisions are made on the shop floor also helps to make it flexible and to avoid machine downtime. As production-planning, programming, machine-setting, and maintenance tasks are assumed by the group, process continuity is ensured. Moreover, existing qualifications of the work force can normally be used and few new ones are required. There is also a reduction of throughput time (Brödner, 1982).

Concern has been expressed that CIM will lead to the social isolation of the relatively few workers remaining on the shop floor to mind the system. The fact that more and more communication takes place through computer terminals, diminishing opportunities for social contact, may well have an adverse effect on individuals and the working atmosphere in plants. Ultimately such dissatisfaction could have negative consequences on the production process overall. System designers need to keep this aspect in mind and provide opportunities for social contact as a contribution to the quality of working life.

Since CIM technology in industry is in its infancy, much of what can be said about its impact on work organization and working conditions must necessarily remain conjecture. However, the opinion of experts is a valuable indicator of future trends. A survey conducted by experts in the Federal Republic of Germany in 1987 shows an interesting pattern of changes following the introduction of CIM technologies (*Figure 5.3*).

5.5   Hazards in the New Working Environment

Although the new job requirements in CIM systems are gradually becoming better known, there is still much uncertainty about the new occupational safety and health hazards they may pose. It stands to reason that physical risks are diminished because fewer workers are in direct contact with production equipment, and most production takes place without direct human intervention. However, the pace of work is usually faster and the amount of shift work and overtime greater; all these factors tend to increase fatigue and the risk of accidents. It has been found that work at computer terminals can be very stressful, particularly in the case of CAD, and it is not unusual for designers or drafting engineers to work six to eight hours a day at CAD work stations. Similar work intensity has been observed in other computerized occupations. The computer, which was designed to facilitate work, ends up exhausting its users. In the long term this can lead to negative health effects.

The findings of a comparative International Labour Office study on CAD applications are instructive. In general, CAD users experienced high job satisfaction because their work was more challenging than traditional drafting. Nevertheless, the report cautions:

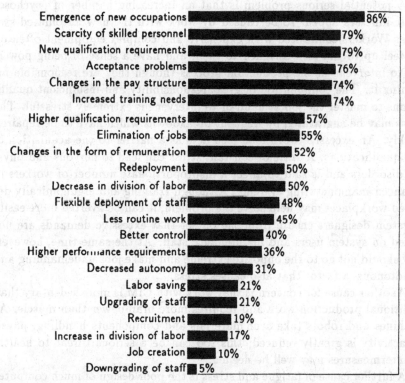

Figure 5.3. Changes in work organization and working conditions, in percent. Source: Köhl *et al.*, 1989, p. 131.

CAD users have also remarked that screen work has something of a "hypnotic" effect which leads to mental exhaustion. Some say that they feel completely drained after working for long hours with a system that exerts a kind of "horrible fascination" as the dialogue with the computer drives the operator on at a fast pace.

Display terminal work can result in eyestrain and back and neck problems. However, the extent of such complaints depends on whether ergonomic factors were taken into account in the design of the equipment and on the amount of time spent at the terminals. In this respect much progress has been made in recent years, and screen displays, particularly in color, now tend to cause less eyestrain. Modern work stations are usually well designed and easy to operate. (Ebel and Ulrich, 1987)

A potential serious problem is that an increasing number of psychosomatic illnesses appear to be caused by work with the new automated systems. Workers confronted with expensive and complex equipment often do not feel up to the task assigned to them and have a sense of being powerless to intervene in the production process though they are responsible for running it. The combination of great responsibility and insufficient qualifications to master the job at hand or to intervene is extremely stressful. The stress may be aggravated by frequent breakdowns which have to be repaired quickly. An excessive work load also acts as a barrier to the acquisition of new qualifications. A state of constant stress can lead to nervous and physical disorders and is said to affect a disproportionate number of workers in advanced manufacturing systems. Although training and ergonomically designed workplaces may help solve the problem, it can be averted more easily if system designers ensure from the outset that excessive demands are not placed on system users and maintenance staff. At the same time, however, they should not go to the opposite extreme and make jobs undemanding and monotonous, a factor that also causes stress.

Another cause for concern is that work with CIM is more sedentary than traditional production work and requires more brainpower than muscle. As machines and robots take over materials and components handling, physical activity is greatly reduced; this too can be a serious threat to health. Countermeasures may well be needed.

A further cause of fatigue and stress is the poor design of much computer software. Much of it is not user friendly and is ill adapted to actual workplace requirements. This can make the interaction between workers and machines very cumbersome. "Cognitive" ergonomics addresses these problems. However, it is a relatively new science; improvements in software design that take into account research findings are only gradually forthcoming.

Another area of continuing concern is the impact of industrial robotics, an important component of CIM, on safety and health. There is wide agreement supported by case studies and accident statistics that the use of robots has significant beneficial effects as it eliminates much unpleasant, monotonous, repetitive, fatiguing, dangerous, and heavy work. In particular, robots can be used in such environmental conditions as cold, heat, intense light, darkness, radiation, vibration, noise, toxic emissions, fire hazards, isolation of workers, and inaccessibility of work stations. Nevertheless, like all machinery, robots can have harmful effects unless handled with the necessary caution. Sources of hazards include errors in the software of the control system and electrical interference from the mains or from radiation. The electrical, hydraulic, or pneumatic controls may malfunction, causing unexpected results. It appears that most recorded accidents with robots have occurred during maintenance and repair work.

The ILO has recently reviewed the basic principles and practices in the safe use of robots applied in various countries (ILO, 1989). The review contains a catalog of recommended measures and technical standards referring, among other things, to guarding, trip devices, reliability of control systems, diagnosis of malfunctions, maintenance rules, robot operation, and safety training and organization.

Despite some overall improvements in the physical working environment, there is one highly automated industry in which physical health hazards have become a cause for concern: the semiconductor industry. Its initial reputation as a "clean" industry has faded. It operates, in fact, with a multitude of cancer-causing, mutagenic, and highly toxic chemicals to which it is feared that workers are exposed on an increasing scale because of negligence, insufficient protection, lack of safety training, and deficiencies in hazard control. Reports have brought to light a high incidence of headaches, respiratory diseases, nausea, skin rashes, allergic reactions, and miscarriages among electronics workers. Moreover, the handling of microscopic components tends to lead to eye damage (*AFL-CIO News*, 1989a; Gassert, 1985).

Finally, even in today's most advanced manufacturing systems many operations cannot be automated. This is the case particularly of assembly tasks, where the work pace is frequently dictated by machines. Workers in such jobs are exposed to cumulative trauma disorders (CTD), i.e., soft-tissue injuries in tendons, muscles, and nerves caused by repetitive motions which may lead to disabilities.

System designers, who usually have an exclusively technical or scientific background, tend to overlook such considerations when planning the installations. However, it is then that preventive measures have to be taken and environmental and ecological concerns have to be addressed. The increase this implies in planning and investment costs is insignificant compared with the cost of rectifying ergonomic and environmental mistakes once a system is installed (Wobbe, 1987).

The high work intensity and the continuous pressure resulting from an ever-increasing volume and complexity of work in information technology jobs have often meant that employees are obliged to do without breaks and to work overtime; the solution would, of course, be adequate staffing. Enterprises ought to have a strong interest in preserving the capacity and efficiency of their work force over the long term by refusing to tolerate conditions of work that are detrimental to physical and mental health.

The principal objective here should be the creation of humane working conditions, for only workers who are treated first and foremost as responsible human beings will be prepared to commit themselves to company goals. A definition of humane work that is apposite to the new technology is the following:

> Work is called humane if it does not damage the psycho-physical health of the worker, does not ... impair his psycho-social well-being, meets his requirements and qualifications, allows him to exercise individual and/or collective control over working conditions and systems of work, and is able to contribute to the development of his personality in activating his potential and furthering his competences. (Martin *et al.*, 1987)

5.6   Preparation of the Work Force for CIM

If people are the key to successful CIM, obviously much hinges on their preparation for the new systems. In all industrialized countries there is a shortage of professional, technical, and managerial personnel able and qualified to mastermind the implementation of CIM. This shortage may well be aggravated in the future by reductions in working time, a major demand of the metal trade unions in most industrialized countries. The lack of adequate computer hardware or software is not the only major constraint; at the shop-floor level the necessary skills also are mostly in short supply. The recruitment difficulties experienced by enterprises, despite much unemployment and the high initial salaries paid to capable young engineers and technicians in this field, are a case in point. This skill shortage may well explain many of the failures reported earlier. Often management does not appear to have a clear idea of where it is going, and workers' representatives are seldom aware of the intricacies and possible social consequences of CIM. There is clearly a case for taking particular care in recruiting capable and motivated staff and for making production work attractive to them.

CIM technologies are new and unfamiliar to the work force and management alike, and training institutions often lag behind the latest developments. However, this is not the only reason for the skill shortage. There is a good deal of evidence that employers have paid too little attention to further training of their work force in new information technology skills. In the United States, for instance, it was found that in 84% of plants using programmable machines the employers provided no systematic operator training for the new functions. Only machine vendors organized some training for their equipment (Kelley and Brooks, 1988).

Such reluctance of employers to impart costly training in such skills was particularly noticeable during economic recessions. Even in the Federal Republic of Germany, with its extensive further training network, it was observed that so far further training in CIM technologies was ad hoc and haphazard, particularly in small- and medium-sized enterprises engaged in small-batch production, which constitute about 90% of all metalworking firms. Only recently have efforts been undertaken to systematize training in this field and to determine specific qualification requirements and training needs. An interesting example is the new occupation, "CAM organizer," which has just been recognized by authorities. A curriculum, comprising 436 hours of practical and theoretical upgrading instruction in information technology subjects for qualified CNC machinists, including examination regulations, has been approved (Fraunhofer-Institut für Arbeitswirtschaft und Organisation, 1989).

There is no easy solution to the problem. But one way of tackling it is through systematic training and further training of the work force based on a strategy specifically designed for the purpose and endorsed by management and workers' representatives. Such training is required before and during the installation of new equipment and should emphasize not only specialized technical competence, including computer literacy, but above all system knowledge, planning, organizational and communication skills, and group dynamics. Ideally, there should be a symbiosis of system and product design and training. Such an approach would also serve to familiarize the work force with the new equipment and thus help to overcome anxiety. This training needs to be carried out mainly by the enterprises themselves in cooperation with system suppliers, since CIM systems are tailor-made to the specific requirements of enterprises and training institutes rarely have the necessary expertise in leading-edge technology. If and when flexible automation becomes more generalized, it may well become necessary to reform training curricula at all levels and make existing training infrastructures more responsive to changing occupational requirements.

Another path lies in the widest possible use of expert advice at the planning stage of CIM and an open discussion of choices among all those concerned – including the workers' representatives who far too often find themselves faced with a *fait accompli*. A thorough discussion of the economic, technical, organizational, and personnel requirements and of the objectives

of a proposed innovation would facilitate an informed assessment of the social consequences and the negotiation of working conditions and training
obligations. Both management and the work force are usually moving into
uncharted territory during the process of CIM introduction and ought to
recognize the fact.

Suppliers of technology and consulting firms specializing in system integration can play a key role in assisting enterprises in the introduction of CIM.
Their contribution tends to be all the more crucial if the enterprise is small
and inexperienced. Firms may need help with financial and strategic planning, feasibility studies, system planning, human resources development, and
project management (the coordination of different suppliers). Whereas many
of these services can of course be contracted, overdependence on suppliers,
such as computer firms, software houses, machine vendors, and management
and engineering consultants, has its drawbacks. It is in the interest of CIM
users to remain in control and build in-house expertise.

Additional in-plant training for CIM, administered to a highly qualified
work force, is claimed to be the most effective way of bridging skill gaps.
At the same time, it must be recognized that the cost of such training can
be prohibitive, particularly for small- and medium-sized enterprises. There
is, therefore, a need for collaboration with industry, employers' and workers'
organizations, governments, professional societies, and research and training
institutions to ensure the funding and the organization of such training. This
is clearly the price that has to be paid for a more rapid diffusion of CIM
technologies.

5.7   The Impact of CIM on Industrial Relations

In the real world the transition to CIM systems, even when it is well planned
and prepared, will rarely be accomplished without creating tension and conflict (whether open or concealed) in organizations. The work force has good
reason to be worried since evidence shows that its interests may not be taken
sufficiently into account or may simply be neglected. Far too often technology is placed before people to cope with as they can, without having been
properly trained to handle it or given a say in its choice. Responsibility is
taken away from the workers, and "de-skilling" occurs; the work is robbed
of its meaning and becomes boring. Small wonder that systems fail.

Workers fear pay losses as a result of reduced overtime, redundancy,
and fewer promotion prospects; they also fear lower staffing levels, machine

pacing and intensification of work, expropriation of know-how through data bases and expert systems, higher stress because of responsibility for expensive capital goods, more intensive shift work, individual performance monitoring by the computer system, and de-skilling. They are apprehensive about the unknown, and concerned about having to adjust to new working patterns, including possible redeployment and the loss of rights. These fears exist, but by applying a strategy that puts people first and seeks genuine consultation at all levels such fears can be overcome.

The positive aspects of introducing advanced systems and the new opportunities they offer will more readily be accepted; these include safer and less physically taxing jobs, enhanced learning and training opportunities, greater responsibility and more interesting assignments, better remuneration, more flexible working hours, generally better working conditions, or greater job security in a more competitive enterprise. The use of new systems can help to overcome the mismatch between personal aspirations and actual work requirements, bring down the rate of absenteeism, and enhance productivity without undue intensification of work. Innovations can, in fact, exert a beneficial influence on industrial relations if more emphasis is placed on consultation at all levels and less play is given to "the arrogant expertise of technologists."

In this respect the Informatics Society of the Federal Republic of Germany observes:

> The machine perspective ignores the fact that the working person possesses strong, albeit tacit work motives whose fulfillment he or she looks on as a sort of existential necessity: the demand for self-reliant expertise and an unexposed scope of action, and endeavors to achieve an informational advantage (particularly over superiors). Every sort of preprogrammed work is apt to jeopardize fulfillment of precisely these motives and to produce resistance or evasive action. This is the essence of the acceptance problem. (IFAC, 1989)

The positive aspects can carry the day only in an atmosphere of social dialogue and goodwill at the enterprise level. Adversarial industrial relations, coupled with arbitrary exercise of management power, could easily spell the failure of CIM projects. Their success presupposes reconciling the interests of management and workers and introducing flexibility in work rules, redeployment measures, and adequate training. Dialogue between the social partners is thus essential for achieving product and process innovation and higher productivity and flexibility in manufacturing.

Unfortunately, the technocentric approach to CIM is widespread and in fact entails an attempt by employers to reassert their control over the work force. On the one hand, most managers resent workers' control over production systems and see it as an encroachment on their prerogatives, particularly in an adversarial climate of industrial relations. On the other hand, unions are often skeptical about greater involvement of workers in the enterprise and distrust managerial strategies encouraging identification with the company. In their view greater responsibility of workers for the production process blurs the lines between "them" and "us" and threatens union identity and solidarity (Turnbull, 1988).

There is indeed evidence that in an adverse climate of industrial relations management tends to resort to an excessive division of labor as a means of restricting the influence of trade unions. In such circumstances management tends to override union jurisdiction and avoids entrusting blue-collar workers with more autonomy and control (e.g., the programming of NC tools) to circumvent rules and constraints imposed by collective-bargaining agreements. In this situation unionization – whenever it leads to restrictive and inflexible work rules and job control, such as manning requirements and seniority rules enforced by the unions – has a perverse effect; it inhibits skill acquisition and upgrading by blue-collar workers (Kelley, 1989). The blue-collar unions are in any case in a difficult situation because CIM technology is bound to accelerate the decline of their influence as a result of membership losses caused by the falling proportion of unskilled and semiskilled workers in advanced manufacturing systems.

However, the greatest threat to workers' autonomy and to job satisfaction stems from centralized control, accompanied by unmotivating rigidity and formal procedures in the production process. Decentralized systems coupled with a maximum of decision making on the shop floor are well suited to small-batch or customized production, tend to enrich jobs and qualifications, reduce machine downtime through flexibility in work assignments, better scheduling, and improved maintenance, and, therefore, enhance productivity. They often prove to be economically superior to rigidly centralized systems with an excessive division of labor (Manske, 1984). These factors should be taken into account when collective agreements are drawn up.

Consensus and cooperation are indeed necessary if CIM systems are to work smoothly, though these requirements do not exclude a resolute defense of the workers' rights and interests. It must not be forgotten that highly skilled workers and technicians and their representatives in integrated manufacturing are in a strong position and cannot easily be replaced. Enterprises installing CIM depend on the quality and commitment of their work force; qualified personnel are needed to maintain the complex and costly equipment and to keep the system working. Moreover, advanced manufacturing systems are vulnerable to industrial actions such as strikes, by a small proportion of their work force; responsible management will therefore be well advised to seek the social dialogue and collective agreements needed to provide a proper framework for their operation. An unorganized work force kept in check by management prerogatives and arbitrariness, subdued by authoritarian supervision and antiunion policies, could easily jeopardize the success of CIM. Industrial relations based on mutual confidence and respect will be far more conducive to success (Kador, 1988).

A telling example of the vulnerability of modern integrated plants to industrial action was reported by *Business Week* (1986). The highly robotized Waterloo tractor plant of John Deere and Company was paralyzed for a considerable time by a dispute with the United Autoworkers' Union which claimed redundancy protection in a new collective agreement. At the Waterloo plant it was impossible to shut down one part of the operation without bringing the rest of the plant to a standstill. This induced the company to rearrange the plant into smaller and more flexible "islands of automation."

The possible undermining of control systems and management objectives by a discontented work force in the advanced technological environment has been described in these terms:

> All [management control systems] have proved susceptible to what might be termed worker subversion. Once subordinates have come to understand the operation of the control system, they may, through superior operating understanding and sheer ingenuity, find the way to weaken its force and even, as in the "capture" of payment-by-results systems, turn it against managerial objectives.... If computerization promises to provide management with a wealth of control-relevant information about the performance of subordinates, there may be little to guarantee that the information itself will be correct. A little distortion goes a long way in the interdependent sets of data stored by computers. Employees may falsify the information which they key in, whether to improve their own compensation or to ease the burden of hierarchical control.... Computer utilization induces some people, and particularly creative people whose jobs underuse their skills, to play with the system. Computerized systems may be even more readily undermined and sabotaged by disgruntled employees than is the assembly line, which was supposed to chain the worker to the job but which proved intermittently vulnerable to sabotage and shutdowns. (Ryan, 1987)

However, it should also be borne in mind that as hierarchical structures change and as middle management is threatened by CIM the role of unions and workers' representatives in enterprises may be weakened. Autonomous groups of highly qualified staff may be able to exert a more direct influence on the determination of their working conditions and thus may feel less need for union representation and intermediaries in their dealings with management. As the minority of core workers come to depend less on union intermediaries, the unions will find it increasingly difficult to organize the underprivileged workers on the periphery such as less-skilled part-time workers, employees in small subcontracting firms, and others in precarious employment conditions. With widespread introduction of CIM technologies, the position of the unions can be expected to weaken.

Since unions tend to recognize the need for enterprises to be competitive, they have generally raised little opposition to technological change, and have often shown a positive attitude. Nevertheless, there have been situations when conflicts among unions arose because CIM technologies cut across narrow traditional craft demarcation lines and challenged established work practices. Such conflicts were not conducive to the adoption of advanced technologies. In the past decade many unions have been involved in technology bargaining and new technology agreements have often eased the transition to advanced technology, as the basis of an understanding between management and workers' representatives. However, such bargaining is often hampered by an understandable lack of technical expertise on the part of union negotiators. Unions have an interest in training their representatives and building sufficient expertise in their ranks to deal with the new social issues raised by the introduction of CIM; some have, in fact, issued guidelines for the conclusion of such agreements to assist their rank and file in negotiations with management.

Given the flexible nature of CIM technologies, there will need to be more recourse to decentralized plant bargaining since many issues are raised that can only be resolved at the plant level. In enterprises introducing CIM, the social dialogue obviously cannot be limited to questions of remuneration and benefits. At any rate, payment-by-results systems may well have to be redesigned since the success of the system and higher productivity depend essentially on the built-in quality-assurance system and reduction of downtime of the automated equipment, and not on the output of individual machine operators. Higher skills and system improvements must be rewarded. The dialogue will have to address issues related to the implementation of the new technology, such as more flexible working-time arrangements, adjusting working conditions to teamwork, developing no-redundancy clauses, establishing operating levels, skill upgrading, and considering plans in the event of plant closures.

Since technology agreements could ease the tensions associated with the introduction of CIM, it is useful to recall the main content of such agreements. The ILO has found that the following issues were usually covered:

- Advance notice and information disclosure to workers or their representatives.
- A requirement for consultation and negotiation on any or all of the aspects of introducing new technology, including the establishment of procedures for such consultation or negotiation.
- Access to outside expertise by workers and their representatives, or by joint committees, to help them participate in the procedures leading to the introduction of new technology, including consultation or negotiation.
- Training for workers' representatives.
- Training or retraining for workers, especially those whose jobs are eliminated or changed.
- Proposals for work organization and job design negotiations and for discussion on job content and job satisfaction.
- Monitoring of workers' performance with new technology.
- Negotiation over job grading and evaluation procedures, where jobs are created or changed by the introduction of new technology.
- The sharing of benefits arising from the use of new technology, such as higher pay or shorter working hours.
- Provision of rest pauses and maximum hours for workers using visual display units or other equipment.
- Negotiation on working time and related issues, such as reduction or rearrangement of weekly working hours or shift work associated with the introduction of new technology.
- Occupational safety and health and ergonomics.

5.8 The Impact on Employment

There is encouraging evidence that by and large the aggregate level of employment in industrial societies is not greatly affected by the introduction of new technologies. The long-term trend of declining manufacturing employment observed in industrialized countries is certainly continuing, owing partly to technological innovations that eliminate unskilled work. It has, for instance, been found that FMS forming part of CIM saves from 50% to 75% of direct labor when compared with conventional systems (IIASA, forthcoming). On the whole, however, job displacement and redeployment of workers in the course of innovation and rationalization appear to balance each other. Where technological change goes along with strong economic growth and expansion of markets and investment, technology even tends to induce positive employment effects through the revitalization of the economy. Japan's technology drive and growth pattern is a case in point. However, it would certainly be vain to pin exaggerated hopes on the introduction of CIM and other high technology and their spin-offs as an employment creation device. "Reindustrialization through high technology" is a misleading concept (Wobbe, 1987; Ebel and Ulrich, 1987b).

Only a very small proportion of the labor force of highly industrialized countries – some 2% to 5% – are engaged in this advanced sector; if past trends and experience are any guide, the proportion will rise only slowly, if at all. Large enterprises can fairly easily muster the human and material resources required for CIM; however, the rather slow diffusion of advanced technology, particularly for small- and medium-sized enterprises, which in industrial societies employ the majority of workers, prevents the effects from being more far-reaching. It was found, for instance, that in 1987 in the United States three out of four machine operators' jobs had not been directly affected by computerized automation (Kelley and Brooks, 1988). In this context it is instructive to look at the distribution of the labor force by level of automation and mechanization in the Federal Republic of Germany (*Figure 5.4*). In 1985–1986 only about 7% of the labor force (1.5 million employees) used programmable means of production such as data-processing equipment, PCs, NC/CNC machines, industrial robots, and computerized medical equipment, while more than 50% continued to use simple tools. The previous survey was conducted in 1979, and since then only a slight increase in the proportion of employees using programmable equipment had been observed. Advanced technology was obviously affecting workers only gradually.

It is clear that most manufacturing workers will not experience radical change as a result of CIM systems in the near future. Nevertheless, it must be expected that CIM will accentuate the already existing labor-market segmentation. A core of highly skilled workers and qualified technicians, engineers, and professional workers manage and operate such systems and are increasingly indispensable. They are generally well paid, and their working conditions are stable thanks to their position as knowledgeable workers

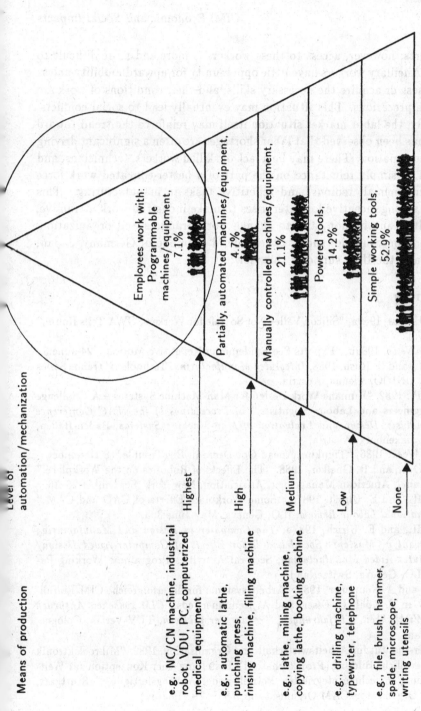

Figure 5.4. Use of means of production by employees in the Federal Republic of Germany in 1985–1986, by level of automation and mechanization. Source: *Mitteilungen aus der Arbeitsmarkt- und Berufsforschung*, 1988, p. 28.

and experts; however, access to these workers is more and more difficult to achieve. Ancillary workers have little opportunity for upward mobility unless they possess or acquire the necessary skills, and their conditions of work are also more precarious. This situation may eventually lead to social conflicts.

Finally, the labor market situation itself may reinforce the trend toward CIM. It has been observed that labor shortages are often a significant driving force of automation. There may be a lack of skilled workers, technicians, and engineers or simply reluctance on the part of a better-educated work force to perform menial, tedious, and repetitive tasks in manufacturing. This acts as a strong incentive for enterprises to make industrial work attractive, to introduce automation, and to change work practices and organization. Evidence from countries such as the Federal Republic of Germany, Japan, Sweden, and the USA supports this view.

References

AFL-CIO News, 1989a, "Silicon Valley Not So Safe for Workers, CWA Tells House," 8 July:3.

AFL-CIO News, 1989b, "Experts See Epidemic in Repetitive Motion," 24 June:3.

Bessant, J., and H. Rush, 1988, *Integrated Manufacturing*, Technology Trends Series No. 8, UNIDO, Vienna, Austria.

Brödner, P., 1982, "Humane Work Design for Man-Machine Systems – A Challenge to Engineers and Labour Scientists," in *Proceedings of the IFAC Conference on Analysis, Design and Evaluation of Man-Machine Systems*, Baden-Baden, IFAC, Laxenburg, Austria.

Business Week, 1986, "Thinking Ahead Got Deere in Big Trouble," 8 December.

Crocker, O.L., and R. Guelker, 1988, "The Effects of Robotics on the Workplace," *Personnel*, American Management Association, New York, September:26–36.

Ebel, K.-H., and E. Ulrich, 1987a, "Some Workplace Effects of CAD and CAM," *International Labour Review*, ILO, Geneva, May-June:365.

Ebel, K.-H., and E. Ulrich, 1987b, *The Computer in Design and Manufacturing – Servant or Master? Social and Labor Effects of Computer-Aided Design/Computer-Aided Manufacturing*, Sectoral Activities Programme Working Paper, ILO, Geneva, Switzerland.

Esser, U., and A. Kemmner, 1989, "Arbeitssysteme für die erfolgreiche CIM-Einführung," in E. Köhl, U. Esser, and A. Kemmner, eds., *CIM zwischen Anspruch und Wirklichkeit – Erfahrungen, Trends, Perspektiven*, TUV-Verlag, Cologne, Germany.

Fraunhofer-Institut für Arbeitswirtschaft und Organisation, 1989, "Mikroelektronik und berufliche Bildung (Projektphase II): Erstellung einer Konzeption zur Weiterbildung in rechnerintegrierten Betrieben der Auftragsfertigung," Stuttgart, CAM Organisator/CAM Organisatorin, *IBV*, **29** (19 July).

Gassert, T., 1985, *Health Hazards in Electronics: A Handbook*, Asia Monitor Resource Centre, Hong Kong.

Handke, G., 1982, "Design and Use of Flexible Manufacturing Systems," in *Proceedings of the Second International Conference on Flexible Manufacturing Systems*, IFS Publications, Kempston, Bedfordshire, UK.

Haywood, B., and J. Bessant, 1987, *The Integration of Production Processes at Firm Level*, Mimeograph, Brighton Polytechnic, Brighton, Sussex, UK.

Hodson, R., and J. Hagan, "Skills and Job Commitment in High Technology Industries in the US," in *New Technology, Work and Employment*, Oxford University Press, Oxford, UK.

IFAC (International Federation of Automatic Control), 1989, "Committee on Social Effects of Automation: Computers and Responsibility," *INFO Pack*, No. 5, August:3.

IIASA, forthcoming, CIM study, Laxenburg, Austria.

ILO, 1989, *Safety in the Use of Industrial Robots*, Occupational Safety and Health Series No. 60, Geneva, Switzerland.

Kador, F.J., 1988, "Das Soziale in High-tech-Unternehmen," *Der Arbeitgeber*, 3(40):94–95.

Kelley, M.R., 1989, "Unionization and Job Design Under Programmable Automation," in *Industrial Relations*, Spring, Berkeley, CA.

Kelley, M.R., and H. Brooks, 1988, *The State of Computerized Automation in US Manufacturing*, Mimeograph, Carnegie-Mellon University, Pittsburgh, PA.

Kidd, P.T., and J.M. Corbett, 1988, "Towards the Joint Social and Technical Design of Advanced Manufacturing Systems," *International Journal of Industrial Ergonomics* 2:307.

Köhl, E., U. Esser, and A. Kemmner, 1989, *CIM zwischen Anspruch und Wirklichkeit – Erfahrungen, Trends, Perspektiven*, TUV-Verlag, Cologne, Germany.

Köhler, C. *et al.*, 1989, "Alternativen der Gestaltung von Arbeits- und Personalstrukturen bei rechnerintegrierter Fertigung," in Kernforschungszentrum Karlsruhe: *Strategische Optionen der Organisations- und Personalentwicklung bei CIM*, Forschungsbericht KfK-PFT 148, August:61–118, Karlsruhe, Germany.

Manske, F., 1984, "Social and Economic Aspects of Alternative Computer-Aided Production Systems in Small and Medium Batch Runs," in *IFAC: Design of work in automated manufacturing systems*, Proceedings of IFAC Workshop, Karlsruhe, Germany, 7–9 November 1983, Pergamon Press, Oxford, UK.

Martin, T., E. Ulrich, and H.-J. Warnecke, 1987, "Appropriate Automation for Flexible Manufacture," in R. Isermann, ed., *Tenth World Congress on Automatic Control*, Munich, July, Vol. 5, IFAC, Laxenburg, Austria.

Mitteilungen aus der Arbeitsmarkt- und Berufsforschung, 1988, Nuremberg, No. 1:28.

Roberts, R., and A. Hickling, 1989, "Computer Integrated Management?" *Multinational Business* 2:18–25.

Ryan, P., 1987, "New Technology and Human Resources," in G. Eliasson and P. Ryan, *The Human Factor in Economic and Technological Change*, Part II, OECD Educational Monographs No. 3, OECD, Paris, France.

Schultz-Wild, R., *et al.*, 1989, *An der Schwelle zu CIM*, RKW-Verlag, Verlag TUV/Rheinland, Cologne, Germany.

Taylor, F.W., 1907, "On the Art of Cutting Metals," *Transactions of the American Society of Mechanical Engineers*, 28:31–350.

Turnbull, P.J., 1988, "The Limits to 'Japanisation' – Just-in-time Labour Relations and the UK Automotive Industry," *New Technology, Work and Employment*, 3(1):7–20.

Warnecke, H.-J, 1987, CIM in Europe, Unpublished manuscript, Stuttgart, Germany.

Wobbe, W., 1987, "Technology, Work and Employment: New Trends in the Structural Change of Society," in *Vocational Training Bulletin* (European Centre for the Development of Vocational Training CEDEFPOP), 1:3-6.

Chapter 6

CIM and Employment: Labor Substitutability

Robert U. Ayres and Shunsuke Mori

6.1 Preliminary Remarks

To estimate the influence of technological progress on the occupational structure of employment, it is necessary to summarize the diversity of workplaces into groups that are comparatively influenced by technological progress. For this purpose it is helpful to define similar tasks or occupations. Tasks are generally more descriptive of the actual work content of a job (Warnken, 1986). On the other hand, the subdivision of the labor force by occupation has the advantage that it establishes a direct connection with educational planning.

To estimate the influence of the technological progress on the level and the structure of employment and to infer the consequences for education, it would be very useful to have data on the occupational composition by sectors and by tasks as well as the task composition by sectors. Such a detailed data base is – to our knowledge – available only for the Federal Republic of Germany (*Figure 6.1*). However relationships between tasks and occupations are likely to be reasonably similar in countries of a comparable level of economic development. Task-by-sector matrices are available only

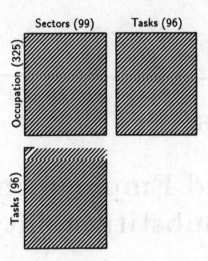

Figure 6.1. System of labor matrices in the Federal Republic of Germany.

for a few countries. Furthermore, occupation-by-sector matrices are available for many more countries.

The following indicators have to be considered to estimate the impact of CIM on employment by occupations.

(1) The fractional share of the workers in a certain occupation potentially affected by the application of a certain CIM technology (e.g., robotics or CAD).
(2) The fractional share of workers actually displaced.
(3) The resulting increase of labor productivity attributable to this technology.

Data about the replacement potential of certain CIM technologies by different occupations and sectors can best be determined on the basis of engineering analysis. This approach is illustrated in Section 6.2. Data on the number of machine tools in use, by category and by type of control, are collected every five years by the *American Machinist* for each metalworking sector (SIC 33–38). The thirteenth survey was published in 1983 and the fourteenth in 1990.

Miller (1983) classified all machine tools into four categories, as shown in *Table 6.1*. A detailed allocation is given in the Appendix to this chapter. He also estimated the percentage of all machine tools in the USA that could, in principle, be operated by level I robots (1982 technology) and by level II

Table 6.1. Low and high estimates of the distribution of metal-cutting machine tools by category.

Category	Percent of machines	
	Low estimate	High estimate
1. Machines designed for low-volume production	39.4	68.2
2. Machines designed for fully automatic operation	12.0	19.9
3. Machines designed for very large or heavy workpieces	1.1	1.7
4. Machines designed for medium- to large-batch production	9.4	46.7

Table 6.2. Estimates of the percent of metal-cutting machine tools that could be operated by level I and level II robots.

Machine types	Percent in metalworking industries
Category 4	9.4
Categories 4 and 2	6.3
Subtotal	15.7 – Maximum for level I robot
Categories 4 and 1 and Categories 4 and 3	37.3
Total, Category 4 (exclusively and jointly)	46.7 – Maximum for level II robot

robots (1990s technology), *Table 6.2*. Combining the results in graphic form yields *Figure 6.2*. This suggests that the upper limit for numerical control (and robotic operation) is about 48% of the existing machine tool population, which would also be about the upper limit of machine operator displacement. This compares well with an earlier industry survey – admittedly limited in scope – carried out at Carnegie-Mellon University (Ayres and Miller, 1983) which suggested that respondents thought that 39.5% of operatives could be replaced by a level II robot (but only 13.6% could be replaced by a level I robot).

The above results can be regarded as a crude sort of validation for the survey methodology. A more far-reaching survey (474 respondents) was carried out in 1984 by the Japan Industrial Robot Association (JIRA, 1985).

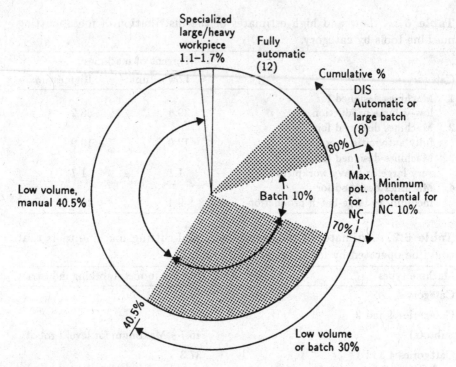

Figure 6.2. Classification of machine tools by use and control. Source: Miller, 1973.

The JIRA study focused on the number of workers replaceable by industrial robots by tasks and by sectors. Based on this, the potential labor displacement matrix for the whole Japanese manufacturing industry can be estimated. It must be noted that the JIRA survey covers only a small part of Japanese industry, although it is much more comprehensive than the Ayres and Miller survey. JIRA results for Japan are summarized in *Table 6.3* (columns 1 and 2). Assuming JIRA's substitutability data to be similar to that of the US manufacturing industry, the potential for labor substitutability of the USA is also estimated (*Table 6.3*, columns 3 and 4). We finally compare the results with the estimates by Ayres and Miller (remaining columns).

The classification of occupations in the JIRA survey is rather different from the survey of Ayres and Miller (1983); nevertheless, it can be concluded that the results are basically consistent. It is noteworthy that the potential displacement ratio estimated from JIRA's survey is roughly within the range for level I and level II robots given by Ayres and Miller (1983).

Another approach taken in the JIRA survey also deserves discussion. The 474 respondents were asked (in effect) how much they would be willing to pay in capital costs to reduce the total number of workers by one. This can be interpreted as the marginal capital value of one robot system per worker replaced. Data are presented in *Figure 6.3* for various tasks in terms of ratios between the average marginal capital value of a single replaced worker (as perceived by managers or entrepreneurs) and the average cost of a robot. It is noteworthy that for most tasks the ratio is greater than unity, implying that, *ceteris paribus*, robots were economically justified in Japan (1984) if they could displace only a single worker. In most cases, the observed displacement ratio is closer to one worker per shift, or two workers per robot.

It is already clear that not all workers are substitutable, even for the most routine tasks. Thus, the marginal willingness-to-pay data presented in *Figure 6.3* might be regarded also as a measure of distance from equilibrium. If all justifiable robots were actually in place, the theoretical ratio should be 0.5 ± 0.1. A high ratio suggests that the potential for substitution is much higher than the current level of penetration. Conversely, a low ratio suggests a very low potential for substitution. This procedure allows one to get an "impression" of the range of labor substitutability due to CIM. This procedure is not necessary if detailed engineering surveys about the potential labor substitutability by sectors and occupations becomes available and more careful computations can be made.

6.2 Procedure for the Estimation of Potential Substitution in Japan and the USA

The objective is to estimate the potential labor substitutability in the USA and Japan attributable to CIM. A survey of 414 companies in Japan in 1985 reported the ratio between potential substitutable workers by industrial robots and existing process workers by task and by industry sector (JIRA, 1985). Based on the data, the potential labor replacement matrix for Japanese manufacturing industry, which lists tasks in the columns and industry sectors in the rows, can be estimated.

Unfortunately, a comparable labor matrix by industry sector and task is not available for the USA. We can only compare the occupation-by-sector

Table 6.3. Comparison of labor displacement estimates in metalworking (in 1000 workers).

Activity or process	Japan[a]		USA[a]		USA[b]			
	SIC34-37	SIC33-38	SIC34-37	SIC33-38	SIC34-37 [I]	[II]	SIC33-38 [I]	[II]
Casting	14.9 [41.8] (35.6%)	17.7 [75.5] (23.4%)	7.9 [35.9] (22.0%)	10.5 [46.5] (22.6%)	1.1 [7.0] (15.7%)	2.8 (40.0%)	3.7 [31.8] (11.6%)	9.3 (29.2%)
Die casting	18.0 [28.5] (63.2%)	21.4 [40.7] (52.6%)	59.4 [154.5] (38.4%)	77.3 [174.6] (44.3%)	.4 [6.5] (6.2%)	1.0 (15.4%)	2.0 [11.7] (17.1%)	4.7 (40.1%)
Plastic forming	22.4 [63.7] (35.2%)	28.2 [75.0] (37.6%)	21.7 [21.7] (100%)	21.7 [21.7] (100%)	7.2 [37.2] (19.4%)	18.2 (48.9%)	8.2 [42.0] (19.5%)	20.7 (49.3%)
Heat treatment	23.0 [113.0] (20.4%)	27.4 [166.3] (16.5%)	14.5 [56.6] (25.7%)	18.1 [78.2] (23.1%)	2.8 [21.3] (13.1%)	11.1 (52.1%)	6.2 [42.9] (14.5%)	23.0 (53.6%)
Forging	11.2 [22.2] (50.5%)	14.0 [54.3] (25.8%)	4.3 [16.8] (25.6%)	4.9 [17.6] (27.8%)	1.2 [7.5] (16%)	5.3 (70.7%)	1.5 [10.0] (15.0%)	7.0 (70.0%)
Press and shearing	54.3 [215.6] (25.2%)	64.6 [254.9] (25.3%)	31.5 [58.2] (54.1%)	37.9 [63.9] (59.3%)	32.9 [202.7] (16.2%)	146.2 (72.1%)	33.6 [221.4] (15.2%)	152.7 (69.0%)
Welding	112.0 [344.1] (32.5%)	134.5 [366.6] (36.7%)	86.2 [253.9] (34.0%)	96.4 [271.6] (35.5%)	86.1 [319.0] (27.0%)	156.3 (49.0%)	93.0 [344.3] (27.0%)	168.7 (49.0%)
Painting	64.9 [180.9] (35.9%)	77.2 [211.2] (36.6%)	18.4 [60.1] (30.6%)	22.2 [66.5] (33.4%)	32.7 [74.4] (44.0%)	49.1 (66.0%)	34.6 [78.5] (44.1%)	51.8 (66.0%)

Plating	25.6 [82.2] (31.1%)	30.5 [112.2] (27.2%)	11.2 [51.8] (21.6%)	14.2 [55.8] (25.4%)	19.8 [61.3] (32.3%)	50.3 (82.0%)	16.4 [66.1] (24.8%)	57.1 (86.4%)
Grinding, machining, etc.	70.3 [241.8] (29.1%)	83.6 [305.4] (27.4%)	151.5 [918.5] (16.5%)	199.1 [990.4] (20.1%)	139.8 [764.1] (18.3%)	363.2 (47.5%)	155.3 [861.8] (18.0%)	397.4 (46.1%)
Assembly	87.0 [372.2] (23.4%)	102.0 [431.3] (23.6%)	297.6 [1097.3] (27.1%)	338.5 [1230.3] (27.5%)	118.3 [1182.7] (10.0%)	354.8 (30.3%)	131.9 [1318.8] (10.0%)	395.6 (30.0%)
Loading and packaging	69.4 [237.1] (29.3%)	82.6 [305.2] (27.1%)	103.0 [338.8] (30.4%)	134.3 [412.1] (32.6%)	10.7 [73.6] (14.5%)	28.1 (38.2%)	14.1 [95.8] (14.7%)	37.0 (38.6%)
Inspection	72.9 [275.3] (26.5%)	86.7 [356.3] (24.3%)	47.9 [325.5] (14.7%)	63.0 [371.1] (17.0%)	33.8 [280.0] (12.1%)	86.2 (30.8%)	40.1 [332.8] (12.5%)	112.0 (33.7%)
Subtotal	645.9 [2218.4] (29.1%)	770.4 [2754.9] (28.0%)	855.1 [3389.6] (25.5%)	1038.1 [3800.3] (29.3%)	486.8 [3037.3] (16.3%)	1272.6 (42.7%)	540.6 [3457.9] (16.0%)	1437.0 (42.5%)
Others	98.8 [103.4] (95.6%)	121.0 [164.5] (73.6%)	645.5 [2070.4] (31.2%)	848.4 [2411.0] (35.2%)	16.7 [154.6] (10.8%)	43.5 (28.1%)	36.2 [258.7] (14.0%)	58.6 (22.7%)
Total	744.7 [2321.8] (32.1%)	891.4 [2919.4] (30.5%)	1500.6 [5460.0] (29.8%)	1886.5 [6211.3] (30.4%)	503.5 [3191.9] (16.0%)	1316.1 (42.0%)	576.8 [3716.6] (15.8%)	1495.6 (41.4%)

Upper, potential displacement of workers; middle, [total employment]; lower, (potential displacement percentage). SIC33–38: primary metal, fabricated metal products, general machinery, electric machinery, transportation machinery, and precision machinery; SIC34–37: fabricated metal products, general machinery, electric machinery, and transportation machinery.

[a] Source: JIRA substitutability data.
[b] Source: Ayres and Miller, 1983.

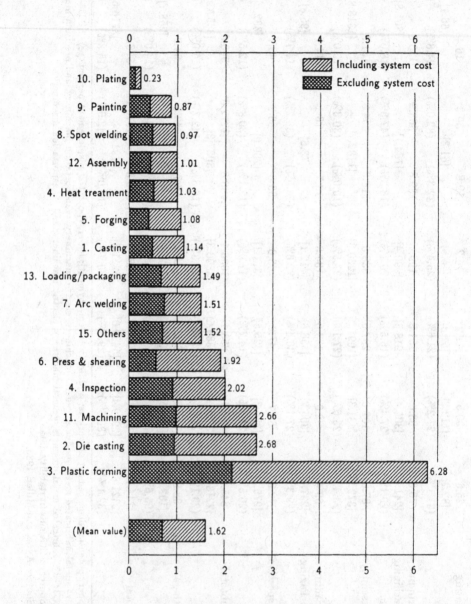

Figure 6.3. Entrepreneurs' willingness to invest to replace one worker (average robot price).

matrix for the USA with that for Japan. To compound the difficulty, tables to convert the data on the national occupational classification systems for the USA and Japan to the ISCO system currently are not available. This makes comparisons difficult.

We describe the three steps necessary to estimate potential labor substitutability in the USA:

(1) We aggregate the detailed occupational labor matrix for the USA to the nearest approximation to the classification of Japan. We then equate it to the task labor matrix. The result is shown in *Table 6.4*.

(2) We then aggregate Japanese occupation-by-industry labor matrix into the same classification as JIRA's task-by-industry labor matrix, say A_j. Hereafter, this aggregated occupation-by-industry labor matrix is denoted by B_j. Let X_j denote the distribution of occupations among tasks, that is, the conversion matrix from B_j to A_j:

$$A_j = B_j X_j \tag{6.1}$$

$$X_j = B_j^{-1} A_j \tag{6.2}$$

where B_j^{-1} is the inverse matrix of B_j, namely,

$$B_j^{-1} = B_j^t (B_j B_j^t)^{-1} . \tag{6.3}$$

(3) We now aggregate US occupation-by-industry labor matrix to a level similar to that of the Japanese table. This aggregated matrix is denoted by B_{US}. Assuming that the conversion matrices of Japan and the USA are the same, we can calculate a task-by-industry labor matrix for the USA, *viz.*,

$$A_{\text{US}} = B_{\text{US}} X_j . \tag{6.4}$$

Next, let us describe the contents of applicable data in JIRA's report and the procedure to estimate the potential labor displacement of the whole manufacturing industry. Let W_i be the total number of workers in sector i and W_{ij} denote the number in the j-th occupational category. Let f_{ij} be the fraction of workers in sector i classified in the j-th occupation category. Similarly, f_{ik} is the fraction of workers in section i employed to perform the k-th task. Thus, by definition

$$W_{ij} = f_{ij} W_i \tag{6.5}$$

$$W_{ik} = f_{ik} W_i , \tag{6.6}$$

where W_{ik} is the upper limit of labor that can be replaced by robots (or other machines) capable of performing the k-th task. Data for f_{ij} are available from the JIRA survey (JIRA, 1985), but f_{ik} must be estimated. An estimate is obtained as follows:

$$f_{ik} = u_{ij} / \sum_{j=1}^{M} u_{ij} \; , \qquad\qquad (6.7)$$

where

$$u_{ik} = \begin{cases} 0 & \text{if task } j \text{ is not required in sector } i \\ 1 & \text{if task } j \text{ is required in sector } i \; . \end{cases} \qquad (6.8)$$

This method is used only if task-by-industry data are not available. The estimate for f_{ik} can be improved by noting two constraints: summing over occupations we have

$$\sum_{j=1}^{M} W_{ij} = W_i \; , \qquad\qquad (6.9)$$

and summing over tasks we have

$$\sum_{k=1}^{N} W_{ij} = W_i \; . \qquad\qquad (6.10)$$

Using (6.8), (6.10) will not (in general) be satisfied. However, by an iterative process of redistributing the errors, successively better estimates can be made. The procedure converges in about five steps.

Tables 6.4 through *6.6* give occupational data and substitutability estimates for the USA. *Tables 6.7* through *6.9* display similar data for Japan.

Table 6.4. US aggregated occupational labor matrix, 1984. Errors in total are due to rounding.

Industry	Casting	Die casting	Plastics forming	Heat treatment	Forging	Press and shearing	Arc welding	Spot welding	Painting
Food and textile	1.15	0	4.97	41.53	0	53.57	0	0	6.06
Wood and paper	0.19	1.26	0	7.59	0	23.85	2.76	0.59	21.90
Chemistry	0	0	165.42	4.12	0	0	4.85	0	6.87
Rubber and cement	0	16.60	0	18.61	0.13	7.88	6.58	0	16.30
Iron and steel	9.68	15.98	0	21.37	0.83	4.32	11.98	2.29	1.98
Nonferrous metals	23.43	11.11	0	23.49	0.42	3.50	2.08	0.90	1.41
Metal products	3.48	34.99	0	9.69	5.70	31.34	21.77	50.19	13.85
General machinery	3.86	60.99	0	8.23	1.66	9.74	28.19	49.90	8.65
Electric machinery	3.52	19.57	16.94	7.03	1.31	7.92	6.77	14.63	10.13
Car and motorcycle	0.81	15.27	3.17	5.19	6.97	3.46	19.04	21.35	8.97
Other transportation	0.77	12.52	1.57	2.89	0.75	2.26	31.25	7.25	17.21
Precision machinery	0.94	4.16	0	0.32	0	1.35	2.83	0.59	4.30
Others	1.77	2.66	11.22	0.96	0	156.57	2.90	0.61	13.79
Total	49.60	195.11	203.29	151.01	17.76	305.75	141.58	148.21	131.47

Industry	Gilding	Grinding	Light assembly	Heavy assembly	Loading	Inspection	Others	Total
Food and textile	0	84.79	0	0	420.83	101.06	2149.44	2863.40
Wood and paper	0	64.81	0	0	211.90	28.44	653.96	1017.24
Chemistry	0	50.09	0	0	121.85	20.00	447.34	655.12
Rubber and cement	0.91	86.27	0	0	167.70	58.48	508.51	1053.40
Iron and steel	2.40	34.31	0	4.30	48.91	15.90	212.28	386.50
Nonferrous metals	1.16	46.04	0	12.80	27.32	13.67	488.71	656.03
Metal products	31.35	246.77	3.93	0	112.16	43.89	472.02	1081.13
General machinery	2.50	353.46	108.36	155.53	58.23	67.81	333.76	1250.88
Electric machinery	13.79	96.98	221.18	331.03	83.34	116.76	344.76	1295.55
Car and motorcycle	1.00	95.30	22.30	147.60	44.50	45.86	200.90	641.70
Other transportation	2.03	80.22	56.79	37.78	13.29	37.50	230.32	534.99
Precision machinery	1.26	37.31	67.11	61.60	24.32	29.74	128.31	364.14
Others	1.77	44.74	20.61	23.89	139.44	17.57	625.19	1063.56
Total	58.16	1321.09	500.28	774.54	1478.30	596.68	6797.07	12865.40

Table 6.5. US potential substitutable workers, 1984. Errors in total are due to rounding.

Industry	Casting	Die casting	Plastics forming	Heat treatment	Forging	Press and shearing	Arc welding	Spot welding	Painting
Food and textile	1.2	0	5.0	11.8	0	28.7	0	0	6.1
Wood and paper	0.2	1.3	0	4.1	0	10.1	2.8	0.6	5.3
Chemistry	0	0	0	3.3	0	0	4.9	0	4.2
Rubber and cement	0	15.6	36.6	3.8	0.1	7.9	6.6	0	4.9
Iron and steel	0.4	2.6	0	0.6	0.2	1.5	1.5	2.3	0.8
Nonferrous metals	1.5	11.1	0	2.7	0.4	3.5	2.1	0.9	1.4
Metal products	2.4	17.3	0	4.2	1.3	10.3	9.9	15.7	5.4
General machinery	1.8	12.6	0	3.1	0.9	7.5	7.2	11.5	3.9
Electric machinery	2.1	15.1	16.9	3.7	1.1	7.9	6.8	13.8	4.7
Car and motorcycle	0.8	8.5	3.2	2.1	0.6	3.5	4.8	7.7	2.6
Other transportation	0.8	5.9	1.6	1.4	0.4	2.3	3.4	5.4	1.8
Precision machinery	0.7	4.2	0	0.3	0	1.4	2.8	0.6	1.6
Others	1.8	2.7	11.2	1.0	0	12.7	2.9	0.6	6.7
Total	13.6	96.6	75.6	42.0	5.1	96.6	55.4	59.0	49.4

Industry	Gilding	Grinding	Light assembly	Heavy assembly	Loading	Inspection	Others	Total
Food and textile	0	84.8	0	0	81.2	39.0	525.8	796.0
Wood and paper	0	43.3	0	0	28.5	13.7	184.7	299.1
Chemistry	0	34.1	0	0	22.5	10.8	145.5	228.6
Rubber and cement	0.9	39.7	0	0	26.2	12.6	169.5	327.6
Iron and steel	0.6	6.6	0	4.3	4.3	2.1	28.1	56.8
Nonferrous metals	1.2	28.1	0	12.8	18.5	8.9	119.7	215.8
Metal products	4.1	44.1	3.9	0	29.0	13.9	188.0	354.6
General machinery	2.5	32.1	54.1	38.0	21.1	10.2	136.9	347.7
Electric machinery	3.6	38.6	65.0	45.6	25.4	12.2	164.5	432.7
Car and motorcycle	1.0	21.6	22.3	25.5	14.2	6.8	91.9	220.1
Other transportation	1.4	15.1	25.4	17.8	13.3	4.8	64.2	166.9
Precision machinery	1.2	12.9	21.8	15.3	8.5	4.1	55.1	132.2
Others	1.8	44.7	20.6	23.9	36.0	17.6	233.1	424.4
Total	18.1	443.9	212.6	182.8	328.7	156.3	2107.1	4194.1

Table 6.6. US percentage of potential substitutability 1984.

Industry	Casting	Die casting	Plastics forming	Heat treatment	Forging	Press and shearing	Arc welding	Spot welding	Painting
Food and textile	100.0	0	100.0	28.4	0	53.6	0	0	100.0
Wood and paper	100.0	100.0	0	54.5	0	42.3	100.0	100.0	24.3
Chemistry	0	0	0	79.1	0	0	100.0	0	61.0
Rubber and cement	0	93.9	22.2	20.4	100.0	100.0	100.0	0	29.9
Iron and steel	3.8	16.2	0	3.0	23.1	35.6	12.3	100.0	40.9
Nonferrous metals	6.6	100.0	0	11.4	100.0	100.0	100.0	100.0	100.0
Metal products	69.8	49.4	0	43.4	22.5	32.7	45.3	31.4	39.1
General machinery	45.8	20.6	0	37.2	56.2	76.7	25.5	23.0	45.5
Electric machinery	60.4	77.3	100.0	52.4	85.6	100.0	100.0	94.2	46.7
Car and motorcycle	100.0	55.4	100.0	39.7	9.0	100.0	25.4	36.1	29.5
Other transportation	100.0	47.1	100.0	49.7	58.3	100.0	10.6	74.2	10.7
Precision machinery	75.8	100.0	0	100.0	0	100.0	100.0	100.0	36.9
Others	100.0	100.0	100.0	100.0	0	8.1	100.0	100.0	48.7
Total (Average)	27.4	49.5	37.2	27.8	28.9	31.5	39.1	39.8	37.6

Industry	Gilding	Grinding	Light assembly	Heavy assembly	Loading	Inspection	Others	Total (Average)
Food and textile	0	100.0	0	0	19.3	38.6	24.5	27.8
Wood and paper	0	66.8	0	0	13.5	48.2	28.2	29.4
Chemistry	0	68.1	0	0	18.5	54.0	32.5	34.9
Rubber and cement	100.0	46.1	0	0	15.6	21.5	33.3	31.3
Iron and steel	25.3	19.2	0	100.0	8.9	13.1	13.3	14.7
Nonferrous metals	100.0	61.0	0	100.0	67.7	65.0	24.5	32.9
Metal products	13.0	17.9	100.0	0	25.9	31.8	39.8	32.8
General machinery	100.0	9.1	49.9	24.4	36.3	15.0	41.0	27.8
Electric machinery	25.8	39.8	29.4	13.8	30.5	10.5	47.7	33.4
Car and motorcycle	100.0	22.6	100.0	17.3	31.9	14.9	45.8	34.3
Other transportation	68.3	18.8	44.7	47.1	100.0	12.7	27.9	31.2
Precision machinery	94.5	34.6	32.5	24.8	35.0	13.8	43.0	36.3
Others	100.0	100.0	100.0	100.0	25.8	100.0	37.3	39.9
Total (Average)	31.1	33.6	42.5	23.6	22.3	26.2	31.0	32.6

Table 6.7. Japanese aggregated occupational labor matrix, 1984. Errors in total are due to rounding.

Industry	Casting	Die casting	Plastics forming	Heat treatment	Forging	Press and shearing	Arc welding	Spot welding	Painting
Food and textile	0	0	34.6	34.6	0	0	0	0	0
Wood and paper	0	15.6	0	15.6	0	31.2	15.6	31.2	31.2
Chemistry	0	0	9.7	0	0	0	0	0	4.9
Rubber and cement	0	0	75.9	10.8	0	10.8	7.2	0	50.6
Iron and steel	18.1	2.6	2.6	38.7	28.4	20.6	10.3	2.6	15.5
Nonferrous metals	13.5	5.4	4.5	6.3	2.7	7.2	2.7	1.8	5.4
Metal products	6.6	5.0	3.3	34.9	5.0	59.8	43.2	31.6	38.2
General machinery	13.7	4.6	3.0	30.5	9.1	45.7	71.6	27.4	61.0
Electric machinery	11.3	8.4	42.2	22.5	0	67.6	46.5	40.8	47.9
Car and motorcycle	8.2	9.8	13.1	19.6	7.4	34.4	31.1	34.4	23.7
Other transportation	2.0	0.7	2.0	5.4	0.7	8.1	10.1	7.4	10.1
Precision machinery	2.1	4.2	4.2	8.3	1.0	11.5	6.3	9.4	9.4
Others	8.7	8.7	32.6	30.4	2.2	43.5	58.7	30.4	58.7
Total	84.2	65.0	227.7	257.6	56.5	340.4	303.3	217.0	356.6

Industry	Gilding	Grinding	Light assembly	Heavy assembly	Loading	Inspection	Others	Total
Food and textile	0	0	34.6	34.6	415.8	277.2	658.4	1490.0
Wood and paper	15.6	31.2	15.6	31.2	109.2	62.4	156.0	561.4
Chemistry	0	4.9	9.7	14.6	48.6	38.9	72.9	204.2
Rubber and cement	7.2	21.7	36.2	3.6	72.3	101.3	61.5	434.0
Iron and steel	18.1	36.1	2.6	7.7	36.1	43.9	36.1	319.9
Nonferrous metals	3.6	10.8	5.4	2.7	15.3	16.2	13.5	117.0
Metal products	23.3	54.9	36.6	24.9	68.2	61.5	23.3	520.3
General machinery	9.1	83.8	45.7	62.5	59.4	74.7	21.3	623.4
Electric machinery	35.2	64.8	94.3	52.1	67.6	91.5	33.8	726.4
Car and motorcycle	13.9	30.3	24.6	18.0	31.1	36.8	15.6	352.0
Other transportation	0.7	8.1	5.4	8.1	10.8	10.8	9.5	100.1
Precision machinery	8.3	16.7	29.2	11.5	16.7	20.9	11.5	159.5
Others	30.4	67.4	58.7	52.2	67.4	73.9	63.0	686.9
Total	165.5	430.6	398.6	323.8	1018.5	909.9	1176.3	6295.0

Table 6.8. Japanese potential substitutable workers, 1984. Errors in total are due to rounding.

Industry	Casting	Die casting	Plastics forming	Heat treatment	Forging	Press and shearing	Arc welding	Spot welding	Painting
Food and textile	0	0	34.1	25.8	0	0	0	0	0
Wood and paper	0	4.2	0	5.3	0	12.6	14.5	11.4	15.0
Chemistry	0	0	5.0	0	0	0	0	0	4.9
Rubber and cement	0	0	6.4	4.9	0	10.8	7.2	0	13.7
Iron and steel	0.9	1.1	1.9	1.5	1.0	3.4	4.0	2.6	4.1
Nonferrous metals	0.8	0.9	1.6	1.2	0.8	2.8	2.7	1.8	3.3
Metal products	3.4	4.2	3.3	5.3	3.8	12.6	14.5	11.4	15.0
General machinery	3.6	4.3	3.0	5.5	4.0	13.0	15.0	11.8	15.5
Electric machinery	4.9	5.9	10.0	7.6	0	17.8	20.6	16.1	21.3
Car and motorcycle	2.4	2.9	4.9	3.7	2.7	8.7	10.1	7.9	10.4
Other transportation	0.6	0.7	1.2	0.9	0.7	2.2	2.6	2.0	2.7
Precision machinery	1.1	1.4	2.3	1.7	1.0	4.1	4.7	3.7	4.9
Others	5.3	6.5	11.0	8.3	2.2	19.5	22.5	17.6	23.3
Total	23.1	32.1	84.8	71.7	16.3	107.6	118.6	86.4	134.1

Industry	Gilding	Grinding	Light assembly	Heavy assembly	Loading	Inspection	Others	Total
Food and textile	0	0	34.6	26.6	77.6	81.4	131.4	414.2
Wood and paper	6.0	16.3	14.6	5.5	16.1	16.9	27.2	165.1
Chemistry	0	4.9	9.7	3.9	11.3	11.9	19.2	71.3
Rubber and cement	5.5	14.8	13.3	3.6	14.6	15.4	24.8	135.0
Iron and steel	1.6	4.4	2.6	1.5	4.4	4.6	7.4	47.0
Nonferrous metals	1.3	3.6	3.2	1.2	3.6	3.7	6.0	38.5
Metal products	6.0	16.3	14.6	5.5	16.1	16.9	23.3	170.7
General machinery	6.2	16.8	15.1	5.7	16.6	17.4	21.3	173.3
Electric machinery	8.5	23.0	20.7	7.8	22.7	23.9	33.8	242.6
Car and motorcycle	4.2	11.3	10.2	3.8	11.2	11.7	15.6	120.7
Other transportation	0.7	2.9	2.6	1.0	2.8	3.0	4.8	31.2
Precision machinery	2.0	5.3	4.7	1.8	5.2	5.5	8.8	57.9
Others	9.3	25.2	22.7	8.5	24.9	26.1	42.2	274.1
Total	51.5	144.7	169.4	76.4	227.1	238.4	364.6	2052.2

Table 6.9. Japanese percentage of potential substitutability, 1984.

Industry	Casting	Die casting	Plastics forming	Heat treatment	Forging	Press and shearing	Arc welding	Spot welding	Painting
Food and textile	0	0	98.5	74.4	0	0	0	0	0
Wood and paper	0	26.9	0	34.3	0	40.3	93.3	36.5	48.2
Chemistry	0	0	51.2	0	0	0	0	0	100.0
Rubber and cement	0	0	8.5	44.8	0	100.0	100.0	0	27.0
Iron and steel	5.2	44.3	74.8	3.8	3.7	16.6	38.5	100.0	26.5
Nonferrous metals	5.7	17.2	34.8	18.8	31.4	38.6	100.0	100.0	61.6
Metal products	51.9	84.0	100.0	15.3	76.8	21.0	33.6	36.1	39.3
General machinery	26.0	94.7	100.0	18.1	43.3	28.4	21.0	42.9	25.5
Electric machinery	43.3	70.2	23.7	33.5	0	26.3	44.3	39.5	44.4
Car and motorcycle	29.3	29.6	37.5	18.9	36.2	25.4	32.5	23.0	44.0
Other transportation	30.0	100.0	61.6	17.4	100.1	27.4	25.3	27.1	26.2
Precision machinery	53.6	32.6	54.9	20.7	100.0	35.5	75.3	39.3	51.9
Others	61.5	74.7	33.6	27.2	100.0	44.8	38.4	58.0	39.7
Total (Average)	27.4	49.5	37.2	27.8	28.9	31.6	39.1	39.8	37.6

Industry	Gilding	Grinding	Light assembly	Heavy assembly	Loading	Inspection	Others	Total (Average)
Food and textile	0	0	100.0	76.7	18.7	29.4	20.0	27.8
Wood and paper	38.7	52.1	93.9	17.7	14.7	27.1	17.5	29.4
Chemistry	0	100.0	100.0	26.6	23.3	30.5	26.3	34.9
Rubber and cement	75.9	68.2	36.9	100.0	20.2	15.2	40.3	31.1
Iron and steel	9.1	12.3	100.0	19.4	12.2	10.5	20.6	14.7
Nonferrous metals	37.1	33.3	60.0	45.2	23.3	23.1	44.7	32.9
Metal products	25.9	29.6	40.0	22.1	23.6	27.4	100.0	32.8
General machinery	68.1	20.0	33.1	9.1	27.9	23.3	100.0	27.8
Electric machinery	24.2	35.5	22.0	15.0	33.7	26.1	100.0	33.4
Car and motorcycle	30.1	37.3	41.4	21.3	35.9	31.8	100.0	34.3
Other transportation	100.0	35.4	47.8	12.0	26.3	27.6	50.8	31.2
Precision machinery	23.4	31.6	16.3	15.6	31.2	26.2	76.9	36.3
Others	30.7	37.4	38.7	16.4	37.0	35.4	66.9	39.9
Total (Average)	31.1	33.6	42.5	23.6	22.3	26.2	31.0	32.6

Appendix

Table A.1. Categories of metal-cutting machine tools in the *American Machinist* 12th inventory.

Machine	Not NC controlled	NC controlled
Turning Machines		
Bench	1	4
Engine and toolroom < 8-inch swing	1	4
Engine and toolroom 9- to 16-inch swing	1	4
Engine and toolroom 17- to 23-inch swing	1	4
Engine and toolroom 24-inch swing and over	1	4
Tracer lathe	1	4
Turret lathe, ram type	1,4	4
Turret lathe, saddle type	1,4	4
Auto chucking vertical and horizontal; single-spindle	2,4	2,4
Auto chucking vertical and horizontal; multispindle	2,4	2,4
Automatic between centers chucking	4	4
Automatic bar (screw) machinery; single-spindle	2,4	2,4
Automatic bar machinery; multispindle	2,4	2,4
Vertical turning and boring mills (VTL < VBM)	3	3
Other (forging, axle, spin, shell)	1,4	4
Boring		
Horizontal boring, drilling, milling (bar machinery); table and planer type	1	4
Horizontal boring, drilling, milling (bar machinery); floor type	1	4
Precision, horizontal, and vertical	1	4
Jig bore, horizontal, and vertical	1	4
Other (not boring lathes)	1	4
Drilling		
Sensitive (hand feed), bench	1	4
Sensitive (hand feed), floor and pedestal	1	4
Upright: single-spindle	1,4	4
Upright: gang	4	4
Upright: turret, not NC	1,4	4
Radial	1	4
Multispindle cluster (adjusted and fixed center)	2,4	2,4
Deep hole (gun drill)	1,2,4	3,4
Other (not unit head and way)	1,4	4

Table A.1. Continued.

Machine	Not NC controlled	NC controlled
Milling		
Bench type (hand or power feed)	1	4
Hand	1	4
Vertical ram type (swivel head and turret)	1	4
General purpose, knee or bed: horizontal (pin, universal, and ram)	1,4	4
General purpose, knee or bed: vertical	1,4	4
Manufacturing, knee or bed	1,4	4
Planer type	1,3	3,4
Profiling and duplicating (die, skin, spar)	4	4
Thread millers	2,4	2,4
Others (spline, router, engraving)	1	4
Tapping Machines	4	4
Threading Machines	2,4	2,4
Multifunction NC Machines (Machining Centers)		
Drill-mill-bore, manual tool change, vertical and horizontal	4	4
Drill-mill-bore, indexing turret	4	4
Drill-mill-bore, auto tool change; vertical	4	4
Drill-mill-bore, auto tool change; horizontal	4	4
Special Way Type and Transfer Machines		
Single-station (several operations on one part)	2	2
Multistation: rotary transfer	2	2
Multistation: in-line transfer	2	2
Broaching Machines		
Internal	4	4
Surface and other	4	4
Planing Machines		
Double column	1,3	3,4
Openside and other	1,3	3,4
Shaping Machines (not gear)		
Horizontal	1	4
Vertical (slotters and keyseaters)	1	4
Cutoff and Sawing Machines		
Hacksaw	2	2
Circular saw (cold)	2	2
Abrasive wheel	2	2
Bandsaw	2	2
Contour sawing and filing	1	4
Other (friction)	2	2

Table A.1. Continued.

Machine	Not NC controlled	NC controlled
Grinding Machines		
External; plain center type	1,4	4
External; universal center type	1,4	4
External; centerless (incl. shoe type)	4	4
External; chucking	1,4	4
Internal; (chucking, centerless shoe type)	1,4	4
Surface; rotary table, vertical and horizontal	1,4	4
Surface; reciprocating, horizontal, manual	1	4
Surface; reciprocating, vertical, horizontal, power	1	4
Disk grinders, not handheld	1	4
Abrasive belt (excl. polishing)	1,4	4
Contour (profile)	4	4
Thread grinders	4	4
Tool and cutter	1	4
Bench, floor and snap	1,4	4
Other (jig)	1	4
Honing Machines		
Internal (combined bore-hone)	1,4	4
External	1,4	4
Lapping Machines		
Flat surface	1	1
Cylindrical	1	1
Other (combined hone-lap)	1	1
Polishing and Buffing Machines		
Polishing stands (bench and floor)	1	
Abrasive-belt, disk, drum (not grinding)	4	
Other (speed lathes and multistation type)	2	2
Gear Cutting and Finishing Machines		
Gear hobbers	2	2
Gear shapers	2	2
Bevel-gear cutters (incl. planer type)	2	2
Gear-tooth finish (grind, lap, shave, etc.)	2	2
Other gear cutting and finishing	2	2
Electrical Machining Units		
Electrical discharge machines (EDM)	2	2
Electro-chemical machines (ECM)	2	2
Electrolytic grinders (ECG or ELG)	2	2

Automatic assembly machines and "other" metal-cutting machines are omitted.

References

Ayres, R.U., and S.M. Miller, 1983, "Robotic Realities: Overview of Prospects and Problems," *Annals of the American Academy of Political & Social Sciences*, Fall [Special Issue: Future Factories, Future Workers].

JIRA (Japan Industrial Robot Association), 1985, *JIRA Report*, Tokyo, Japan (in Japanese, extracted by S. Mori).

Miller, S.M., 1983, *Potential Impacts of Robotics on Manufacturing Costs within the Metalworking Industries*, PhD Thesis, Carnegie-Mellon University, Pittsburgh, PA.

Warnken, J., 1986, "Zur Entwicklung der 'internen' Anpassungsfähigkeit der Berufe bis zum Jahre 2000," *MittAB 1986* 1:120.

Chapter 7

Scientific, Training, and Socioeconomic Aspects of CIM in the USSR

Felix I. Peregudov and Iouri M. Solomentsev

7.1 Research in the USSR

The dynamic development of modern economies is defined to a great extent by their machine-building sectors, including such basic branches as machine tool manufacturing. At present, this branch produces many kinds of equipment, e.g., CNC machine tools, automatic machine tools, groups of automated machinery, and flexible manufacturing systems (FMS). In the USSR, machine tool building, instrument making, and electronic and computing machinery are referred to as priority sectors.

The development of flexible manufacturing both in the USSR and elsewhere is conditioned by societal requirements. In the USSR, as in many countries, small-batch production represents about 70% to 80% of total production. This requires the development of equipment that can be quickly and easily readjusted. In turn, social considerations define scientific tasks connected with the development of flexible automated manufacturing at the level of separate production sites and in the systems at such sites, workshops, and automated plants.

In several countries, such as the USA, Japan, Germany, and Italy, research programs have been developed for solving such important and complex problems. The Soviet government also has given very close attention to these issues. The USSR State Committee for Science and Technology has evolved major programs for "technology, machines, and production of the future" and "information technologies." These are aimed at resolving problems associated with computer integrated manufacturing (CIM). Other programs exist in different regions (Moscow and Leningrad) and ministries (the Ministry of Machine-Building Industry and the Ministry of Electronics Industry). The USSR State Committee for Public Education also has developed a research program on CIM-related problems.

Several higher educational establishments and universities in the USSR have sufficient potential to solve scientific and technical problems associated with CIM. These educational establishments are responsible for training specialists for this "scientist-consuming" field (not less than 5,000 specialists are needed per year).

In 1987, the USSR State Committee for Science and Technology announced a competition for the development of automated plant concepts. Sixty-nine organizations from different branches of the USSR Academy of Sciences and higher educational establishments took part in this competition. Three working collectives jointly secured first place, including the Moscow Machine Tool and Small Tool Institute. As a result of this competition, the USSR State Committee for Science and Technology has chosen several plants to undergo basic reconstruction to create a wide range of flexible automated manufacturing.

The leading higher educational establishments, such as the Moscow Machine Tool and Small Tool Institute, the Moscow Higher Technical Educational Establishment (Bauman), the Leningrad Polytechnic Institute, and the Moscow Aircraft Technological Institute, are participating in the intensive work being conducted in this field. They also are contributing funds to these projects.

The application of the latest scientific principles in higher educational establishments, of course, raises their prestige. However, even before the development of such state programs, higher educational establishments had already made great contributions to CIM concepts in the USSR, from the point of view of the development of new technologies, tools, equipment, robotics, computer automated design systems, and so on. These developments are currently being implemented in industrial production.

7.2 CIM and Personnel Training in the USSR

CIM is needed to increase profitability. Competitiveness in external markets cannot be achieved by simply decreasing the number of the staff or by increasing the volume of labor. Increased productivity can mainly be achieved by increasing the efficiency of operations, e.g., by reducing the lead times of new products and inventories of work-in-process. Such manufacturing should be based on the most advanced techniques. (To set up CIM with outdated techniques is a recipe for failure.) CIM demands that all manufacturing subsystems must create, process, and control information dealing with organization, design, technical development, planning, scheduling, process control, and inventory control. This information must be integrated in the CIM system by means of automated functions and automated data acquisition and communication. This "central nervous system" links different manufacturing functions and substitutes a unified "organism" for a collection of independent activities. It forms the basis for optimal functioning of the organism as a whole.

CIM should also be regarded as a strategy for current enterprise activity which provides opportunities to change or intensify the contents of the system's separate elements, in accordance with changes imposed by the environment and which are aimed at manufacturing efficiency. CIM projects also envisage a structural tuning of each automated function within the configuration and specific features of the technological system that is being automated.

The first CIM goal is to organize manufacturing as a single unit using integrated data systems to automate information, material, and control flows among various manufacturing subsystems. These subsystems carry out predetermined functions at the various manufacturing levels. These functions ensure the following:

- Organization of production linked to receipt of external orders.
- Decision making concerning the manufacture of new equipment, production updating, and its integration with external criteria.
- Integration of design and the production process, e.g., machine (machine part) design and technical developments.
- Development of schedules and lists of supplies.
- Control over manufacturing processes.
- Coordination of all operations during the development and execution of the planned task.

- Equipment control and manipulation of control devices (drives, automatic electric devices).

In CIM structures each function is determined as a subsystem with specified functions at each manufacturing level of a plant, shop, work site, or module. Information data in CIM are accessible to every subsystem. The manufacturing system, from organizational subsystem to equipment control subsystems, provides feedback on the state of production. This feedback allows for readjustments of the preplanned control functions. In this way the design and manufacturing processes within CIM can be regarded as a whole.

Integration of the design process with planning and manufacturing provides numerous advantages:

- Quick readjustment of production processes for the manufacture of new products.
- Decreased technical design costs.
- Reduced overall time for the introduction of new products.
- Improved product quality.
- Increased productivity of engineers.
- Increased productivity of assembly operations.
- Increased productivity of equipment.
- Decreased work-in-progress.
- Decreased labor costs (reduced numbers).

CIM includes technological equipment, staff, design and manufacturing software, hardware, and so on. The problem of integration is accordingly the integration among the staff and between staff and the hardware/software, the staff and the equipment, and the equipment and hardware/software. The solution to integration problems in CIM, in general, is based on the development of standards to deal with such problems. The Soviet CIM project also carries out feasibility studies to assess which equipment (such as hardware and software) is the most appropriate to develop an integrated manufacturing strategy.

The classic concept of traditional manufacturing systems consists of several discrete parts and subsystems. Since these have distinctive limits and tasks, and in part can be regarded as "information islands" or "operations islands," the next logical step of integrated manufacturing relies on improvements on the basis of function automation. The complete execution of CIM creates the opportunity to bridge the gap between these "islands" and provides an opportunity to design and create automatic plants. Naturally this creates problems regarding new tasks and staff training. These include identifying CIM problems; acquiring the most advanced techniques, equipment, tools, robots, and robot systems; obtaining information and measuring devices; and maintaining control systems, CAD systems, economic structures, and manufacturing organizations.

CIM constitutes a large system, and its subcomponents should be completely integrated. However, in general, staff training can be arranged so that each subcomponent is matched with particular specialization and skills. Since it is practically impossible to train a universal CIM specialist, this problem has to be solved with a team of technologists, designers, robot specialists, managers, CAD experts, economists, among others. Thus, we suggest the need for an interdisciplinary team of specialists who work closely with each other, during both the training process and the manufacturing process, and who can collectively "cover" the whole of CIM problems.

Here we recognize an important characteristic of CIM, *viz.*, that training should be conducted on a single methodological foundation comprising the whole process of system development and its operation through its service life. This fact translates into a program-oriented principle of lifelong staff training.

In the general scheme of machine life cycles and machine tool development and production, organization is very important. This is so regardless of the type of product that society needs. But it is on the basis of machine tools that automated manufacturing is necessarily organized. The methods of machine tool development may be broken down into the following stages:

- A theoretical study of each stage of machine tool development.
- Simulation of each stage and the provision of information exchange between the stages.
- A design of an algorithm for the machine tool development process.
- Formation of software/hardware complexes to conduct design and management at all levels.

Thus, we can think of information techniques in automated manufacturing. This defines several scientific tasks for design and technological informatics, from data acquisition, processing, storage, and communication in machine tool building to the technological development processes at the design and production stage (as well as during its exploitation), in order to adopt optimal solutions. The training of CIM specialists should be conducted on the basis of this methodology.

The main principles of specialist training are as follows:

(1) *Fundamental education.* This does not mean an increase in the amount of classroom hours. The fundamental sciences (for example, mathematics) should be presented in all subjects during the whole process of education. This not only allows deeper fundamental training of students, which is especially important for modeling of types of production, but also calls for the tutorial staff to stress the physical and mathematical standards.

(2) *Computerization of education.* This plays a principal role in the training of CIM specialists, because all stages of machine tool development eventually lead to problem-solving and optimization tasks. The program determines the whole computer-training process starting from the first term to the postgraduate level. At the leading higher educational establishments, students have 350 to 450 hours of hands-on computer time. Students who take courses in the field of CIM management and computer aided design spend at least 600 hours with the computer (in some cases more than 1,000 hours). The curriculum of the Moscow Machine Tool and Small Tool Institute might serve as a model for other establishments regarding this process.

(3) *Team training of specialists.* This concept is the direct application of program methods of training for special problems. As noted above, the necessary set of specializations for CIM support is being undertaken. This approach does not exclude any speciality from the team or "module." In practice, the Moscow Institute conducts work on complex CIM problems. It is also important to note that work placement at different enterprises according to this modular approach is undertaken by teams of specialists from different fields.

(4) *Flexibility in staff training.* This means a relatively rapid reorientation in future specialists training that could be determined by production requirements or by new scientific and technological developments. Flexibility is achieved through teaching methods aimed at problem solving by means of continuity of subjects, modules of appropriate courses, and fundamentalization of education.

(5) *Education system.* The education process is developed according to the following scheme: theoretical mastering of certain application fields, modeling, working out algorithms, and developing the task for hardware and software for CIM management and CAD systems.

(6) *Independent knowledge acquisition.* The number of compulsory classes and examinations has been reduced, and the role of the students' creative activities has been increased. Most class work (laboratory exercises,

design activities, or practical training) is conducted as a research project that must be defined, solved, and concluded.
(7) *Close ties with manufacturing.* In this area, closer research ventures with the customer have been established. This includes appointments or "institute chairs" at enterprises and research institutes. These appointees collaborate on the use of equipment, facilities, and computer hardware development and support scientific developments. They also can be used to retrain specialists.

Currently, support for CIM specialists is complex. Engineers and scientists (postgraduates and doctoral candidates), technicians, programmers, operators, and other qualified personnel need training. Also, and more important, constant retraining and qualification improvement courses of personnel involved in all stages of CIM development and maintenance are required. This should be done at the leading higher educational establishments, R&D institutes, and advanced manufacturing centers.

Usually staff support for enterprises (including enterprises dealing with CIM) is carried out according to agreements between higher educational establishments and customers. The main terms defined in these agreements include specialists' salaries (partial compensation of the costs besides the main source of financing – the state budget). The cost of CIM specialist training is composed of the following:

• Wages of professors, tutorial staff, research staff, and service staff (30%) and student grants (15%).
• Expenses for facilities (equipment, tools, computers, etc.) and amortization expenses (35%).
• Amortization expenses for the institute; construction and maintenance (20%).

We believe that currently the cost of training one CIM specialist during the whole period of education is in excess of 60,000 rubles, equivalent to a professor's salary for 10 years. This figure is determined by the availability of highly qualified professors, tutorial staff, and researchers and by the high cost of technological equipment, especially computers. The physical teaching environment also is important.

However, the cost for a highly qualified CIM specialist working at an enterprise that implements the CIM concept should be returned in three years. This is true for specialists in management, software, and technology designs. Presently in the USSR, the state underwrites most of the expenses for specialist training; the customer then partially compensates for these expenses by paying for wages or salaries or by purchasing equipment, computers, tools, etc. This process has only recently been implemented, and the process will be mainly determined by future economic changes in the Soviet Union.

7.3   CIM and Science in Higher Education

The staff-training approach described in Section 7.2 is possible on the basis of a general reorganization of scientific activities within the higher educational establishments, which requires an intensive development of science in these educational establishments.

Experience and practice, both in the USSR and elsewhere, have shown that one criterion (besides staff support, material support, and educational methods) for any higher educational establishment is knowledge of and access to state-of-the-art technologies. The linkage of R&D with higher educational establishments has become a world tendency. At most of the leading world universities, the ratio of teaching staff to research workers is about 1:2.5 or 1:3.

Today, structural changes in teaching at higher educational establishments can be carried out that afford the opportunity to direct the efforts of professors, tutors, researchers, and engineering and technical staff toward resolving the overall problems posed by multispeciality higher educational establishments. For example, the Moscow Machine Tool and Small Tool Institute has set up a Design and Technological Informatics Center, with the main goal of concentrating the scientific potential of the Institute toward solving CIM problems and exploring potentials. The Center has a science and technological council, and all issues from ideas to financing are discussed during its meetings.

The structure of the Institute is determined by the nature of CIM problems themselves. It can be regarded as a mechanism for uniting scientists and specialists working on separate projects. Professionals working in the field head the Center, its departments, and the "interchair" task groups. They recruit temporary task groups from the permanent chairs and departments of the Institute after the specific theme of scientific development has been defined.

The setting up of science–training–engineering centers (STEC) is the next stage of development at higher educational establishments. STECs are organized, in essence, toward flexible principles. We already have several examples of such STECs: Taganrog Radio Technical Institute, Novotcherkask Polytechnical Institute, Higher Technical College (Bauman), and Moscow Machine Tool and Small Tool Institute. STECs should consist of a training department, including all forms of training and retraining; a scientific department to conduct research and development; an engineering department; and a pilot manufacturing plant.

This structure necessarily requires a wide exchange of specialists between different STEC departments (staff rotation). STECs raise the quality of staff training by such mechanisms as the following:

- Research can be transferred into independent practical projects.
- In-house training allows specialists to learn future techniques, new principles of machine design, advanced systems design, and computer control techniques.
- Money earned by "selling" science is easily channeled to strengthen the material basis of any STEC department.
- Within STEC, the upgrading of staff training will raise the professional skills needed to train and retrain students.

7.4 CIM and International Cooperation in Higher Educational Institutions

When working on complex problems like CIM, contact with foreign specialists is extremely important not only to exchange scientific information, but to establish joint activities leading to real projects on a contract basis. Since a real need exists to develop international projects in the field of CIM, specialists in advanced capitalist countries have welcomed this idea, especially where the projects are science-oriented and directly financed by the government.

Since CIM development is a science-consuming activity, international cooperation is welcomed by the USSR State Committee on Education. The creation of joint ventures can serve as an example. This provides an opportunity to immerse personnel in new techniques and closely follow strategic trends of science and technology. This calls for increasing the number of students attending universities throughout the world.

7.5   Conclusions

(1) The scientific potential of higher educational establishments should become involved in the intricate and science-consuming problems of CIM.

(2) Such establishments should develop a wide-range of scientific training courses.

(3) Modern higher educational institutions should be scientific manufacturing training centers comprising departments within which the staff – scientific, professorial, and tutorial staff – and students can learn from each other.

(4) The emergence of complicated CIM problems demands the integration of higher educational establishments, technical colleges, secondary schools, vocational schools, and retraining centers. This may not entail physical integration, but the motivation for educational restructuring should be curriculum development.

(5) There is an urgent need for the development of programs aimed at differentiated training – having in view selection of the most capable people (so-called elite groups without any negative connotation) who later will generate new ideas.

(6) Wide integration with developed countries in training and research is urgently needed. We are concerned not only with the training of staff internationally in the Soviet higher educational schools and universities, but also with the development of complicated science-consuming work (CIM included) on a contract basis or financed by the governments of interested countries.

Part III

Meso Effects:
Industrial Structures

Chapter 8

The Economic Impact of Technological Change

Rumen Dobrinsky and Shunichi Furukawa

The analysis and forecast of the economic impacts of CIM can be regarded in the broader context of the socioeconomic consequences that are brought about by technological change. This research requires a major effort in theoretical and empirical analysis to identify and describe the phenomenon by an appropriate model capable of providing the relevant information.

The limited resources that were available within the CIM Project did not allow us to construct the sophisticated environment which would be required to approach the problem in detail. The Project relied on external collaboration, and many of the studies presented in this volume were prepared outside IIASA through the efforts of the Project's collaborating network.

These studies reflect a variety of approaches to the problems of social and economic impacts of CIM diffusion, emphasizing different aspects of this technological development or applying different methodological background. The advantages and disadvantages of different methods for estimating the sectoral effects of technological changes such as CIM have been discussed repeatedly in the literature (Brooks, 1985; Friedrich and Roenning, 1985).

The first attempt to use an input–output model to analyze multisectoral economic impacts of microelectronic applications was made by Fleissner *et al.* (1981). Leontief and Duchin (1986) subsequently published a study in which the impact of computer-based automation on employment was analyzed using an input–output model for the USA. An input–output model similar to the Leontief model was used to calculate the relative industry and occupational effects of alternative levels of the use and production of industrial robots in the USA (Howell, 1985). A multinational and multisectoral econometric model, which covered eight regions linked through an international trade matrix, was used to analyze the impacts of industrial robots on employment (Kinoshita and Yamada, 1989).

As these examples suggest, the main advantage of the input–output approach is that it reflects not only the productivity and employment effects of CIM application in a given sector but also the indirect effects on other sectors of the economy. However, even with the help of input–output models, not all of the important effects of CIM applications can be estimated. Some methodological and practical limitations of these studies should be mentioned:

(1) There is no straightforward way that effects such as changed work content and work environment, which are conditioned by CIM application, can be reflected in the model. In the CIM literature these (hardly quantifiable) effects are especially emphasized (Ayres, 1988).

(2) Input–output tables for most countries are out of date. Fleissner *et al.* (1981) used the input–output tables from 1970 and 1976. Leontief and Duchin used tables from 1967, 1972, and 1977. This led to severe problems in parameter forecasting (see Friedrich and Roenning, 1985).

(3) At the microeconomic level, the effects of CIM application are likely to be quite marked in comparison with traditional technology. But these major effects will not immediately be "transferable" to the sectoral and the macroeconomic levels to the same extent (see also Ayres and Miller, 1983).

The impact of CIM has so many aspects that it is unlikely to be completely reflected by any model, and thus several complementary approaches are needed. The recent studies reported in this volume have tried to reduce these limitations. Some are based on models; others rely on economic analysis and expertise. The variety of methods actually reflects the interdisciplinary approach to study of the socioeconomic impacts of CIM which was adopted within the Project.

In this brief survey we concentrate on some recent econometric studies which were carried out within the IIASA CIM Project and which followed

the pioneering works mentioned above. We shall try to compare the methodological approaches to the problem, the assumptions made by the different
authors, and their main results. We focus on the studies of Yamada (Chapter 10, referred to as the Y model), the Polish team (Chapter 11, P model),
Dobrinsky (Chapter 14, D model), and Mori (Chapter 4, M model). Of
course, when we consider specific topics we shall refer to the other studies
in the volume which are not based on econometric models.

8.1   The Methodological Framework

The econometric studies of the impact of CIM technologies were based on
different methodological approaches, but they have one feature in common:
they address these issues from a macro perspective, i.e., at a national or
sectoral level. The only exception is the M model which is a combination of
macro and micro performances. Another common feature is that they focus
on the diffusion of CIM technologies mainly in the metalworking industries,
which obviously are and will be the main recipients of this technological development. The Y model and the P model are based on input–output models in a Leontief–Keynesian framework (demand-driven, with an econometric
block of final demand); the D model relies on a supply-driven neoclassical
growth model; and the M model, on a specific type of production function.

Figures 8.1 to *8.4* present a simplified picture of the main causal relationships as they are reflected in the four models. It should be noted that
these pictures refer only to the treatment of CIM in the models and the way
it influences the model performance.

The Y model and the P model are rather similar in structure and actually
were built on the same data base for the countries analyzed (Japan, the USA,
and the FRG in the case of Y model and Japan and the USA in the case
of P model). The input–output tables have been specially designed for this
exercise by aggregating original input–output tables into 21-sector tables but
with a disaggregated representation of the five metalworking sectors.

There are, however, some important differences in introducing the links
between the CIM diffusion process and the economic performance. The Y
model (*Figure 8.1*) is more comprehensive in this sense as it considers the
impact of CIM diffusion on both the input structure (the technological coefficients of the I–O matrix) and the final demand structure (the consumption and investment coefficients). The changes in final demand drive the
technologically changed I–O model to define the corresponding output. In

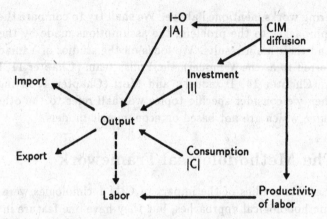

Figure 8.1. Causal relations in Y model.

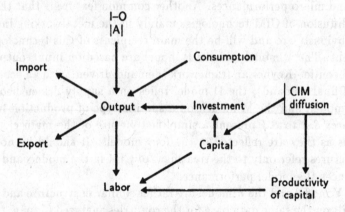

Figure 8.2. Causal relations in P model.

addition, the Y model relates technological change directly to labor productivity. Output and labor productivity are used to determine labor demand; two possibilities are given for this: through production functions or through labor demand functions. We cannot specify precisely which type of function was actually used in simulations.

The P model (*Figure 8.2*) does not consider CIM-induced changes in the A-technological matrix; consumption is modeled by demand functions which are not directly linked to the CIM diffusion process. The impact of CIM is reflected in two ways: through the labor substitution effect and change in

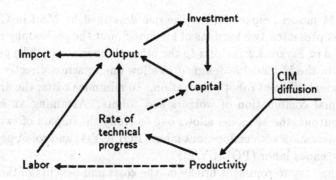

Figure 8.3. Causal relations in D model.

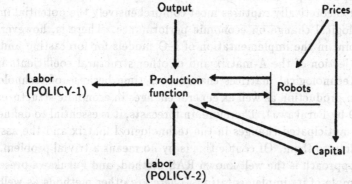

Figure 8.4. Causal relations in M model.

capital productivity and through the capital requirements for new equipment. The latter is transformed into changes of investment demand, and through final demand this influences the level of output. On the other hand, changed capital, changed labor, and changed output through a production function determine labor demand.

The D model (see *Figure 8.3*) is essentially a supply-driven neoclassical growth model and leaves less space for feedback maneuvers. The CIM diffusion is described in detail as changes in labor productivity in the five metalworking industries. This, in turn, is translated into changes in the level of technical progress which causes changes in output. An additional loop (output-investment-capital-output) produces an additional spillover effect. There exists the possibility to link labor productivity (labor and technical progress) to capital turnover, but this was not done in the actual simulations.

The M model (*Figure 8.4*) is the one described by Mori in Chapter 4. (Mori has presented two versions of his model, but the philosophy of the one illustrated in *Figure 8.4* is closer to the other three models.) The production function in the M model is designed to allow one to assess directly the labor enhancement effect of robot penetration. To minimize costs, the firms select the optimal combination of workers and robots. Assuming an exogenous level of output, the M model allows one to study the impact of two policies: robot penetration with endogenous labor (POLICY-1) and robot penetration with exogenous labor (POLICY-2).

We shall try to comment briefly on the costs and benefits of the methodological frameworks and the specific assumptions of the different authors.

The input–output framework is undoubtedly the most consistent and the one which theoretically captures most comprehensively the potential impact of technological change on economic performance. There is, however, one crucial point in the implementation of I–O models for forecasting and that is the projection of the A-matrix and of other structural coefficients in the future. Technological matrices change over time because of technological changes in production as well as general changes in economic structures (see Chapter 9 by Furukawa). Therefore, in forecasts, it is essential to define precisely the anticipated changes in the technological matrix and the assumptions leading to them. Of course this is by no means a trivial problem. The classical approach is the well-known RAS method, and Furukawa presents a good example of its implementation. There are other methods as well. For example, Leontief is more inclined to apply expert knowledge in forecasting the technological coefficients (see Leontief and Duchin, 1986); other more sophisticated versions of the RAS technique have also been developed recently (Snower, 1990).

Of the two I–O-based studies, only the study by Yamada tackles this problem assuming a time trend for the A-coefficients based on past observations. This approach does not distinguish precisely which future changes in A are actually due to technological change (and, going further, which are caused by CIM) and which are due to other factors. In addition this approach does not consider changes in the consumption and investment coefficients. This might be the reason for a couple of somewhat contradictory outcomes in the simulation results in his otherwise excellent study.

The Polish team tried to incorporate explicitly the capital structure and labor substitution effects in the metalworking industries in the CIM diffusion scenarios in its model. The researchers ignored, however, the other important links to the input and final demand structures.

In general, the input–output approach is a time-consuming and expensive one. Of course the effort invested always pays off, but the limited resources of the Project did not allow us to extend this study beyond three countries.

The neoclassical growth framework applied by Dobrinsky in Chapter 14 is rather limited in scope as compared with the I–O one. The supply-driven approach is less flexible with respect to analyzing the economic interrelations. On the other hand, his approach allows one to describe precisely the CIM diffusion scenarios and to assess the pure effect of this technological development. The models are quite simple, and it was possible to cover 11 countries in this study.

The M model is more specific as compared with the other three as it addresses developments at a sectoral level. Its most advantageous feature is that it enables one to assess the direct labor substitution effect of the new technology.

The ideal methodological framework to analyze the economic impacts of CIM would be an input–output model that considers structural changes in input and final demand as in the Y model, the effect on capital and productivity as in the P model, the CIM diffusion scenarios as in the D model, and the labor substitution effects as in the M model. In addition, however, it should allow one to distinguish explicitly the structural change due to technological change, in particular, to CIM diffusion.

8.2   The Economic Impacts of CIM in Simulations

8.2.1   Control solutions

The first problem to be solved in the practical implementation of the models is a methodological one. Is it possible to analyze only one technological development such as CIM apart from the overall technological change? Is it possible to distinguish precisely which changes in economic performance are due only to CIM? These are very difficult questions to answer, and there are good reasons to answer them negatively. In fact, do we observe in reality an isolated technological development and are the statistics on its impact taken separately?

Economists have invented an approach which they claim can help them give a positive answer to these questions (or, at least, to avoid the direct answer). These are the conditional (what if ...) forecasts and the evaluations of relative performance. That is, we design scenarios under alternative assumptions for the future and measure the difference in the economic performance. This difference is supposedly related to the difference in the initial exogenous assumptions. This is actually how this problem is tackled by the econometric studies within the CIM Project.

Now the next problem arises: How to design the reference or control scenario to serve as a basis for the evaluation of the model performance under a CIM diffusion scenario? Paradoxically (and unfortunately) there is no good solution to this problem. If one chooses a "normal" growth path as a reference scenario the danger arises that this "normal" path already contains some of the technological change that we would like to treat separately. On the other hand, a "no-technological-change" control scenario is an unlikely and unrealistic development pattern to be considered as a reference point. Anyway, the second option usually is considered as the "lesser evil," and this is what has actually been adopted by our econometric studies. Again the M study is an exception in the sense that it does not make use of a reference scenario but rather compares the performance of the model under the two formulated employment policies and different exogenous assumptions.

Table 8.1 gives a brief overview of the reference scenarios in three of the CIM-related econometric studies: the Y model, the P model, and the D model. Though not always directly comparable the assumptions of the three studies are more or less consistent in the sense that they reflect the general notion of "no-technological-change" development pattern of the economy. The Y model and the P model do this by assuming an unchanging A-matrix and no structural change on the final demand side. In the D model this is done by fixing the exogenously set rate of technical progress to zero.

There are, however, some differences in the performance of the models, and these are due to the difference in their driving forces. The Y model and the P model are demand-driven, and their performance is to a large extent determined by the exogenously set demand items (basically these are the public expenditures). This is especially true for the growth rate of real GNP, real consumption, and real investment in the reference scenarios. The D model is supply-driven, and its performance is determined by the exogenously set labor input and investment ratio. These are the causes for the differences in the growth paths projected by the reference scenarios of the three models.

In spite of these differences, all reference scenarios in the three models are characterized by slow growth, which by any measure is below a realistic

Table 8.1. Exogenous assumptions and main results for Japan, the USA, and the FRG.

	Japan			USA			FRG	
	Y model 1990–2000	P model 1990–2000	D model 1985–2000	Y model 1990–2000	P model 1990–2000	D model 1985–2000	Y model 1990–2000	D model 1985–2000
Exogenous assumptions								
Exchange rate (.../$)	120	–	–	–	–	–	1.8	–
Population (million)	107.6[a]	–	–	208.2[b]	–	–	58.95[c]	–
Discount rate (%)	4	–	–	–	10.8[f]	–	4.7	–
Money supply (%)	–	–	9.0	–	–	9.0	–	6.5
Public consumption (%)	3.5	}3.5[d]	–	1.5	}1.5[d]	–	2.8	–
Public investment (%)	2.0		–	1.5		–	2.0	–
Labor input (%)	–	–	0.4	–	–	1.4	–	–0.7
Investment ratio (%)	–	–	0.314	–	–	0.183	–	0.203
Technological change	No change	No change	Zero rate	No change	No change	Zero rate	No change	Zero rate
Results								
Real GNP (%)	2.64	2.39	1.17	1.39	2.70	1.57	2.06	0.43
Consumption (%)	1.65	1.47	–	0.27	2.65	–	1.40	–
Housing investment (%)	1.36	–0.45	}1.43[e]	–4.65	7.09	}1.10[e]	0.36	}0.66[e]
Business investment (%)	1.35	2.09		4.19	3.61		2.96	
Exports (%)	7.16	5.33	–	6.75	2.31	–	4.19	–
Imports (%)	4.45	2.85	4.16	4.05	2.74	2.72	4.34	3.37
Employment (%)	0.80	0.64	–	1.10	0.77	–	0.53	–
Rate of unemployment (%)	–7.16	1.52	–	–1.65	–2.30	–	–8.26	–
GNP deflator (%)	6.36	3.22	8.85	4.68	7.45	4.70	2.83	5.68
Wage rate (%)	9.24	4.73	–	5.17	9.14	–	4.67	–

[a]Population over 15 years in 2000. [b]Population over 16 years in 2000. [c]Total population in 2000. [d]Total government expenditures.
[e]Total investment. [f]Prime rate. [g]Labor force in 2000 (million).

forecast for these countries. These control scenarios only serve as the reference point to feature the effect induced by a technological change such as CIM.

8.2.2 Modeling the CIM diffusion

Once the reference point is fixed, the next step is to design the scenarios featuring the diffusion of CIM technologies. In the macroeconomic framework of the econometric models the conventional diffusion models are irrelevant and require additional adaptation. The traditional diffusion models yield figures such as absolute numbers of production units or level of penetration (population density) of the new technology. This might be a necessary but not a sufficient condition for the incorporation of a diffusion process into the econometric model. What is actually needed for a macroeconomic model is a transformation of these changes into changes of the model parameters, i.e., an additional block or submodel performing this transformation.

As already mentioned the modeling teams have adopted different approaches to the problem of transformation. In the Y model no formal transformation block exists; instead the CIM diffusion is depicted implicitly as changes in the A-matrix and in the final demand structure. These changes, however, are only described as time trends and do not necessarily reflect a specific CIM diffusion process. Then a scenario with technological change is one in which these parameter changes are exogenously set.

The P model contains a CIM diffusion submodel of the changing structure of fixed assets (traditional equipment versus CIM equipment); these changes are transformed (through the labor substitution effect implied in the production function) into a changed level of labor demand. The A-matrix and the other structural coefficients, however, are not sensitive to the changes.

The D model contains a detailed submodel describing the CIM diffusion process of the changing capital structure (traditional equipment and three levels of automated technologies). This in turn is transformed into changing structures of labor and output.

The M model describes the diffusion process as exogenously fixed levels of robot penetration; two such cases (high and low) have been evaluated.

Due to the rather different nature of the models, it is practically impossible to compare the CIM diffusion assumptions directly. It should be noted though that three of the modeling teams (Y model, P model, and D model) have based their CIM diffusion scenarios at least partly on the IIASA FMS data base. In addition the Y model adopts as an exogenous input the CIM-induced labor productivity changes resulting from the D model. So there is at least some qualitative consistency in the assumptions regarding the CIM diffusion in the different studies.

8.2.3 The macroeconomic impact of CIM: Comparison of results

One of the major findings of the IIASA CIM Project is the revolutionary effect that CIM is expected to have on virtually all aspects of economic, social, and even political life. This is highlighted by the results presented in the volumes of the final report. Ayres and Bodda (Chapter 3) briefly summarize the direct benefits of CIM as labor saving, capacity augmenting/capital saving, capital sharing/saving, product quality improvements; and acceleration of product performance improvement. The combination of these effects results in an overall reduction of pay-back time for CIM equipment and, in the final run, in a general acceleration of the capital turnover and an increase in capital productivity.

It is not realistic to expect that any model could address in detail the rich variety of effects and their interrelations. Turning back to the econometric models we can see that they are rather modest in scope and treat only some specific macroeconomic issues of CIM-induced changes in economic performance. As the models, which were built by different teams, are different in nature so are the issues that they address in the simulations. The results of these simulations are well described and analyzed in the papers contained in this volume, so we shall concentrate on the comparison of the different findings as well as on the general conclusions concerning the expected socioeconomic impact of CIM.

Table 8.2 presents an overview of some of the important simulation results of three modeling efforts: the Y model, the P model, and the D model. These results describe on a comparative basis the future economic performance of Japan, the USA, and the FRG in a scenario with CIM diffusion in the five metalworking sectors as well as the "pure" CIM-induced effect, i.e., the difference between the CIM and the reference scenarios. It should be added that for this comparison we have selected Case 3 of the Y model (technological change only in the five metalworking sectors), which is the closest among the three Y scenarios to the CIM scenarios of the P and D models.

Table 8.2. Comparison of simulation results for Japan, the USA, and the FRG.

	Japan			USA			FRG	
	Y model 1990–2000	P model 1990–2000	D model 1985–2000	Y model 1990–2000	P model 1990–2000	D model 1985–2000	Y model 1990–2000	D model 1985–2000
Average annual growth rates, %								
CIM scenarios								
Real GNP	3.20	2.65	1.82	1.50	2.77	1.94	2.51	0.85
Consumption	1.88	1.55	–	0.23	2.84	–	1.25	–
Housing investment	1.42	−0.24	}2.08^a	−4.80	7.89	}1.47^a	0.26	}1.08^a
Business investment	2.85	2.04		4.06	3.17		4.32	
Exports	7.85	6.37	–	6.96	2.47	–	4.63	–
Imports	4.85	3.18	4.58	3.62	2.69	3.01	4.05	3.54
Employment	0.80	0.67	–	1.13	0.63	–	0.49	–
Rate of unemployment	−6.59	0.30	–	−1.97	0.45	–	−4.03	–
GNP deflator	5.56	3.02	8.29	4.39	7.08	5.58	1.97	5.34
Wage rate	8.70	4.62	–	5.08	8.92	–	3.62	–
CIM scenario vs. reference scenario								
Real GNP	0.56	0.25	0.65	0.11	0.08	0.37	0.45	0.42
Consumption	0.23	0.08	–	−0.04	0.19	–	−0.15	–
Housing investment	0.06	0.22	}0.65^a	−0.15	0.80	}0.37^a	−0.10	}0.42^a
Business investment	1.50	−0.05		−0.13	−0.44		1.36	
Exports	0.69	1.04	–	0.21	0.17	–	0.44	–
Imports	0.40	0.28	0.42	−0.43	−0.05	0.28	−0.29	0.16
Employment	0.00	0.08	–	0.03	−0.14	–	−0.04	–
Rate of employment	0.57	−1.21	–	−0.32	2.75	–	4.23	–
GNP deflator	−0.80	−0.20	−0.56	−0.29	−0.37	0.87	−0.86	−0.35
Wage rate	−0.54	−0.11	–	−0.09	−0.22	–	−1.05	–

^{a}Total investment.

In spite of attempts to use the same (or similar) assumptions in the CIM scenarios of the three studies, it was not possible to make them completely unified. Besides the models possess different properties and may manifest different performance even if the exogenous assumptions are identical.

To our judgment we could assume that the differences in the absolute figures depicting the economic performance under the CIM scenario (the upper part of *Table 8.2*) to a large extent are attributed to the underlying philosophy of the models. Of greater importance is the sensitivity of the models to the shock of the CIM diffusion, and it is described by the discrepancy levels shown in the lower part of the table.

Although there is not a full synchronism in the reactions of the models, some remarkable similarities are symptomatic of the expected impact of CIM.

The first important symptomatic result is that in all experiments GNP responds positively to the CIM diffusion, i.e., the CIM diffusion leads to a higher economic growth in the countries. The evaluated quantitative effect on GDP in the different studies is of comparable magnitude and scale, and the cross-country differences can be attributed to the different weight of the metalworking industries in the economies (higher in Japan and the FRG and lower in the USA).

Another identical response is the positive change of exports. The interpretation of this phenomenon could well be the fact that this technological innovation increases the export potential of the countries.

A third symptom is the negative response of the GNP deflator to the CIM diffusion (the USA in the D model is the only exception due to a specific function for the velocity of money). The most important factor for this (together with other factors which we shall consider later) is the reduction of unit costs caused by the CIM equipment.

A fourth similar reaction is a positive change of real wages (with the exception of the FRG in the Y model). Although the nominal wage rate in general responds negatively, the absolute level of its change is smaller than that of the price deflator which results in a positive change of real wages. This in turn leads to an increase in consumption which in a demand-driven model further increases output and GNP.

Actually in the CIM results of the two demand-driven input–output models, one can trace something very similar to the the effect of a Salter cycle: the more productive CIM technologies introduced into the metalworking sectors cause reduction in labor demand. This reduces labor costs and leads to price decreases; the latter stimulates demand and induces an increase in output. In Chapter 10 Yamada analyzes the Salter causality chain in quantitative terms.

At the same time there are cross-country as well as cross-model differences in the response to the CIM shock. Thus the rate of unemployment in Japan responds positively in the Y model and negatively in the P model. The rate of unemployment in the USA manifests exactly the opposite reaction in the two models – negative in the Y model and positive in the P model. There are differences in the responses of consumption, investment, and imports as well.

There are three main reasons for these differences. The first one is connected with the differences in the modeling methodology (see the causality chains in *Figures 8.1* to *8.4*). The second factor is the difference in some behavior equations in the model. As a third factor we mention the country-specific characteristics. The combination of these factors affects to a large degree the labor demand (and subsequently, the level of unemployment), which is subject to various interrelated influences. In the Y model the CIM diffusion is translated into changes of labor productivity, and this in turn is used (through the inverse of a production function or through a labor-demand function) to derive employment. In the P model the technological change is expressed in changes of capital productivity, which is thence fed to the inverse of a production function to determine labor demand. Due to the different responses from the two cases, the overall reaction of the economy can also be different from what we observe in the results of the two models.

This might be the reason for the different estimations in the three studies of the effect of CIM diffusion on labor employment in the metalworking sectors (*Table 8.3*). The results of the Y and P models are the directly evaluated figures for the decrease in labor demand in the five sectors in the CIM scenario as compared with the reference scenario. The D model, which is supply-driven, does not allow for a direct estimate of the changes in labor demand, so the figures presented are only indirect estimations. The estimation is made under the assumption that every newly created work place associated with CIM technologies (these figures are calculated in the model) causes an equivalent (by labor productivity) decrease in labor demand in the corresponding sector. The ranges given in the table correspond to the productivity ratios ranging from one to the actual CIM productivity ratios (see Chapter 14). The table also contains the estimated effects on labor due to robot penetration in Japan according to the results of the M model.

The results for the effect on employment present a broad variety of responses to the CIM diffusion. Some of the disparity in the responses may be

Table 8.3. CIM effect on employment in metalworking industries.

Decrease in labor demand in 2000: CIM vs. reference scenario (thousands)	ISIC 381 Final metal products	ISIC 382 Non-electrical machinery	ISIC 383 Electrical machinery	ISIC 384 Transport equipment	ISIC 385 Scientific equipment	ISIC 38 Metalworking total
Japan						
Y model	269	1,384	1,078	1,058	213	4,002
P model	551	461	731	466	126	2,335
D model[a]	265+477	406+725	568+1,000	319+578	82+153	1,640+2,933
M model[b]			9%+15%	−8%+12%		
USA						
Y model	656	1,035	818	1,280	325	4,114
P model	140	210	350	580	200	1,480
D model[a]	179+341	543+1,015	721+1,363	654+1,207	272+503	2,369+4,429
FRG						
Y model	44	294	573	522	88	1,521
D model[a]	56+110	257+475	152+292	187+359	25+47	677+1,283

[a]The D model estimates are made under the assumption that new CIM work places cause an equivalent (by labor productivity) decrease in labor demand in the corresponding sector.
[b]2000 vs. 1985; decrease is only due to industrial robot penetration.

attributed to differences in the sectoral employment figures actually used by the different studies. The estimations of the Y model, in general, seem a bit overestimated as the changes in labor demand approach levels comparable with the actual employment levels in these sectors at the present time. However, as stated above, this also might be due to a difference in the statistical base.

In any case these results as a whole reveal the dramatic structural change brought about by CIM, which is occurring and which will deepen substantially in the future. The sectoral impacts on the level of output (see Chapters 10 and 14) are also impressive; CIM might lead to as much as 100% increases in the sectoral output of the metalworking industries by the year 2000.

8.3   Conclusions

Of course in the end we have to answer the question What lessons have we learned from the rather expensive and time-consuming exercise of CIM econometric modeling?

First we have to reemphasize that the models are capable of catching only some of the socioeconomic aspects of technological change. One should be aware of this and should not expect from the models more than they can give. One obvious lesson is that traditional econometric models, such as the ones which were used within the CIM Project, are not relevant to capture the complexity of the phenomenon we are studying. However, it seems that no better models exist now. The development of better models will clearly be a challenge for future research.

However, within the scope of what the models *can do*, there are some very interesting and encouraging results. The most important of these results is the presence in the different studies of a stable set of identical, or at least similar, indicators describing the economic performance of the countries under CIM diffusion (the "symptomatic" results which we discussed in Section 8.2). The fact that these results were arrived at in independent studies increases the reliability of the findings.

We theorize that the existence of symptomatic results might be a good reason to extend their validity beyond the boundaries of the countries for which they were estimated and thus to consider them as a global characteristic of the economic impact of CIM. This is a speculation, but there is no evidence from the studies rejecting such a hypothesis.

At the same time there are several areas where the expected impact of CIM is not clearly defined. The overall effect on employment and the impact on consumption, investment, and imports may well be considered as country-specific, but they clearly deserve future study.

It is a pity that some interesting issues such as the cross-sector CIM sensitivity of the models were not analyzed separately in any of the econometric studies (this is well within the boundaries of the analytical power of the models). Such an analysis might be helpful for evaluating the pure spillover effect of the CIM diffusion. This is another possible area for future research.

Acknowledgment

Many valuable suggestions and comments by Robert Ayres are gratefully acknowledged.

References

Ayres, R., 1988, "Complexity, Reliability, and Design: Manufacturing Implications," *Manufacturing Review* 1(1).

Ayres, R., and S.M. Miller, 1983, *Robotics: Applications and Social Implications*, Ballinger Publishing Company, Cambridge, MA.

Brooks, H., 1985, "Automation Technology and Employment," *ATLAS Bulletin*, November, United Nations, New York, NY.

Fleissner, P., R. Dell'mour, and P.P. Sint, 1981, "Makroökonomische Aspekte der Mikroelektronik," in *Mikroelektronik – Anwendungen, Verbreitung und Auswirkungen am Beispiel Österreichs*, Springer-Verlag, Vienna, Austria.

Friedrich, W., and G. Roenning, 1985, Arbeitsmarktwirkungen moderner Technologien, Institut für Sozialforschung und Gesellschaftspolitik, Cologne, Germany.

Howell, D.R., 1985, "The Future Employment Impact of Industrial Robots," *Journal of Technological Forecasting and Social Change* 28(4):297–310.

Kinoshita, S., and M. Yamada, 1989, "The Impact of Robotization on Macro and Sectoral Economies within a World Econometric Model," *Journal of Technological Forecasting and Social Change* 35(2–3):211–230.

Leontief, W., and F. Duchin, 1986, *The Future Impact of Automation on Workers*, Oxford University Press, New York, NY.

Snower, D., 1990, "New Methods for Updating Input-Output Matrices," *Economic Systems Research* 2(1):27–38.

Chapter 9

Input–Output Structural Changes and CIM

Shunichi Furukawa

Industrial structures have been changing rapidly since the 1970s. This period was a turning point for energy systems and production technology. Since the mid-1970s, energy-saving technologies have been emphasized to reduce energy inputs per unit of production. For that purpose, major investments were needed to replace old production facilities with new ones. In this process, those industries which strengthened their international competitiveness through technological innovations, have not only survived but also expanded their market shares in the world. In particular, newly industrializing economies (NIEs) in Asia have expanded their market shares by increased international competitiveness in both labor inputs and energy inputs.

A second turning point was the worldwide economic recession in the early 1980s. In this period, consumption patterns changed from more volume to more varieties, from emphasis on quantity to emphasis on quality, from large scale to more compact, from heavy to light, and from a single purpose to a multipurpose. This has accelerated innovation in production technologies and input structures. At the same time, electronics and communication equipment have penetrated more deeply into various production systems and products. The computer integrated manufacturing (CIM) system can be considered to be a composite of several production information technologies that are brought together through managerial and organizational adjustments. The CIM system will mainly be regarded as a flexible production system, a labor-saving production system, and a capital-intensive system. It has been argued that CIM is the core of a new industrial transformation (Ayres, 1989).

This chapter analyzes the implications of long-term structural and technological changes and their effects, using an input–output approach. Three countries (Japan, the USA, and the Federal Republic of Germany) were selected for a comparative study. A time series of input–output tables are used for the analysis, and input structures are predicted based on these time-series data. The input–output tables used in the study are for the years 1973 to 1981. In addition, 1985 annual data are used for Japan (MITI, 1977, 1982), supplementary 1984 and 1985 data are used for the USA (US Department of Commerce, 1984, 1985), and biennial tables from 1978 to 1986 for the FRG (Statistisches Bundesamt, 1970–1986). An aggregation from large-scale tables to 53 industrial sectors was carried out (see Appendix I and Appendix II). In addition, the input coefficients and capital formation for the three countries for the year 2000 are predicted.

The chapter consists of three sections: 9.1 Structural Changes in Inputs; 9.2 Changes in Investment Pattern, and 9.3 Effects of Structural Changes. The method of projection of input–output tables forecasts for the year 2000 are described in this chapter.

9.1 Structural Changes in Inputs

An input coefficient is defined as follows:

$$a_{ij}^* = q_{ij}/Q_j \; , \tag{9.1}$$

where Q_j is the output of industry j and q_{ij} is an intermediate input from industry i, which industry j uses for producing its output. The a_{ij}^* can be measured in either quantity terms or value terms. The expression a_{ij}^* indicates the number of units of input i that are used for one unit of product j. In this case, a_{ij}^* is called a technological input coefficient. Input coefficients are relatively stable as long as technologies do not change. This can be assumed for a short period, but it is not necessarily valid in the long term. Three major factors can change a_{ij}^*: changes in material inputs; changes in production efficiency; and changes in production mix.

These factors are not independent of each other. Coefficient changes in an input–output table show us the results of such technological changes but not the causes. In observing input coefficients in time series, it is possible to trace how technological input coefficients have changed over time. If input coefficients in quantity terms were available as time series, it would be possible to analyze the trends and infer the factors that affected each structural change. However, it is not always possible to obtain the input–output table in terms of quantity. In some cases there are records of the amounts of major materials used for certain products such as iron and steel. But, it is impossible to list every input item. In most cases, the data are recorded in value terms of necessity, due to aggregation of activities and commodities within sectors, and so on. Therefore, most countries have published input–output tables in value terms. An input coefficient in value terms, a_{ij}, can be formulated as

$$a_{ij} = x_{ij}/X_j \tag{9.2}$$
$$= p_i q_{ij}/p_j Q_j \ , \tag{9.3}$$

where X_j is the value of the output of industry j, x_{ij} is the intermediate input from sector i which industry j uses for producing the output, and p_i is the price of product i. Equation (9.3) shows that a_{ij} depends not only on input q_{ij} and production Q_j but also on prices, p_i and p_j. Therefore, a_{ij} is always affected by the relative price level, which is not always stable. In a time-series analysis, we can minimize this problem by using input–output tables at constant prices, where both tables (a base-year table and a current-year table) are valued at the same price levels. The constant price input–output tables are available from 1978 to 1986 biennially for the FRG and from 1973 to 1981 for Japan. In addition, an input–output table for the USA based on 1982 constant prices was estimated. But, those for the USA are not published. For an international comparison and to use the most recent tables, it was not possible to use input–output tables in constant prices.

9.1.1 Intermediate input trends

The total intermediate input ratio of industry j, w_j, is defined as

$$w_j = \sum_i x_{ij}/X_j \tag{9.4}$$

or

$$w_j = \sum_i a_{ij} \; . \hspace{4cm} (9.5)$$

The intermediate input ratio indicates to what extent a production activity depends on intermediate inputs or on goods and services produced by other sectors (including the sector itself). *Table 9.1* shows the total intermediate input ratios (TIIR) of three countries in the latest year (1985 for the USA and Japan, and 1986 for the FRG). *Table 9.2* shows a deviation of intermediate input ratio for the latest year from the average. From the tables, two implications can be drawn regarding the changing pattern of total intermediate input ratios among the three countries.

(1) Total intermediate input ratios of the services industries (Sectors 43 to 52) are distributed between 0.1 and 0.6 in Japan, 0.2 and 0.5 in the USA, and 0.1 and 0.9 in the FRG. Two sectors in Japan, the transportation and warehousing sector (Sector 46) and the business-services sector (Sector 51), use more intermediate inputs than those of the USA, but the other five Japanese services industries, the real estate sector (Sector 45), the communication sector (Sector 47), the financial-services sector (Sector 44), the health, education, and social services sector (Sector 50), and the other personal-services sector (Sector 52) use less intermediate inputs than their counterparts in the USA.

(2) The ratios of machinery industry (Sectors 29 to 38) are distributed between 0.57 and 0.72 in Japan, between 0.48 and 0.72 in the USA, and between 0.44 and 0.72 in the FRG. Japanese machinery industries, except the electronics and communication-equipment sector, use slightly more intermediate inputs than those of the USA and the FRG. This implies that the Japanese machinery industry is a low value-added sector compared with the other two countries.

9.1.2 Changes in total intermediate input ratio

Changes in total intermediate input ratio are quite different among the three countries in the long-term trend. We classify the sectors into five categories: (1) the sectors which have increased intermediate input ratios in both Japan and the USA; (2) the sectors which have reduced ratios in both Japan and the USA; (3) the sectors which have an increased ratio only in Japan; (4) the sectors which have an increased ratio only in the USA; and (5) the sectors which have an increased ratio only in the FRG, *Table 9.3*.

Table 9.1. Total intermediate input ratios (TIIR).

Sector	Japan 1985	USA 1985	FRG 1986	Sector	Japan 1985	USA 1985	FRG 1986
1	0.3037	0.5276	0.5706	27	0.7307	0.7167	0.8294
2	0.7638	0.7495		28	0.5748	0.5666	0.5553
3	0.4932	0.4717		29	0.5986	0.4758	0.5740
4	0.4558	0.4987	0.4612	30	0.5830	0.4811	
5	0.3543	0.5999	0.5575	31	0.6571	0.6638	0.7164
6	0.5639	0.4298		32	0.6932	0.6428	0.4819
7	0.4557	0.5078	0.5504	33	0.6107	0.7181	
8	0.3443	0.2212	0.1541	34	0.6118	0.4751	
9	0.7235	0.7564	0.7562	35	0.6245	0.5439	
10	0.2067	0.4590	0.2036	36	0.7186	0.6219	0.6357
11	0.7099	0.7564	0.6404	37	0.6288	0.5428	0.5765
12	0.6792	0.6659	0.6295	38	0.5663	0.5227	0.4385
13	0.7060	0.6366	0.5986	39	0.6084	0.5725	0.5084
14	0.5907	0.5805		40	0.5675	0.5804	0.5053
15	0.7688	0.6495	0.7546	41	0.6267	0.4450	0.5071
16	0.6571	0.6463	0.6767	42	0.4473	0.5304	0.5496
17	0.5134	0.5439	0.4802	43	0.3291	0.3466	0.3314
18	0.7991	0.6765	0.7049	44	0.2484	0.4713	0.8826
19	0.7652	0.6718	0.6243	45	0.1281	0.1992	0.2052
20	0.5957	0.6118		46	0.5447	0.4550	0.4894
21	0.7503	0.8114	0.7548	47	0.2125	0.2580	0.1114
22	0.6193	0.6314	0.5048	48	0.4586	0.4358	
23	0.5962	0.6128	0.6030	49	0.2993	0.4259	0.3216
24	0.5475	0.5299	0.5816	50	0.3428	0.3867	0.6498
25	0.6266	0.5722	0.5765	51	0.4319	0.3090	
26	0.7975	0.6083	0.7823	52	0.4002	0.4749	0.3585

The Japanese machinery sectors, except for household electrical appliances, automobiles, and other transport equipment, have reduced intermediate input ratios. But in the USA the trends have been almost reverse. The electrical machinery industries, such as the electronics and communication-equipment sector and the office and service machinery sector, have increased their total intermediate input ratios. The automobile sector and the other transport-equipment sector have reduced ratios. In general, it can be stated that the Japanese machinery sector has been changing from a low value-added industry to a high value-added industry. On the other hand, the USA has already reached high value-added industrialization, and intermediate inputs have been shifting from material suppliers to the service sectors.

Table 9.2. Deviations of the latest TIIR from the average.

Sector	Japan 85–AVG	USA 85–AVG	FRG 86–AVG	Sector	Japan 85–AVG	USA 85–AVG	FRG 86–AVG
1	0.0216	0.0489	0.0135	27	−0.0309	−0.0255	0.0128
2	0.0405	−0.0734		28	0.0142	0.0006	−0.0003
3	0.0817	−0.0158		29	−0.0226	−0.0248	0.0080
4	0.0926	0.0418	0.0576	30	−0.0381	−0.0208	
5	0.0271	0.0527	−0.0817	31	−0.0197	0.0466	0.1191
6	0.1212	−0.0138		32	0.0239	0.0270	0.0024
7	0.0705	0.0532	−0.0506	33	−0.0197	0.1079	
8	0.0555	−0.0412	−0.0598	34	−0.0241	−0.0123	
9	−0.0401	0.0099	0.0063	35	−0.0315	0.0120	
10	−0.0721	−0.0635	0.0082	36	0.0733	−0.0068	0.0240
11	−0.0093	0.0256	0.0211	37	0.0377	−0.0102	−0.0075
12	−0.0072	0.0169	0.0186	38	−0.0033	0.0331	0.0333
13	−0.0105	−0.0046	0.0015	39	−0.0213	0.0031	−0.0008
14	−0.0320	0.0046		40	−0.0076	−0.0302	0.0193
15	0.0386	0.0089	0.0164	41	0.0540	−0.0232	0.0181
16	−0.0705	0.0108	0.0109	42	−0.0172	−0.0389	0.0451
17	−0.0044	0.0073	0.0078	43	0.0277	0.0318	0.0033
18	0.0218	−0.0247	−0.0066	44	0.0281	0.0432	0.0127
19	0.0050	−0.0170	0.0119	45	−0.0552	−0.0065	−0.0010
20	−0.0136	−0.0087		46	0.1323	−0.0030	0.0269
21	0.0175	−0.0274	0.0785	47	0.0524	0.0314	0.0060
22	−0.0080	0.0375	0.0060	48	−0.0452	0.0106	
23	−0.1016	0.0169	0.0435	49	−0.0436	0.0235	0.0107
24	−0.0219	0.0289	0.0264	50	0.0425	0.0133	−0.0072
25	0.0021	−0.0002	0.0145	51	−0.0194	0.0006	
26	−0.0055	−0.0267	0.0049	52	0.0025	−0.0101	−0.0042

9.1.3 TIIR excluding intrasectoral input

Table 9.4 shows the total intermediate input ratios minus the intrasectoral transaction for the entire period (average), the latest year (1985 or 1986), and the deviation of the latest year from the average. This can be defined as

$$w_j^* = \sum_i a_{ij} (i = j)$$

or

$$w_j^* = \sum_i a_{ij} - a_{ii} \; , \tag{9.6}$$

Table 9.3. Changes of intermediate input ratios.

Sector code	Sector name

(1) Sectors which have increased intermediate input ratios by more than 0.02 in both Japan and the USA.

1	Agriculture
4	Forestry and fishery
5	Metallic ore mining
7	Coal mining
32	Household electrical appliances
43	Wholesale and retail trade
44	Financial services
47	Communication

(2) Sectors which have reduced intermediate input ratios by more than 0.02 in both Japan and the USA.

10	Food and beverages
27	Basic nonferrous metals
29	General industrial machinery
30	Special industrial machinery

(3) Sectors which have an increased intermediate input ratio by more than 0.02 only in Japan.

2	Livestock
3	Agricultural services
6	Nonmetallic ore mining
8	Crude petroleum
18	Basic chemicals
37	Other transport equipment
41	Construction repair

(4) Sectors which have an increased intermediate input ratio by more than 0.02 only in the USA.

22	Rubber and plastic products
33	Electronics and communication equipment

(5) Sectors which have an increased intermediate input ratio by more than 0.02 only in the FRG.

21	Petroleum refinery
23	Leather products
42	Electricity, gas, and water supply
33	Electronics and communication equipment

Table 9.4. Total intermediate input ratios less diagonal elements.

Sector	Average during the period			1985	1985	1986	Deviations from average		
	Japan	USA	FRG	Japan	USA	FRG	Japan	USA	FRG
1	0.265	0.461	0.433	0.282	0.508	0.425	0.018	0.048	−0.008
2	0.722	0.643		0.736	0.602		0.014	−0.041	
3	0.397	0.475		0.493	0.473		0.096	−0.002	
4	0.221	0.471	0.409	0.298	0.521	0.438	0.077	0.051	0.030
5	0.340	0.490	0.318	0.354	0.540	0.317	0.015	0.050	−0.000
6	0.462	0.411		0.562	0.427		0.100	0.015	
7	0.375	0.313	0.350	0.456	0.347	0.357	0.081	0.033	0.008
8	0.272	0.200	0.193	0.314	0.189	0.188	0.042	−0.011	−0.004
9	0.611	0.577	0.519	0.551	0.584	0.518	−0.060	0.007	−0.001
10	0.272	0.307	0.193	0.199	0.271	0.187	−0.073	−0.036	−0.006
11	0.392	0.435	0.340	0.400	0.430	0.335	0.007	−0.004	−0.005
12	0.644	0.491	0.533	0.659	0.516	0.551	0.015	0.025	0.018
13	0.630	0.347	0.419	0.590	0.354	0.427	−0.040	0.006	0.007
14	0.594	0.572		0.561	0.578		−0.033	0.005	
15	0.335	0.467	0.454	0.423	0.471	0.440	0.087	0.004	−0.014
16	0.731	0.589	0.534	0.644	0.563	0.519	−0.087	−0.026	−0.015
17	0.399	0.452	0.426	0.393	0.432	0.436	−0.006	−0.020	0.011
18	0.426	0.493	0.346	0.448	0.491	0.323	0.022	−0.002	−0.023
19	0.652	0.661	0.496	0.735	0.696	0.488	0.083	0.035	−0.008
20	0.563	0.559		0.500	0.550		−0.063	−0.009	
21	0.702	0.762	0.573	0.678	0.745	0.454	−0.024	−0.016	−0.119
22	0.536	0.552	0.459	0.455	0.558	0.428	−0.081	0.006	−0.031
23	0.557	0.407	0.365	0.434	0.412	0.379	−0.124	0.005	0.014
24	0.520	0.438	0.410	0.451	0.450	0.417	−0.069	0.012	0.007
25	0.513	0.451	0.397	0.491	0.435	0.393	−0.023	−0.017	−0.004
26	0.267	0.447	0.257	0.267	0.417	0.265	0.000	−0.030	0.009

where a w_j^* is a total intermediate input ratio except intrasectoral transaction and a_{ii} represents an input coefficient for an intrasector i or a diagonal element of a_{ij}. In this table, interindustrial linkages between sectors can be observed. Japan and the USA exhibit different changing patterns. For instance, the electronics and communication-equipment industry (Sector 33) in Japan has increased intermediate inputs including intrasectoral inputs during the period, while that of the USA has a slightly declining total intermediate input ratio due to sharply decreasing intrasectoral inputs, despite increased inputs from the other sectors. In effect the US industry is becoming less vertically integrated. The automobile industry of Japan has

Table 9.4. Continued.

Sector	Average during the period			1985	1985	1986	Deviations from average		
	Japan	USA	FRG	Japan	USA	FRG	Japan	USA	FRG
27	0.408	0.397	0.386	0.334	0.420	0.384	−0.074	0.022	−0.002
28	0.517	0.500	0.480	0.523	0.494	0.474	0.006	−0.007	−0.006
29	0.413	0.385	0.384	0.438	0.373	0.387	0.024	−0.012	0.003
30	0.471	0.435		0.433	0.429		−0.037	−0.006	
31	0.591	0.501	0.496	0.465	0.548	0.559	−0.126	0.047	0.062
32	0.552	0.614	0.357	0.552	0.625	0.343	−0.000	0.011	−0.014
33	0.395	0.421		0.475	0.465		0.080	0.044	
34	0.561	0.407		0.533	0.399		−0.028	−0.008	
35	0.561	0.487		0.424	0.482		−0.137	−0.005	
36	0.396	0.398	0.427	0.326	0.397	0.438	−0.070	−0.001	0.012
37	0.514	0.424	0.466	0.433	0.407	0.428	−0.081	−0.017	−0.038
38	0.386	0.473	0.386	0.398	0.498	0.411	0.011	0.025	0.025
39	0.589	0.526	0.469	0.562	0.535	0.463	−0.027	0.009	−0.006
40	0.578	0.600	0.464	0.567	0.570	0.470	−0.010	−0.030	0.006
41	0.578	0.457	0.492	0.626	0.466	0.492	0.048	0.009	−0.000
42	0.453	0.349	0.409	0.421	0.339	0.423	−0.032	−0.010	0.014
43	0.277	0.317	0.296	0.310	0.332	0.293	0.033	0.015	−0.003
44	0.198	0.233	0.245	0.198	0.251	0.258	0.000	0.018	0.013
45	0.146	0.127	0.207	0.126	0.117	0.207	−0.020	−0.010	−0.000
46	0.306	0.321	0.332	0.440	0.322	0.315	0.134	0.001	−0.017
47	0.150	0.221	0.071	0.190	0.240	0.068	0.040	0.019	−0.003
48	0.498	0.421		0.459	0.410		−0.039	−0.011	
49	0.328	0.439	0.302	0.299	0.545	0.312	−0.029	0.107	0.010
50	0.293	0.360	0.446	0.338	0.362	0.451	0.045	0.002	0.004
51	0.392	0.226		0.318	0.219		−0.074	−0.008	
52	0.393	0.440	0.255	0.390	0.426	0.230	−0.003	−0.014	−0.025

shifted in the opposite direction, using less intermediate inputs from the other sectors and more intrasectoral inputs. The increasing importance of intrasectoral inputs implies that the Japanese automobile industry is becoming more vertically integrated. The other machinery industries have changed the intrasectoral inputs in parallel with the total intermediate inputs, i.e., no change in vertical integration.

In general, most industries have been increasing inputs from services industries as shown in *Table 9.5*. But, among manufacturing industries, it is interesting that Japanese textile industries and electrical machinery industries, except the electronics and communication-equipment sector, have

Table 9.5. Services input ratios by industry.

Sector	Japan 1981	Japan 1985	USA 1982	USA 1985	FRG 1982	FRG 1986
1	0.0511	0.0544	0.1030	0.1754	0.0603	0.0663
2	0.0869	0.0948	0.0864	0.0800		
3	0.0995	0.1221	0.1427	0.1553		
4	0.0488	0.0737	0.0655	0.0692	0.0675	0.0874
5	0.1170	0.1321	0.0884	0.1053	0.0706	0.0906
6	0.1358	0.4398	0.1118	0.1269		
7	0.0652	0.1364	0.0806	0.1012	0.0590	0.0678
8	0.0692	0.1368	0.0775	0.1009	0.0296	0.0460
9	0.0549	0.0686	0.0797	0.0843	0.0762	0.0900
10	0.0177	0.0272	0.0747	0.0701	0.0767	0.0820
11	0.0926	0.0811	0.0703	0.0770	0.0778	0.0817
12	0.1009	0.0607	0.0800	0.0860	0.1236	0.1449
13	0.0767	0.0883	0.0669	0.0660	0.1051	0.1117
14	0.0769	0.0891	0.1174	0.1289		
15	0.0918	0.0909	0.0849	0.0878	0.0887	0.0903
16	0.0670	0.0835	0.0756	0.0697	0.1048	0.1148
17	0.1074	0.1166	0.1763	0.1707	0.1050	0.1153
18	0.0838	0.0958	0.1120	0.1128	0.1209	0.1286
19	0.1039	0.0849	0.0888	0.0974	0.1053	0.1077
20	0.1602	0.1558	0.2140	0.2233		
21	0.0555	0.0493	0.0656	0.0790	0.0269	0.0509
22	0.0681	0.0788	0.0885	0.0923	0.0887	0.0924
23	0.0314	0.0579	0.0963	0.1070	0.0803	0.0915
24	0.1158	0.1047	0.0944	0.1021	0.1042	0.1211
25	0.1283	0.1385	0.1441	0.1359	0.1503	0.1680
26	0.0502	0.0659	0.0779	0.0662	0.0621	0.0656

actually reduced service input ratios. Those industries exported a great deal in the 1970s, but now they are losing out to some of the newly industrializing economies (NIEs). By contrast, the Japanese nonelectrical machinery industries have increased service inputs sharply, and these industries are now exporting their products not only to developing countries but also to developed countries. In the USA, two industries have increased levels of the service inputs. One is the office and service-machinery industry, and the other is the electronics and communication-equipment industry. In the FRG, only the office and service-machinery industry has increased service inputs significantly.

Table 9.5. Continued.

Sector	Japan 1981	1985	USA 1982	1985	FRG 1982	1986
27	0.0765	0.0830	0.0725	0.0790	0.0423	0.0445
28	0.0677	0.0892	0.0795	0.0823	0.1027	0.1125
29	0.0653	0.0908	0.0869	0.0914	0.1105	0.1165
30	0.0696	0.0964	0.0622	0.0646		
31	0.0884	0.0936	0.1021	0.1381	0.1542	0.2086
32	0.1230	0.0903	0.0981	0.1036	0.1047	0.1009
33	0.1092	0.1133	0.1393	0.1613		
34	0.1224	0.1095	0.0811	0.0843		
35	0.1004	0.0940	0.0932	0.0976		
36	0.0630	0.0647	0.0627	0.0615	0.0815	0.0934
37	0.0786	0.0843	0.1057	0.1101	0.0975	0.0974
38	0.1071	0.0990	0.1133	0.1269	0.0923	0.1148
39	0.0722	0.1040	0.1287	0.1411	0.1010	0.1206
40	0.0924	0.1341	0.1869	0.1620	0.1161	0.1321
41	0.0861	0.1145	0.0703	0.0652	0.0789	0.0916
42	0.0762	0.1115	0.0513	0.0519	0.0518	0.0811
43	0.1766	0.2452	0.2280	0.2418	0.2051	0.2089
44	0.1505	0.1902	0.4005	0.4087	0.8405	0.8318
45	0.0806	0.0538	0.1297	0.1300	0.0921	0.0946
46	0.1992	0.2059	0.2579	0.2304	0.2491	0.2777
47	0.0723	0.1447	0.1179	0.1159	0.0698	0.0787
48	0.3657	0.3456	0.3772	0.3657		
49	0.1935	0.1353	0.1183	0.1805	0.1341	0.1463
50	0.0473	0.0951	0.2114	0.2239	0.4320	0.4328
51	0.1845	0.2589	0.2043	0.2010		
52	0.0786	0.1175	0.1821	0.1897	0.1704	0.1843

Assuming at least two input–output tables are available for different points in time, one can study the effects of structural changes during the period at the sector level. Observed changes in input coefficients can be interpreted as industrial changes that have occurred. Denote $A(0)$ as a matrix of input coefficients in the base year and $A(t)$ as that of the current year t. Then, the relation between $A(0)$ and $A(t)$ can be written as

$$A(t) = RA(0)S + U \; , \qquad (9.7)$$

Table 9.6. Coefficients of commodity technology changes (R) in Japan.

Sector	1971	1978	1979	1980	1981	1985
1	1.4683	1.3747	1.2252	1.2393	1.1431	1.0000
2	1.4428	1.3694	1.3565	1.3537	1.3251	1.0000
3	0.8157	0.7125	0.7175	0.7007	0.6257	1.0000
4	1.2267	1.2484	1.2571	1.2855	1.2934	1.0000
5	1.5981	1.4506	1.2153	1.3446	1.4574	1.0000
6	1.4035	1.2678	1.3548	1.4270	1.4023	1.0000
7	1.3760	1.1687	0.9465	0.9341	0.9259	1.0000
8	1.0676	1.0889	1.1128	1.1169	1.1171	1.0000
9	1.0025	0.9196	0.8408	0.8546	0.8296	1.0000
10	0.1155	0.2877	0.3128	0.3030	0.3082	1.0000
11	0.8458	0.8135	0.8181	0.8291	0.8037	1.0000
12	1.7928	1.8777	1.9033	1.9321	1.9008	1.0000
13	1.2256	1.2446	1.2272	1.3048	1.2873	1.0000
14	1.3750	1.3841	1.3170	1.2782	1.2614	1.0000
15	1.1776	1.1679	1.1322	1.1421	1.2158	1.0000
16	0.8002	0.8101	0.8206	0.8304	0.8415	1.0000
17	0.9348	0.9494	0.9713	0.9732	0.9737	1.0000
18	0.8316	0.8434	0.8432	0.8537	0.8818	1.0000
19	1.2940	1.1843	1.2112	1.2498	1.2739	1.0000
20	0.9398	1.0604	1.0675	1.0595	1.0640	1.0000
21	1.2779	1.2882	1.2679	1.2919	1.3859	1.0000
22	0.7607	0.8378	0.8534	0.8545	0.8515	1.0000
23	0.9671	0.9516	0.9764	0.9959	0.9838	1.0000
24	1.2102	1.1631	1.1892	1.1848	1.1874	1.0000
25	1.0968	1.0965	1.0866	1.1234	1.1198	1.0000
26	1.1530	1.1556	1.1867	1.1863	1.1810	1.0000

where

$$A(t) = \begin{bmatrix} a_{11}(t) & \cdots & a_{1j}(t) & \cdots & a_{1n}(t) \\ a_{i1}(t) & \cdots & a_{ij}(t) & \cdots & a_{in}(t) \\ a_{n1}(t) & \cdots & a_{nj}(t) & \cdots & a_{nn}(t) \end{bmatrix}$$

and

$$A(0) = \begin{bmatrix} a_{11}(0) & \cdots & a_{1j}(0) & \cdots & a_{1n}(0) \\ a_{i1}(0) & \cdots & a_{ij}(0) & \cdots & a_{in}(0) \\ a_{n1}(0) & \cdots & a_{nj}(0) & \cdots & a_{nn}(0) \end{bmatrix} .$$

Table 9.6.   Continued.

Sector	1971	1978	1979	1980	1981	1985
27	1.0196	0.9932	1.0014	1.0283	1.0708	1.0000
28	0.8452	0.8308	0.8565	0.8721	0.8919	1.0000
29	1.4078	1.4169	1.4477	1.4713	1.5685	1.0000
30	0.9761	1.0037	1.0147	1.0253	1.0497	1.0000
31	0.6122	0.5454	0.5171	0.4969	0.3916	1.0000
32	1.0536	1.1627	1.1638	1.1582	1.1091	1.0000
33	1.6198	1.5452	1.5379	1.4961	1.4577	1.0000
34	1.1228	0.9876	0.8661	0.8366	0.7244	1.0000
35	0.5674	0.5095	0.5359	0.5108	0.4696	1.0000
36	0.9680	0.8972	0.9270	0.8594	0.7758	1.0000
37	0.8883	0.8986	0.9061	0.9172	0.9282	1.0000
38	1.2491	1.2640	1.2813	1.2701	1.2035	1.0000
39	0.4150	0.4217	0.4225	0.4154	0.4133	1.0000
40	1.0000	1.0000	1.0000	1.0000	1.0000	1.0000
41	0.8432	0.7918	0.8168	0.8343	0.8199	1.0000
42	0.7206	0.8279	0.8421	0.8059	0.9917	1.0000
43	1.1729	1.1452	1.2054	1.1883	1.0435	1.0000
44	0.7687	1.0276	1.1118	1.0993	1.0937	1.0000
45	0.8063	1.2301	1.2802	1.2538	1.2130	1.0000
46	0.5761	0.6275	0.6577	0.6749	0.6742	1.0000
47	1.0346	1.2326	1.3606	1.3079	1.2202	1.0000
48	1.2189	1.2759	1.3559	1.3486	1.3014	1.0000
49	1.0000	1.0000	1.0000	1.0000	1.0000	1.0000
50	0.2405	0.2830	0.2927	0.2774	0.2619	1.0000
51	0.7346	0.6559	0.6685	0.6756	0.6757	1.0000
52	0.6274	0.5918	0.6076	0.6218	0.6319	1.0000

In equation (9.7), R and S represent a diagonal matrix of coefficients of commodity technology change, and a diagonal matrix of coefficients of industrial technology change, respectively:

$$R = \begin{bmatrix} r_{11} & & 0 \\ & \ddots & \\ 0 & & r_{nn} \end{bmatrix} \qquad S = \begin{bmatrix} s_{11} & & 0 \\ & \ddots & \\ 0 & & s_{nn} \end{bmatrix}.$$

The matrix U reflects changes of input coefficients that are due to factors other than R and S:

$$U = \begin{bmatrix} u_{11} & \cdots & u_{1n} \\ & & \\ u_{n1} & \cdots & u_{nn} \end{bmatrix}.$$

Table 9.7. Coefficients of commodity technology changes (R) in the USA.

Sector	1977	1982	1984	1985
1	1.1862	1.0718	1.0331	1.0000
2	1.1017	1.0625	1.0253	1.0000
3	0.7798	0.8014	0.9340	1.0000
4	1.3274	1.1085	1.0676	1.0000
5	1.4274	1.2515	1.0389	1.0000
6	1.0744	1.0845	1.0083	1.0000
7	1.1908	1.0831	1.1244	1.0000
8	0.8785	1.1526	1.0906	1.0000
9	1.1847	1.0660	1.0440	1.0000
10	1.1452	1.0705	1.2204	1.0000
11	1.0992	0.9650	0.9978	1.0000
12	1.2994	1.0253	0.9965	1.0000
13	1.3351	0.9037	0.9858	1.0000
14	1.0791	0.8963	0.9165	1.0000
15	1.0576	0.9918	1.0209	1.0000
16	1.0268	0.9611	0.9887	1.0000
17	0.9438	0.9499	0.9674	1.0000
18	1.0969	1.0382	1.0246	1.0000
19	1.0714	0.9826	1.0255	1.0000
20	0.9934	0.9610	0.9671	1.0000
21	1.0225	1.3389	1.2445	1.0000
22	0.9926	0.8919	0.9523	1.0000
23	0.9678	0.9328	0.9568	1.0000
24	1.1442	1.0122	0.9682	1.0000
25	1.0991	0.9061	0.9538	1.0000
26	1.4954	1.1485	1.0501	1.0000

The matrices R and S are calculated by an iteration method, which was developed by Stone and called the RAS method (Stone *et al.*, 1962).

Tables 9.6 to *9.8* show the results of the calculation of R coefficients of Japan, the USA, and the FRG, respectively. *Tables 9.9* to *9.11* show the S coefficients for those countries. First, we can explore the stability of the coefficients during the period. The variations of coefficients are divided into types. The first type nearly constant in time. The second type changes monotonically with time. The third type is not closely correlated to a time trend (or has large variances within the period). There are several reasons why these coefficients can be unstable. One major factor is that R and S are measured at current prices. Thus, in cases where the prices changed by

Table 9.7. Continued.

Sector	1977	1982	1984	1985
27	1.3918	1.1076	1.0764	1.0000
28	1.1213	1.0113	0.9691	1.0000
29	0.9508	0.9261	0.9214	1.0000
30	1.2488	1.1770	1.0626	1.0000
31	1.0164	1.0066	1.1236	1.0000
32	1.1254	0.9249	1.0076	1.0000
33	0.9281	0.9989	1.0634	1.0000
34	1.0823	1.0559	0.9878	1.0000
35	0.8893	0.8318	0.9407	1.0000
36	0.9955	0.8722	0.9671	1.0000
37	0.7871	0.8630	0.9735	1.0000
38	1.0169	1.0881	1.0534	1.0000
39	1.3942	1.1077	1.0347	1.0000
40	1.0000	1.0000	1.0000	1.0000
41	1.5476	1.0715	1.2180	1.0000
42	0.7649	1.1058	1.0771	1.0000
43	1.0486	0.8725	0.8943	1.0000
44	0.8981	0.9480	0.9436	1.0000
45	0.7346	0.8566	0.9661	1.0000
46	1.0088	1.0738	1.0056	1.0000
47	1.0385	1.2467	0.9783	1.0000
48	0.8239	0.7612	0.9975	1.0000
49	0.8480	0.9139	0.9402	1.0000
50	0.8621	0.9114	0.8454	1.0000
51	0.6876	0.8399	0.8953	1.0000
52	0.9935	1.0893	0.9957	1.0000

a large degree and without any trend, R and S coefficients might be affected by price changes in either the input side or the output side. However, this chapter, does not analyze price factors. This is because few cases include both input–output tables in current prices and constant prices. I have, nevertheless, estimated future input coefficients for the year 2000 to analyze effects of both changes in the input structure and in the investment pattern. The procedure adopted is to multiply the input coefficient matrix for the base year, $A(1985)$ for Japan and the USA and $A(1986)$ for the FRG, by an estimated R and S for the year 2000, $R(2000)$ and $S(2000)$, *viz.*,

$$A(2000) = R(2000)\, A(1985)\, S(2000), \text{ for Japan and the USA}$$

Table 9.8. Coefficients of commodity technology changes (R) in the FRG.

Sector	1978	1980	1982	1984	1986
1	1.27418	1.09252	1.14521	1.06298	1.00000
2	1.00000	1.00000	1.00000	1.00000	1.00000
3	1.00000	1.00000	1.00000	1.00000	1.00000
4	1.40611	1.24901	1.06068	0.97910	1.00000
5	0.96888	1.02383	1.02538	1.07171	1.00000
6	1.00000	1.00000	1.00000	1.00000	1.00000
7	1.00998	1.12958	1.14491	1.11601	1.00000
8	1.45993	1.85826	1.86630	1.70619	1.00000
9	1.34956	1.16730	1.12672	1.08754	1.00000
10	1.28560	1.06493	0.97139	0.96613	1.00000
11	1.10191	1.02868	0.88947	0.89652	1.00000
12	1.23205	1.04400	0.97363	0.94055	1.00000
13	1.41836	1.23639	0.99828	1.02756	1.00000
14	1.00000	1.00000	1.00000	1.00000	1.00000
15	0.88881	0.92079	0.96499	1.03174	1.00000
16	0.97972	0.93670	0.94594	0.91069	1.00000
17	1.12173	1.08363	1.02291	0.98791	1.00000
18	0.93315	0.90696	0.94742	0.99764	1.00000
19	0.85906	0.93715	0.87458	0.94111	1.00000
20	1.00000	1.00000	1.00000	1.00000	1.00000
21	1.17316	1.58928	1.63113	1.54063	1.00000
22	1.00319	1.05863	0.98921	0.95222	1.00000
23	1.28627	1.02617	0.98251	0.91787	1.00000
24	1.10095	1.06950	1.00693	0.96098	1.00000
25	1.10943	1.06296	1.02817	1.00232	1.00000
26	1.21704	1.16766	1.03658	0.98308	1.00000

$$A(2000) = R(2000)\ A(1986)\ S(2000), \text{ for the FRG.}$$

This estimation method shows how the input coefficients, A, would evolve from the base year to the year 2000, if we could assume that the changing patterns of the R and S coefficients are essentially dependent on the time trend. If one could assume that the time trend is an adequate surrogate for technological change, and that the previous trend can be extrapolated into the future, the estimated future input coefficients reflect input structural changes resulting from the accumulation of all technological changes affecting production. Many conceptual and practical problems remain in this simple procedure, but it may still have an advantage in deriving possible future

Table 9.8. Continued.

Sector	1978	1980	1982	1984	1986
27	1.00180	1.16983	1.00146	1.11989	1.00000
28	1.16095	1.07258	1.03113	0.94965	1.00000
29	1.06742	0.96924	0.92197	0.98425	1.00000
30	1.00000	1.00000	1.00000	1.00000	1.00000
31	0.94536	0.91471	1.12037	0.90436	1.00000
32	0.95395	0.91264	0.85388	0.91453	1.00000
33	1.00000	1.00000	1.00000	1.00000	1.00000
34	1.00000	1.00000	1.00000	1.00000	1.00000
35	1.00000	1.00000	1.00000	1.00000	1.00000
36	1.03978	1.04007	0.96697	0.91006	1.00000
37	0.79091	0.93577	0.95709	0.95118	1.00000
38	1.15589	1.11147	0.98616	0.98553	1.00000
39	1.37969	1.38006	1.12236	0.98411	1.00000
40	1.21267	1.17333	0.99521	0.98983	1.00000
41	0.92261	0.88992	0.90339	0.92753	1.00000
42	0.86529	0.84456	0.97905	0.95948	1.00000
43	1.04342	0.99888	0.96874	1.00442	1.00000
44	0.83320	0.83164	0.93742	0.94919	1.00000
45	0.62591	0.69470	0.86225	0.90595	1.00000
46	0.89853	0.91812	0.92295	0.98312	1.00000
47	0.94418	0.88284	0.92048	0.90897	1.00000
48	1.00000	1.00000	1.00000	1.00000	1.00000
49	0.90845	0.92063	0.99797	0.98731	1.00000
50	0.91266	0.93450	0.94453	0.98668	1.00000
51	1.00000	1.00000	1.00000	1.00000	1.00000
52	0.72838	0.76999	0.83317	0.87600	1.00000

patterns directly from past long-term trends. If we can add more specific technological information, then we can obtain more realistic results. *Table 9.12* shows the estimated R and S coefficients of Japan, the USA, and the FRG for the year 2000.

9.2 Changes in Investment Pattern

Tables 9.13 to *9.15* allow one to compare the structural changes of capital formation in three countries. Again, the pattern of structural change in Japan is quite different from that of the USA. For instance, Japanese investment in the construction sector (Sector 40) sharply dropped from 62%

Table 9.9. Coefficients of industry technology changes (S) in Japan.

Sector	1977	1978	1979	1980	1981	1985
1	0.8835	0.8990	0.9308	0.9637	1.0838	1.0000
2	0.9528	0.9989	0.9945	1.0078	1.0531	1.0000
3	0.9029	0.6284	0.6372	0.6482	0.6740	1.0000
4	0.6949	0.6638	0.6743	0.6623	0.7034	1.0000
5	1.1385	1.1456	1.1303	1.1311	1.2488	1.0000
6	1.0537	1.0895	1.0923	1.1002	1.1245	1.0000
7	0.7846	0.7976	0.7825	0.8077	0.8549	1.0000
8	0.7769	0.8537	0.8702	0.8628	0.8165	1.0000
9	0.9499	0.9626	0.9749	0.9870	0.9999	1.0000
10	1.1892	1.1607	1.2075	1.1907	1.2225	1.0000
11	1.0586	1.0699	1.0771	1.0873	1.1068	1.0000
12	1.0502	1.0612	1.0705	1.0836	1.1124	1.0000
13	0.9087	0.9098	0.9111	0.9134	0.9361	1.0000
14	0.9465	0.9631	0.9686	0.9835	1.0093	1.0000
15	0.8852	0.8840	0.8955	0.9078	0.9054	1.0000
16	1.0521	1.0749	1.0919	1.1114	1.1161	1.0000
17	0.9906	1.0231	1.0337	1.0448	1.0596	1.0000
18	1.0387	1.0473	1.0670	1.0746	1.0223	1.0000
19	1.1635	1.1623	1.1820	1.1984	1.1466	1.0000
20	1.1942	1.1558	1.1542	1.1934	1.2117	1.0000
21	0.9390	0.9558	0.9748	0.9825	0.9765	1.0000
22	1.0587	1.0489	1.0486	1.0816	1.0987	1.0000
23	1.2227	1.2375	1.2524	1.2715	1.2888	1.0000
24	0.9653	0.9919	0.9883	1.0071	1.0294	1.0000
25	0.9105	0.9128	0.9406	0.9509	0.9615	1.0000
26	0.9064	0.8997	0.8850	0.8945	0.9076	1.0000

in 1973 to 49% in 1985 due to relatively low growth rates (less than 4% per annum), whereas the construction share in the USA is very stable at about 45%. In the FRG, the share of the construction sector is between that of Japan and the USA, but the actual level is the lowest of the three countries at about 30% in 1986. In addition, Japanese investment in the electronics and communication-equipment industry (Sector 33) expanded at an annual rate of 18%, and the share was between 2% and 7%. The Japanese share exceeds that of the USA (about 5%) in 1985, whereas the USA has also expanded the investment in this sector at a substantially high level of growth rate (about 14%) during the period. It can be concluded from the situation

Table 9.9. Continued.

Sector	1977	1978	1979	1980	1981	1985
27	0.9981	1.0474	1.0261	1.0095	0.9690	1.0000
28	0.9265	0.9256	0.9556	0.9734	0.9645	1.0000
29	0.9263	0.9372	0.9502	0.9620	0.9730	1.0000
30	1.0093	1.0275	1.0340	1.0450	1.0691	1.0000
31	1.1835	1.2072	1.2106	1.2535	1.3404	1.0000
32	1.0038	0.9942	0.9946	1.0173	1.0660	1.0000
33	1.0306	1.1628	1.1506	1.2239	1.2867	1.0000
34	1.0215	1.0959	1.1002	1.1486	1.1901	1.0000
35	1.1869	1.1982	1.2226	1.2319	1.2378	1.0000
36	0.9304	1.0411	1.0164	1.0626	1.1278	1.0000
37	0.9518	0.9642	0.9766	0.9890	1.0014	1.0000
38	0.9081	0.9197	0.9321	0.9438	0.9538	1.0000
39	1.1063	1.0701	1.0874	1.1133	1.1564	1.0000
40	1.0037	1.0244	1.0207	1.0405	1.0630	1.0000
41	0.9526	0.9557	0.9691	0.9749	0.9930	1.0000
42	1.1303	1.0448	1.0546	1.0970	1.0196	1.0000
43	1.1469	0.9756	0.9078	0.9245	0.9497	1.0000
44	0.9048	0.9188	0.9216	0.9353	0.9507	1.0000
45	1.3460	1.8952	1.8080	1.8542	1.8774	1.0000
46	0.8050	0.7754	0.7226	0.7427	0.7484	1.0000
47	0.9159	0.7408	0.7194	0.7533	0.7962	1.0000
48	1.4216	1.2373	1.1340	1.1829	1.2860	1.0000
49	1.2241	1.1415	1.1315	1.1559	1.1901	1.0000
50	0.8102	0.8120	0.7969	0.8050	0.8349	1.0000
51	1.1361	1.1100	1.0657	1.0847	1.1145	1.0000
52	1.0536	1.0505	1.0405	1.0582	1.1017	1.0000

described above that Japanese investments concentrated mainly in electronics and communication equipment during the period, while that of the USA were widely distributed.

Extrapolation on this basis of historical trends implies that the increase in Japanese capital formation will go to the nonconstruction sectors, particularly the electrical and electronics sectors. Conceivably, the Japanese construction share could be decreased to the same level as FRG, while capital formation in the electronics and communication-equipment sector could grow from 7% to 28% of total capital formation by the year 2000.[1] The same sort of extrapolation implies that in the USA about 45% will be absorbed by the construction sector (Sector 40) and only a small part will go to any single nonconstruction sector. In the FRG, the share of nonelectrical

Table 9.10. Coefficients of industry technology changes (S) in the USA.

Sector	1977	1982	1984	1985
1	0.9374	0.8062	0.9633	1.0000
2	1.0770	1.1439	1.0252	1.0000
3	1.0079	1.0434	0.9669	1.0000
4	0.7326	1.0008	0.8919	1.0000
5	0.7862	0.8802	0.9456	1.0000
6	1.0329	0.9813	0.9168	1.0000
7	0.7838	0.8883	0.9519	1.0000
8	1.3764	0.9052	0.9097	1.0000
9	0.9041	0.9994	1.0025	1.0000
10	1.2112	1.2705	1.0225	1.0000
11	0.9134	1.0026	1.0101	1.0000
12	0.8682	0.9879	1.0032	1.0000
13	0.7972	1.1037	0.9853	1.0000
14	0.9098	1.0028	1.0196	1.0000
15	0.9498	1.0147	1.0002	1.0000
16	1.0196	1.0473	1.0650	1.0000
17	1.0642	1.0680	1.0673	1.0000
18	0.9847	1.0219	0.9363	1.0000
19	0.8976	0.9665	0.8954	1.0000
20	1.0875	1.0895	1.0465	1.0000
21	1.1453	0.9615	0.9348	1.0000
22	0.8922	0.9932	1.0158	1.0000
23	0.9521	0.9816	1.0190	1.0000
24	0.9079	0.9615	1.0008	1.0000
25	1.0080	1.0821	1.0426	1.0000
26	0.9585	1.0785	1.0420	1.0000

machinery (Sector 29) is now 14%, the highest among the three countries. If the past trend continues, the share will be 17% by the year 2000.

9.2.1 Percentage share of machinery and equipment

Table 9.16 shows the percentage share of machinery and equipment in three countries. The percentage share of machinery and equipment in capital formation in all three countries is almost the same level – from 36% to 39%. But, the composition in the USA is quite different from the other two countries. The largest share in the USA is automobiles at 9.7%.[2] In Japan and the FRG, investment in the industrial machinery sector is the largest, about 14%. The share of electrical machinery in Japan is the highest among

Table 9.10. Continued.

Sector	1977	1982	1984	1985
27	0.8864	1.0466	0.9856	1.0000
28	0.8706	0.9521	1.0149	1.0000
29	0.9838	1.0132	0.9949	1.0000
30	1.0052	1.0041	1.0327	1.0000
31	0.8448	0.9432	0.9599	1.0000
32	0.8737	0.9922	1.0256	1.0000
33	0.7885	0.9470	1.0185	1.0000
34	1.0001	1.0256	1.0309	1.0000
35	0.9537	0.9898	1.0389	1.0000
36	0.9996	1.0282	1.0272	1.0000
37	1.0925	1.0657	1.0467	1.0000
38	0.8902	0.9695	1.0066	1.0000
39	0.9524	0.9975	1.0188	1.0000
40	1.0553	1.1793	1.0556	1.0000
41	0.9611	0.9852	0.9379	1.0000
42	1.1530	1.1002	0.9494	1.0000
43	0.9752	1.0401	1.0351	1.0000
44	0.9947	1.0427	1.0784	1.0000
45	1.1740	1.0891	1.0139	1.0000
46	0.9982	1.0757	0.9861	1.0000
47	0.7664	0.9906	0.9827	1.0000
48	1.0812	1.0398	1.0616	1.0000
49	0.6385	0.7525	0.6573	1.0000
50	1.1006	1.0390	1.0508	1.0000
51	1.1380	1.0873	1.0552	1.0000
52	1.0571	1.0690	1.0277	1.0000

the three countries, about 13%. The share of office and service machines in the USA is the highest among the three countries, 6.4%. Therefore, one can conclude that Japan and the FRG invested mainly in industrial machinery and electrical machinery, and the USA invested more widely and covered the whole field of machinery and equipment. It seems that the investment pattern of the USA insinuates a direction for Japan and the FRG, so they will probably increase their share of housing and office equipment for social needs in the future.

Table 9.11. Coefficients of industry technology changes (S) in the FRG.

Sector	1978	1980	1982	1984	1986
1	1.27418	1.09252	1.14521	1.06298	1.00000
2	1.00000	1.00000	1.00000	1.00000	1.00000
3	1.00000	1.00000	1.00000	1.00000	1.00000
4	1.40611	1.24901	1.06068	0.97910	1.00000
5	0.96888	1.02383	1.02538	1.07171	1.00000
6	1.00000	1.00000	1.00000	1.00000	1.00000
7	1.00998	1.12958	1.14491	1.11601	1.00000
8	1.45993	1.85826	1.86630	1.70619	1.00000
9	1.34956	1.16730	1.12672	1.08754	1.00000
10	1.28560	1.06493	0.97139	0.96613	1.00000
11	1.10191	1.02868	0.88947	0.89652	1.00000
12	1.23205	1.04400	0.97363	0.94055	1.00000
13	1.41836	1.23639	0.99828	1.02756	1.00000
14	1.00000	1.00000	1.00000	1.00000	1.00000
15	0.88881	0.92079	0.96499	1.03174	1.00000
16	0.97972	0.93670	0.94594	0.91069	1.00000
17	1.12173	1.08363	1.02291	0.98791	1.00000
18	0.93315	0.90696	0.94742	0.99764	1.00000
19	0.85906	0.93715	0.87458	0.94111	1.00000
20	1.00000	1.00000	1.00000	1.00000	1.00000
21	1.17316	1.58928	1.63113	1.54063	1.00000
22	1.00319	1.05863	0.98921	0.95222	1.00000
23	1.28627	1.02617	0.98251	0.91787	1.00000
24	1.10095	1.06950	1.00693	0.96098	1.00000
25	1.10943	1.06296	1.02817	1.00232	1.00000
26	1.21704	1.16766	1.03658	0.98308	1.00000

9.2.2 Annual growth rates of capital formation for machinery and equipment

Table 9.17 shows the annual growth rates of capital formation for machinery and equipment in the three countries, based on the period 1977 to 1985 or 1978 to 1986. The growth rates are measured in current US dollars; therefore, the rates are influence by exchange rate changes.[3] Regarding the total, Japan and the USA maintained high growth rates during the period. In comparison with Japanese and US investments, the FRG investments for machinery and equipment grew more slowly. The growth rate is less than half of Japan and the USA. In Japan and the USA, the growth rates of office and service machinery were highest. Both countries record almost the

Table 9.11. Continued.

Sector	1978	1980	1982	1984	1986
27	1.00180	1.16983	1.00146	1.11989	1.00000
28	1.16095	1.07258	1.03113	0.94965	1.00000
29	1.06742	0.96924	0.92197	0.98425	1.00000
30	1.00000	1.00000	1.00000	1.00000	1.00000
31	0.94536	0.91471	1.12037	0.90436	1.00000
32	0.95395	0.91264	0.85388	0.91453	1.00000
33	1.00000	1.00000	1.00000	1.00000	1.00000
34	1.00000	1.00000	1.00000	1.00000	1.00000
35	1.00000	1.00000	1.00000	1.00000	1.00000
36	1.03978	1.04007	0.96697	0.91006	1.00000
37	0.79091	0.93577	0.95709	0.95118	1.00000
38	1.15589	1.11147	0.98616	0.98553	1.00000
39	1.37969	1.38006	1.12236	0.98411	1.00000
40	1.21267	1.17333	0.99521	0.98983	1.00000
41	0.92261	0.88992	0.90339	0.92753	1.00000
42	0.86529	0.84456	0.97905	0.95948	1.00000
43	1.04342	0.99888	0.96874	1.00442	1.00000
44	0.83320	0.83164	0.93742	0.94919	1.00000
45	0.62591	0.69470	0.86225	0.90595	1.00000
46	0.89853	0.91812	0.92295	0.98312	1.00000
47	0.94418	0.88284	0.92048	0.90897	1.00000
48	1.00000	1.00000	1.00000	1.00000	1.00000
49	0.90845	0.92063	0.99797	0.98731	1.00000
50	0.91266	0.93450	0.94453	0.98668	1.00000
51	1.00000	1.00000	1.00000	1.00000	1.00000
52	0.72838	0.76999	0.83317	0.87600	1.00000

same rate (about 20%). The table shows that the Japanese growth rates for industrial machinery and scientific equipment are much higher than in the other two countries.

9.3 Effects of the Structural Changes

CIM will affect two aspects to the interindustrial structure. As CIM is introduced to a certain industry, that industry needs to adjust the previous investment pattern in terms of both magnitude and allocation structure.

Table 9.12. Estimated coefficients of commodity technology change (R) and industry technology change (S), in year 2000.

Sector	R Coefficients			S Coefficients		
	Japan	USA	FRG	Japan	USA	FRG
1	0.6091	0.8062	0.8508	1.1768	1.8486	1.0852
2	0.5843	0.8531	1.0000	1.0404	0.7357	1.0000
3	1.1556	1.3721	1.0000	2.3492	0.8739	1.0000
4	0.7659	0.7912	0.7233	2.2679	1.0067	1.1604
5	0.5887	0.5872	0.9968	0.7901	1.4041	0.9104
6	0.5846	0.8143	1.0000	0.8980	0.8833	1.0000
7	1.4653	0.9546	0.9723	1.5625	1.4419	1.0153
8	1.5506	0.9695	0.7988	1.4803	0.9026	0.6104
9	1.2547	0.8370	0.7865	1.0911	1.1249	1.1008
10	18.6918	1.2467	0.8060	0.7118	0.6213	1.0006
11	1.4418	0.9492	0.9037	0.9101	1.1943	1.0423
12	0.4264	0.7899	0.8397	0.8866	1.2063	1.0499
13	0.7548	0.9933	0.6934	1.2818	0.9487	1.0741
14	0.5840	0.9214	1.0000	1.0802	1.1841	1.0000
15	0.8174	0.9713	1.0551	1.3105	1.0685	0.9992
16	1.3504	0.9702	1.0517	0.8672	1.1570	1.0517
17	1.0725	0.9783	0.9153	0.9725	1.0782	1.0093
18	1.3464	0.8847	1.0370	0.8951	0.8042	0.9124
19	0.7560	0.9977	1.0704	0.7187	0.8631	0.9285
20	1.2515	0.9431	1.0000	0.7029	0.9458	1.0000
21	0.9947	0.9374	0.8894	1.1009	0.8171	1.1700
22	1.3738	1.0343	0.9914	0.8858	1.2427	0.9312
23	0.8156	0.9574	0.8202	0.6523	1.2494	1.0840
24	0.8537	0.7971	0.9296	1.0212	1.2350	1.0410
25	1.0060	0.9732	0.9258	1.1822	0.9865	0.9912
26	0.7986	0.6792	0.8411	1.2840	0.9640	1.0526

Much of the new investments will be spent on electrical machinery and electronics equipment such as microcomputers, communication networks, programmable controllers, and industrial robots. Also, the industry will spend more on research and development of its proprietary software system. As we have observed the changing pattern of the investment in all three countries, capital formation in the electronics and communication-equipment sector has increased faster than in other kinds of equipment. This is likely to continue. R&D investment will change the input coefficients of industry for the services sectors, such as the business-service sector and the educational-service

Table 9.12. Continued.

Sector	R Coefficients			S Coefficients		
	Japan	USA	FRG	Japan	USA	FRG
27	0.9444	0.7971	0.9330	1.0267	0.9337	0.9812
28	1.2400	0.8109	0.9017	1.1498	1.3187	1.0298
29	0.6762	0.9187	0.9390	1.1639	0.9976	1.0399
30	1.0146	0.7398	1.0000	0.9644	1.1341	1.0000
31	1.6291	1.1270	1.1186	0.6685	1.2530	1.0586
32	1.1224	1.0705	1.0171	0.9839	1.2649	0.9726
33	0.5084	1.1271	1.0000	0.8111	1.5981	1.0000
34	1.0637	0.8199	1.0000	0.8820	1.0707	1.0000
35	1.7276	1.2205	1.0000	0.6974	1.2493	1.0000
36	1.2710	1.1326	0.9736	1.0358	1.1012	1.0533
37	1.1112	1.3185	1.1477	1.0802	0.9953	1.0131
38	0.7043	0.9088	0.8772	1.2100	1.2660	1.0415
39	3.2781	0.7170	0.7323	0.8152	1.1630	0.9690
40	1.0000	1.0000	0.8338	1.0000	0.8216	1.0372
41	1.3757	1.1479	1.0566	1.1227	0.9344	1.0493
42	1.2985	1.1006	1.1075	0.8555	0.6626	1.1532
43	0.8186	0.9129	0.9559	1.0201	1.1031	0.8628
44	1.0907	0.9471	1.1288	1.2326	1.2369	0.8723
45	0.9500	1.3429	1.2954	0.4892	0.8565	0.9905
46	2.4308	0.8291	1.0546	1.8121	0.8712	0.9715
47	1.1276	0.6443	1.0585	1.8353	1.2007	1.1567
48	0.7910	1.8235	1.0000	0.6137	1.0918	1.0000
49	1.0000	1.0287	1.0726	1.0000	1.0444	0.9771
50	7.0795	0.8257	1.0491	1.5041	1.0868	0.9357
51	1.6189	1.1977	1.0000	0.7814	0.9652	1.0000
52	1.5833	0.8150	1.2095	0.8745	0.9508	0.9236

sector. In this analysis, it is assumed that the CIM will strengthen and accelerate past trends in regard to structural changes of input coefficients and capital formation until the year 2000. The changes of input coefficient are reflected by the commodity technology changes (R) and the industry technology changes (S). The changes of capital formation assume that past growth rates will last until the year 2000. From these two assumption, we discuss the effects of the CIM application indirectly.

Some of the effects can be calculated as follows: let

$$X^k = (I - A)^{-1} K , \qquad (9.8)$$

Table 9.13. Capital formation, Japan.

Sector	Values in billion yen			Average annual growth (%)		Composition ratios (%)		
	1973	1985	2000	1973–1985	1985–2000	1973	1985	2000
1	50	75	184	3.5	6.1	0.2	0.1	0.1
2	56	193	108	10.8	-3.8	0.2	0.3	0.1
12	92	106	102	1.1	-0.2	0.3	0.2	0.1
13	7	21	128	10.0	12.8	0.0	0.0	0.1
14	508	866	1,097	4.5	1.6	1.6	1.4	0.6
26	-75	-122	-67	4.1	-3.9	-0.2	-0.2	- 0.0
27	-155	-6	-13	-24.1	5.7	-0.5	-0.0	- 0.0
28	256	302	1,005	1.4	8.3	0.8	0.5	0.5
29	1,215	4,246	14,341	11.0	8.5	3.7	6.7	7.6
30	2,263	5,357	11,226	7.4	5.1	7.0	8.4	5.9
31	289	1,146	10,450	12.2	15.9	0.9	1.8	5.5
32	463	1,340	5,022	9.3	9.2	1.4	2.1	2.6
33	616	4,489	53,880	18.0	18.0	1.9	7.0	28.4
34	555	1,733	8,613	10.0	11.3	1.7	2.7	4.5
35	141	734	2,880	14.7	9.5	0.4	1.2	1.5
36	2,548	3,148	4,645	1.8	2.6	7.9	4.9	2.4
37	1,064	1,479	1,199	2.8	-1.4	3.3	2.3	0.6
38	159	939	9,227	15.9	16.5	0.5	1.5	4.9
39	289	536	1,456	5.3	6.9	0.9	0.8	0.8
40	20,101	31,478	52,546	3.8	3.5	61.9	49.4	27.7
43	1,902	5,143	9,292	8.6	4.0	5.9	8.1	4.9
46	103	364	2,465	11.1	13.6	0.3	0.6	1.3
	32,447[a]	63,711[a]	189,908[a]	5.8[b]	7.6[b]	100.0[a]	100.0[a]	100.0[a]

[a]Total: may not add up due to rounding.
[b]Average.

where X^k denotes total production requirements induced by capital formation and K, the capital formation. Since X^k includes both direct and indirect requirements, we can separate the indirect requirements, X^{*k}, from the total by subtracting the capital formation from the total:

$$X^{*k} = X^k - K$$
$$= (I - A)^{-1}AK . \tag{9.9}$$

The total production requirement can be compared with the initial impact, i.e., total of the capital formation as an exogenous vector (K). *Table 9.18* indicates the total production requirements (both direct and indirect)

Table 9.14. Capital formation, USA.

Sector	Values in billion yen			Average annual growth (%)		Composition ratios (%)		
	1977	1985	2000	1977–1985	1985–2000	1977	1985	2000
5	374	529	880	4.4	3.5	0.1	0.1	0.1
8	116	387	3,219	16.3	15.2	0.0	0.1	0.1
12	892	1,566	3,909	7.3	6.3	0.3	0.2	0.2
13	11	15	23	4.0	3.0	0.0	0.0	0.0
14	5,050	12,768	63,147	12.3	11.2	1.6	1.9	1.9
18	541	769	1,292	4.5	3.5	0.2	0.1	0.1
22	58	94	202	6.2	5.2	0.0	0.0	0.0
26	5	21	269	19.6	18.5	0.0	0.0	0.0
27	106	89	56	-2.2	-3.1	0.0	0.0	0.0
28	4,675	6,362	9,849	3.9	3.0	1.4	1.0	0.9
29	6,772	10,748	22,202	5.9	5.0	2.1	1.6	1.5
30	32,802	48,896	89,798	5.1	4.1	10.1	7.4	6.9
31	10,418	42,510	515,807	19.2	18.1	3.2	6.4	6.8
32	1,607	2,530	5,148	5.8	4.8	0.5	0.4	0.4
33	10,655	33,550	250,394	15.4	14.3	3.3	5.1	5.2
34	5,854	10,871	30,145	8.0	7.0	1.8	1.6	1.6
35	1,588	4,604	29,433	14.2	13.2	0.5	0.7	0.7
36	30,854	64,314	221,483	9.6	8.6	9.5	9.7	9.4
37	11,100	13,785	17,978	2.7	1.8	3.4	2.1	1.9
38	8,758	21,344	98,533	11.8	10.7	2.7	3.2	3.2
39	1,283	2,445	7,117	8.4	7.4	0.4	0.4	0.4
40	150,890	297,601	923,914	8.9	7.8	46.4	44.8	43.3
41	22	17,128	56,795	129.8	8.3	0.0	2.6	5.3
43	24,668	46,718	134,412	8.3	7.3	7.6	7.0	6.8
45	10,747	16,900	34,311	5.8	4.8	3.3	2.5	2.4
46	1,976	3,594	9,585	7.8	6.8	0.6	0.5	0.5
47	3,385	4,178	5,386	2.7	1.7	1.0	0.6	0.6
	325,207[a]	664,316[a]	2,535,289[a]	9.3[b]	9.3[b]	100.0[a]	100.0[a]	100.0[a]

[a]Total: may not add up due to rounding.
[b]Average.

and the ratios to total investment. In the three countries, total production requirements are almost double the direct investments. Japanese multipliers (X^k/K) are typically higher than the US or German multipliers. Those of the FRG are higher than those of the USA. This implies that Japanese capital formation induces more indirect production via interindustrial linkages

Table 9.15. Capital formation, FRG.

Sector	Values in million DM			Average annual growth (%)		Composition ratios (%)		
	1978	1986	2000	1978– 1986	1986– 2000	1978	1986	2000
4	191	252	409	3.5	3.5	0.1	0.1	0.1
11	387	405	643	0.6	3.4	0.2	0.1	0.1
13	4,989	5,969	9,224	2.3	3.2	2.0	1.7	1.6
19	588	1,091	2,015	8.0	4.5	0.2	0.3	0.4
25	14	31	70	10.4	6.0	0.0	0.0	0.0
26	4,499	7,461	12,509	6.5	3.8	1.8	2.1	2.2
27	223	250	702	1.4	7.6	0.1	0.1	0.1
28	13,849	20,024	26,871	4.7	2.1	5.5	5.6	4.7
29	32,653	48,115	97,609	5.0	5.2	12.9	13.5	17.1
31	6,176	15,510	29,885	12.2	4.8	2.4	4.3	5.2
32	21,784	35,123	68,750	6.2	4.9	8.6	9.8	12.1
36	14,920	22,214	28,820	5.1	1.9	5.9	6.2	5.1
37	3,660	4,696	5,953	3.2	1.7	1.4	1.3	1.0
38	2,687	3,274	6,778	2.5	5.3	1.1	0.9	1.2
39	376	467	816	2.7	4.1	0.1	0.1	0.1
40	85,161	105,756	145,261	2.7	2.3	33.7	29.6	25.5
41	38,934	53,906	72,178	4.2	2.1	15.4	15.1	12.7
43	12,562	17,417	31,808	4.2	4.4	5.0	4.9	5.6
46	759	927	2,027	2.5	5.7	0.3	0.3	0.4
52	8,168	14,132	27,343	7.1	4.8	3.2	4.0	4.8
	252,580[a]	357,020[a]	569,670[a]	4.4[b]	3.4[b]	100.0[a]	100.0[a]	100.0[a]

[a]Total.
[b]Average.

Table 9.16. Percent share of machinery and equipment in total capital formation.

Sector	Japan 1985	USA 1985	FRG 1986
Industrial machinery (IMC)	15.07	8.98	13.48
Office and service machinery (OMC)	1.80	6.40	4.34
Electrical machinery (EMC)	13.02	7.76	9.84
Automobiles (AUT)	4.94	9.68	6.22
Other transport equipment (OTQ)	2.32	2.08	1.32
Scientific equipment (SCQ)	1.47	3.21	0.92
Total	38.62	38.11	36.12

Table 9.17. Percent of annual growth rates of capital formation for machinery and equipment, in million US dollars.

Sector	Japan 1977–1985	USA 1977–1985	FRG 1978–1986
Industrial machinery (IMC)	11.00	5.26	3.95
Office and service machinery (OMC)	20.76	19.22	11.11
Electrical machinery (EMC)	13.06	12.78	5.12
Automobiles (AUT)	4.36	9.62	4.08
Other transport equipment (OTQ)	5.90	2.74	2.16
Scientific equipment (SCQ)	26.88	11.78	1.51
Average[a]	10.79	9.73	4.81

[a]Average of all 53 sectors.

Table 9.18. Capital formation and the induced production requirement and the ratio.

	Japan 1981	1985	USA 1982	1985	FRG 1982	1986
Capital form.[a]	252,746	267,085	491,042	664,316	127,738	164,412
Total of X^{ka}	604,927	623,633	1,043,530	1,325,013	277,301	347,951
Total of X^{*ka}	352,182	356,548	568,344	693,726	149,563	183,539
X^k/K^b	2.39342	2.33495	2.12513	1.99455	2.17085	2.11634
X^{*k}/K^c	1.39342	1.33495	1.15742	1.04427	1.17085	1.11634
Ratio	58%	57%	54%	52%	54%	53%

[a]In current US dollars.
[b]Total production requirement.
[c]Indirect production requirement.

than the others. Such linkages are weaker for the USA than for the other countries.

It follows that the indirect production requirement induced in Japan is the highest among the three countries. In addition, it is a common pattern in the three countries that the multiplier for 1985 or 1986 is lower than the one for 1981 or 1982. There may be two main reasons why the multiplier becomes lower over time. One reason is that the industrial structure is changing from manufacturing-oriented to service-oriented. This can be confirmed from the patterns of changes in input structures. Another reason is that material inputs are gradually transferred from domestic products to imported ones; this process is occurring fastest in the USA. Accordingly, interindustrial linkages are becoming weaker. This also suggests that the pattern in Japan and the FRG should eventually correspond to the US pattern.

Table 9.19. Total production requirements induced by capital formation.

	Japan			USA			FRG		
Sector	1981	1985	2000	1982	1985	2000	1982	1986	2000
	Total requirements (billion US $)								
PRM	28	22	118	39	36	52	10	9	21
LTM	34	34	173	60	78	203	13	15	27
HVM	130	124	841	180	204	499	70	82	150
MCH	140	164	1,566	270	367	1,889	58	85	186
CON	142	135	426	241	324	1,010	66	76	125
UTL	14	13	128	30	31	82	6	7	14
SRV	118	133	904	223	286	841	54	74	160
Total	605	624	4,155	1,044	1,325	4,577	277	348	683
	Composition ratios (%)								
PRM	4.6	3.5	2.8	3.8	2.7	1.1	3.8	2.5	3.0
LTM	5.5	5.4	4.2	5.8	5.9	4.4	4.6	4.4	3.9
HVM	21.5	19.9	20.2	17.3	15.4	10.9	25.4	23.4	22.0
MCH	23.1	26.2	37.7	25.9	27.7	41.3	20.8	24.5	27.2
CON	23.5	21.6	10.3	23.1	24.5	22.1	23.9	21.9	18.4
UTL	2.4	2.1	3.1	2.9	2.3	1.8	2.0	1.9	2.1
SRV	19.4	21.3	21.8	21.4	21.6	18.4	19.5	21.3	23.4
Total[a]	100.0	100.0	100.0	100.0	100.0	100.0	100.0	100.0	100.0

[a]Totals may not add up due to rounding.

Table 9.19 shows the total production requirements induced by capital formation for the three countries. It also shows the composition ratios, at aggregated sector level. *Table 9.20* shows the composition ratios of total production requirements induced by capital formation for metal products and machinery, attributable to capital formation at the sectoral level.

In the mid-1980s, the total induced production requirement (due to capital formation) for the machinery sector (MCH) was largest in all three countries. The construction sector (CON) in Japan and the heavy manufacturing industry (HVM) in the FRG were the largest in the early 1980s. Since the machinery industry (MCH) of the USA had the most shares in the early 1980s, it appears that Japan and the FRG have followed the USA. The total requirement for output of the machinery industry has been increasing at higher growth rates than the other sectors as a long-term trend. If this trend lasts to the year 2000, about 40% of the total requirements induced by capital formation would be shared by the machinery industry in both Japan and the USA. The case of the FRG is not as extreme as the other two countries, but about 30% of the total requirement would be shared by the machinery sectors.

The level of requirement for heavy manufacturing industries (HVM), the services sector (SRV), and the construction sector (CON) is not much different from the machinery industry in the 1980s in the three countries. However, the weight of the construction sector has declined, and it can be supposed that this trend will continue in the future. In the case of Japan, the weight of the construction sector is supposed to drop to about 10%.

The total requirement for the service sector has been gradually and steadily increasing as the composition ratio of the mid-1980s is higher than it was in the early 1980s in all three countries. Since the total induced requirement for services is always the same as the indirect requirements, it should be noted that capital formation always has a significant effect on the service sectors. Considering that most service outputs are not tradable and cannot be imported from foreign countries, the growing demand for services will increase their share in domestic production.

The demand for raw materials for machinery production and construction has not increased during the 1980s. Primary sectors (PRM) and light manufacturing industries (LTM) are not as dependent on capital formation, and the induced requirement share is continuously decreasing. The requirement for raw materials from heavy manufacturing industries has not decreased in value terms, but the relative weight of the requirement has fallen during the 1980s.

More than 25% of the production requirements induced by capital formation in the mid-1980s were for machinery and equipment. *Table 9.20* displays more specific observations. Each country has a specific pattern. Japan appears to be specializing both in the electrical machinery and equipment and in industrial machinery; the USA, in the automobile sector and the office and service machinery and equipment sector; and the FRG, in the industrial machinery sector.

As shown in *Table 9.18*, the ratio of indirect production requirements to the total in the mid-1980s was 57%, 52%, and 53% for Japan, the USA, and the FRG, respectively.[4] The ratio of Japan was higher than the others. This means that the investment to capital formation generates indirect domestic production requirements greater than the amount of direct investment. If this ratio increases, then industrial linkages between capital formation and all industries will be deeper and strengthened, and capital formation will create more demands. From the table, it appears that the

Table 9.20. Composition ratios of total production requirements induced by capital formation, for metal and machinery industries, in percent.

Sector	Japan			USA			FRG		
	1981	1985	2000	1982	1985	2000	1982	1986	2000
26	6.8	6.5	5.5	3.2	2.5	1.2	8.1	7.8	6.5
27	1.8	1.9	2.6	2.1	1.8	1.4	1.8	1.6	1.9
28	3.3	3.1	2.5	4.5	4.2	2.6	4.4	4.3	3.6
Metals	11.9	11.5	10.6	9.8	8.5	5.2	14.3	13.7	12.0
29	5.1	5.0	5.5	2.3	2.1	1.3	8.1	9.0	10.8
30	3.5	4.5	2.8	5.0	4.3	2.3	–	–	–
31	0.7	1.0	2.7	3.3	4.1	14.2	1.5	2.5	3.0
32	1.1	1.2	1.4	0.3	0.3	0.2	6.5	7.6	8.8
33	4.6	3.8	12.0	3.8	4.1	10.9	–	–	–
34	1.4	1.6	2.2	1.5	1.5	1.4	–	–	–
35	1.3	2.7	5.7	0.9	1.1	1.4	–	–	–
36	4.2	4.1	2.7	5.0	7.0	6.7	3.4	4.0	3.2
37	0.5	1.4	0.5	1.8	1.3	0.6	0.8	0.8	0.6
38	0.6	0.9	2.1	1.9	1.9	2.4	0.5	0.6	0.7
Machinery	23.1	26.2	37.7	25.9	27.7	41.3	20.8	24.5	27.2

[a]Totals may not add up due to rounding.

ratio will increase in Japan but not in the USA. However, trade imbalance and resulting friction might retard or even reverse this trend.

Table 9.21 shows the indirect production requirement induced by capital formation at the aggregated sectoral level. The table indicates that the indirect production requirement for heavy manufacturing sectors is the greatest in Japan during the 1980s and in the FRG in 1986. The composition ratio is more than one-third of the total in both countries. However, the ratio of heavy manufacturing sectors is decreasing in all three countries. One reason seems to be that the nonelectrical machinery industry and construction sectors are reducing their indirect supply to heavy manufacturing industries. This point will be discussed later.

Next, we note the trend for induced service requirements (SRV). In the USA, the table shows that the induced requirement for services is about one-third of the all industries in the 1980s. This is higher than heavy manufacturing. In Japan and the FRG, the ratio is slightly lower than that in the USA but has been steadily increasing in the 1980s.

The induced requirement for the machinery sectors (MCH) are about 17% in Japan, 16% in the USA, and 14% in the FRG in the mid-1980s.

Table 9.21. Indirect production requirements induced by capital formation.

Sector	Japan			USA			FRG		
	1981	1985	2000	1982	1985	2000	1982	1986	2000
	Indirect requirements (billion US $)								
PRM	27	21	115	38	35	48	10	9	21
LTM	29	27	151	46	61	129	10	12	21
HVM	128	123	834	172	196	487	60	68	127
MCH	56	60	618	85	114	608	18	26	54
CON	1	3	16	9	9	30	2	3	5
UTL	14	13	128	30	31	82	6	7	14
SRV	96	109	813	188	248	658	43	59	126
Total	352	357	2,674	568	694	2,042	150	184	366
	Share (%)								
PRM	7.7	5.9	4.3	6.7	5.1	2.3	6.9	4.8	5.6
LTM	8.2	7.6	5.6	8.2	8.8	6.3	6.9	6.6	5.7
HVM	36.4	34.6	31.2	30.3	28.3	23.9	40.4	37.2	34.6
MCH	16.0	16.9	23.1	15.0	16.4	29.8	11.8	14.1	14.6
CON	0.4	0.7	0.6	1.6	1.3	1.5	1.4	1.4	1.2
UTL	4.0	3.8	4.8	5.3	4.4	4.0	3.7	3.6	4.0
SRV	27.2	30.5	30.4	33.0	35.7	32.2	28.9	32.2	34.3
Total[a]	100.0	100.0	100.0	100.0	100.0	100.0	100.0	100.0	100.0

[a]Totals may not add up due to rounding.

While these figures are lower than for the two sectors mentioned above, they are steadily growing in all three countries.

From these observations, it seems that the service sectors and the machinery sectors will increase their share of all production requirements through new investment and industrial structural changes, which are accelerated with new technologies such as CIM introduction and high-tech industrialization. Such technological innovations will increase demands for electronics and communication equipment, office and service machinery, industrial machinery, as well as service inputs. In contrast, demands for raw materials will be transformed from metal products to chemical products including petrochemical and plastic products.

The discussion so far has been based on the metalworking sector's gross output and to what extent demand would be induced by capital formation; I will now discuss the gross outputs which depend on the capital formation. It is more convenient to define a measure of dependency of gross output on capital formation as a ratio of induced production requirement to gross outputs. *Table 9.22* shows that the dependency of total gross output of each

Table 9.22. Ratios of induced production requirements to gross outputs, in percent.

Sector	Japan 1981	1985	USA 1982	1985	FRG 1982	1986
PRM	32.4	26.7	10.4	10.2	21.0	17.2
LTM	11.3	10.2	9.6	11.0	8.6	8.6
HVM	29.0	28.3	25.7	27.1	31.1	30.4
MCH	36.3	36.1	43.0	44.0	32.8	33.8
CON	60.4	62.1	80.2	81.7	86.7	85.5
UTL	17.9	15.4	10.9	10.2	14.3	13.3
SRV	11.0	10.4	9.3	9.5	9.2	9.2
Average[a]	23.7	21.9	18.9	19.6	21.3	20.5

[a]Averages for all 53 sectors.

country is almost at the same level, 20% more or less. But, the dependencies by sector vary more widely among countries. The construction sector in Japan depends on capital formation at about 60%. This is less than for the USA and the FRG (about 80%). The machinery industry in the USA depends more on capital formation than it does in Japan and the FRG. The degree of dependency of primary sector in Japan and the FRG is higher than the USA. But both countries are reducing this dependency. This reflects the fact that both countries import more primary products than the USA. The other sectoral dependency ratios are almost at the same level among countries. The ratios for the service sector and the light manufacturing industry are about 10%, and those for the heavy manufacturing sector are about 30%.

We have observed production requirements induced by capital formation in details. As mentioned above, induced production requirements combine the effects of two factors – (1) input structures and (2) capital formation structure. To compare such production requirements at two points in time, it is more convenient to separate the effects by factor. The production requirements induced by capital formation in time $t1$ and $t2$ are defined in matrix form as

$$\tilde{X}(t1) = B(t1)F_k(t1) \tag{9.10}$$

$$\tilde{X}(t2) = B(t2)F_k(t2) \ , \tag{9.11}$$

where $\tilde{X}$ is a vector of production requirements induced by capital formation, F_k is a vector of capital formation, and B is an inverse matrix of $(I\text{-}A)$. From equations (9.10) and (9.11), we can derive the following equations:

$$
\begin{aligned}
d\tilde{X} &= \tilde{X}(t2) - \tilde{X}(t1) \\
&= B(t2)F_k(t2) - B(t1)F_k(t1) \\
&= [B(t1) + dB][F_k(t1) + dF_k] - B(t1)F_k(t1) \\
&= B(t1)dF_k + dBF_k(t1) + dF_k dF_k \ ,
\end{aligned}
\tag{9.12}
$$

where d is the increment.

Thus, changes of production requirement can be classified into three categories: (1) changes in capital formation, (2) technological changes, and (3) the cross-product term (Uno, 1989).

Tables 9.23, 9.25, and *9.27* show the composition ratios of production requirement changes due to the three factors mentioned above for the three countries from 1981–1985. During the early 1980s, the three countries developed along somewhat different paths. In Japan, changes in capital formation brought increased production requirements for three sectors: industrial machinery and equipment, services, and heavy manufacturing products. More than 50% of all growth of the induced production requirement is allocated to the machinery sector, and about 20% each to the services sector and heavy manufacturing. In the USA, the allocation to the machinery sector is lower than in Japan but, in contrast, induced production is allocated more to construction. The FRG is very close to the Japanese pattern, but the details are different. For example, the effects of changes in capital formation are larger in the heavy manufacturing industry and services sectors in Japan. In addition, the effects of technological changes reduce induced production requirements for heavy manufacturing industry more in Japan.

Regarding effects of changes in technologies, Japan and the USA have very similar patterns. As a result of technological changes, induced production requirements are reduced in the primary sector, heavy manufacturing sector, and utility sector. One difference is that production requirements in the Japanese services sector have been strongly induced by technological change. In the FRG, induced effects of technological changes are quite minor by contrast.

Finally, we discuss future trends of effects of technological changes and changes in structure of capital formation based on the input–output table projected for the year 2000. *Tables 9.24, 9.26*, and *9.28* show a composition ratio table of changes of production requirement by changing factor. In Japan, technological changes may affect compositional changes of induced

Table 9.23. Changes of production requirement by factor 1981–1985, Japan.

Sector	BdF	dBF	dBdF	dX
	Composition ratio (%)			
PRM	5.6	– 9.9	–1.2	–5.5
LTM	8.7	– 2.7	–2.1	3.9
HVM	20.4	–10.6	–4.4	5.5
MCH	55.2	– 3.1	–0.8	51.3
CON	3.3	1.4	0.2	4.8
UTL	2.7	– 2.1	–0.3	0.3
SRV	21.6	16.9	1.1	39.6
Total[a]	117.5	–10.0	–7.6	100.0

Table 9.24. Changes of production requirement by factor 1985–2000, Japan.

Sector	BdF	dBF	dBdF	dX
	Composition ratio(%)			
PRM	2.0	0.2	0.4	2.6
LTM	3.0	0.1	0.6	3.7
HVM	14.0	1.9	4.6	20.5
MCH	37.3	1.7	3.4	42.4
CON	5.8	0.0	0.1	5.9
UTL	1.9	0.5	1.1	3.5
SRV	13.9	2.3	5.3	21.5
Total[a]	77.8	6.6	15.5	100.0

Table 9.25. Changes of production requirement by factor 1981–1985, USA.

Sector	BdF	dBF	dBdF	dX
	Composition ratio (%)			
PRM	4.1	– 3.9	–1.2	–1.0
LTM	6.3	– 0.3	–0.1	5.9
HVM	18.8	– 8.5	–2.5	7.9
MCH	33.2	– 0.6	–0.1	32.5
CON	28.9	– 0.7	–0.2	28.0
UTL	3.2	– 2.2	–0.7	0.3
SRV	27.5	– 0.9	–0.3	26.3
Total[a]	122.0	–17.0	–5.0	100.0

Table 9.26. Changes of production requirement by factor 1985–2000, USA.

Sector	BdF	dBF	dBdF	dX
	Composition ratio(%)			
PRM	3.3	–0.7	– 2.1	0.5
LTM	7.2	–0.9	– 2.4	3.9
HVM	19.7	–2.7	– 7.7	9.2
MCH	47.5	–0.3	0.2	47.4
CON	21.7	–0.1	– 0.2	21.4
UTL	3.2	–0.4	– 1.2	1.6
SRV	27.8	–3.1	– 8.7	16.0
Total[a]	130.5	–8.2	–22.2	100.0

Table 9.27. Changes of production requirement by factor 1981–1985, FRG.

Sector	BdF	dBF	dBdF	dX
	Composition ratio (%)			
PRM	4.9	–10.9	–1.5	–7.5
LTM	5.0	– 2.1	–0.3	2.6
HVM	31.7	–21.0	–3.0	7.6
MCH	51.8	2.2	1.0	55.0
CON	5.1	0.1	0.0	5.2
UTL	2.8	– 1.3	–0.2	1.3
SRV	27.4	7.0	1.4	35.8
Total[a]	128.8	–26.1	–2.7	100.0

Table 9.28. Changes of production requirement by factor 1985–2000, FRG.

Sector	BdF	dBF	dBdF	dX
	Composition ratio(%)			
PRM	2.5	0.9	0.5	3.8
LTM	4.0	–0.6	–0.3	3.1
HVM	20.2	–0.4	–0.2	19.6
MCH	31.5	0.0	0.0	31.5
CON	12.8	–0.1	0.0	12.7
UTL	1.9	0.4	0.2	2.4
SRV	22.2	2.8	1.7	26.8
Total[a]	95.0	3.1	1.9	100.0

[a]Totals may not add up due to rounding.

production requirement, shifting from machinery and services sectors to raw-material sectors. In particular, the share for induced service demands may not be as high as in the 1980s. By contrast, in the USA the construction sector will decrease its share of induced production, but the machinery sectors will increase their share. In addition, technological changes may reduce the share of demands for services and heavy manufacturing.

Table 9.29 and *Table 9.30* list selected sectors, for which the composition ratios of induced production requirements change by more than 1%, due to changes in capital formation and due to technological changes, respectively. As a result of structural changes in capital formation, the construction sector (Sector 40) in Japan and the FRG will be affected positively compared with the 1980s, while for the USA the impact will be negative. The electronics and communication-equipment industry (Sector 33) both in Japan and in the USA will increase their shares. In addition, office and service machinery (Sector 31) and services requirements (Sectors 43–52), will also increase their relative shares in the year 2000. It seems that these changes are typical of the impacts of CIM.

From the 1980s to the year 2000, the effects of technological changes may not be so strong for manufacturing and services as compared with the early 1980s. In general, induced indirect production requirements for the manufacturing sectors exhibit growth in consequence of capital formation, but a contrary effect is due to changes during the period. However, induced indirect requirements for the service sectors grow because of capital formation and technological change. In the USA and the FRG, this trend may be maintained up to the year 2000, although at a decelerating pace. But, in the case of Japanese manufacturing sectors, the projection indicates almost the reverse situation, *viz.*, technological changes would apparently continue to increase induced indirect requirements for both manufacturing and services. There are two major reasons for this phenomenon. One is that, as a result of input substitutions of raw materials (from lower quality to higher quality), the share of raw material inputs increases relatively. The other reason is that, due to vertical integration in the machinery sector, either intrasectoral input ratios or input coefficients for the manufacturing sector per se are raised. However, it is difficult to conclude from this simple projection for the year 2000 whether Japanese technological changes will bring about effects that are similar to those in the USA or effects that are unique to Japan. We need more information for further discussions.

Table 9.29. Composition ratio of changes of production requirement due to changes in capital formation (difference between 1981–1985 and 1985–2000 by more than 1%).

Japan				USA				FRG			
	A	B	B–A		A	B	B–A		A	B	B–A
33	5.0	15.5	10.5	31	7.3	17.3	10.0	40	5.1	12.8	7.7
40	3.4	5.8	2.4	33	5.6	11.6	6.0	25	0.9	2.2	1.3
28	0.4	1.8	1.4	37	−0.6	0.4	1.0	27	2.8	1.8	−1.0
50	0.3	1.4	1.1	45	4.6	3.0	−1.6	42	2.8	1.9	−1.0
46	2.0	3.1	1.0	36	13.7	7.2	−6.5	21	2.0	1.0	−1.0
38	3.7	2.7	− 1.0	40	30.5	21.4	−9.1	8	1.6	0.6	−1.1
32	2.2	1.2	− 1.0					18	3.9	2.7	−1.2
8	2.3	1.0	− 1.2					43	7.9	6.1	−1.8
13	1.7	0.4	− 1.2					28	4.8	2.8	−2.0
2	1.3	−0.0	− 1.3					29	16.2	13.9	−2.3
44	3.8	2.1	− 1.7					52	12.5	10.1	−2.4
21	3.4	1.5	− 1.9					32	14.1	10.6	−3.5
14	2.8	0.2	− 2.6					26	14.6	7.8	−6.8
27	4.6	2.0	− 2.6					36	9.2	2.1	−7.1
26	8.7	4.0	− 4.7					31	10.8	3.6	−7.2
37	7.3	0.0	− 7.2								
43	11.2	3.8	− 7.4								
29	15.5	4.2	−11.3								
30	14.1	2.0	−12.0								

A = 1981–1985; B = 1985–2000.

Notes

[1] The growth of capital formation in electronics and communication equipment was extremely high and constant in Japan during the base period. Although the growth rates calculated by several methods have been compared, estimates indicate very high growth rate for this sector, with estimated shares between about 22% and 28%. The highest estimate (28%) happens to result from the method that was used to compare the three countries. Recently, Japanese industries have been accelerating information network investment. According to the survey by *Nihon Keizai Shimbun* [Japanese Economic Newspaper], the growth rate of investment related to the information network has been more than 30% in the past three years. The survey report also said that the computer communication network innovation will continue during the 1990s. As these investments aim not only at increasing productivity but also at competing with others, the process will presumably continue until the various proprietary information systems are completed. The survey also indicates that this investment trend will continue

Table 9.30. Composition ratio of changes of production requirement due to technological changes (difference between 1981–1985 and 1985–2000 by more than 1%).

Japan				USA				FRG		
	A	B	B–A		A	B	B–A		A	B	B–A
21	−9.3	0.4	9.7	8	−2.8	−0.4	2.5	21	−7.6	0.9	8.5
29	−9.0	0.5	9.5	21	−2.6	−0.3	2.3	8	−7.6	0.9	8.5
33	−6.9	−0.0	6.9	26	−2.7	−0.6	2.1	26	−7.0	−2.1	5.0
8	−6.2	0.3	6.5	42	−2.3	−0.4	1.9	18	−1.7	0.6	2.4
26	−4.8	0.5	5.3	27	−1.5	−0.3	1.2	27	−2.0	0.3	2.3
43	−5.1	−0.1	5.0	51	0.8	−0.9	−1.8	42	−1.3	0.4	1.7
25	−3.4	0.2	3.6	43	1.1	−0.8	−1.9	7	−1.5	0.1	1.6
44	−2.3	0.5	2.8					25	−1.6	−0.1	1.6
42	−2.2	0.5	2.7					43	−1.4	0.2	1.6
47	−2.1	0.1	2.2					28	−1.6	−0.2	1.4
6	−1.8	−0.1	1.8					9	−1.1	−0.1	1.0
12	−1.7	0.0	1.7					32	2.0	0.0	−1.9
15	−1.5	0.0	1.6					52	7.7	1.6	−6.0
5	−1.3	−0.1	1.2								
31	1.0	0.0	− 1.0								
11	1.2	0.0	− 1.2								
40	1.5	0.0	− 1.4								
22	2.0	0.3	− 1.7								
36	3.6	0.5	− 3.2								
28	3.5	0.1	− 3.4								
51	6.2	0.1	− 6.1								
50	6.9	0.3	− 6.6								
35	8.2	0.5	− 7.8								
46	13.9	1.2	−12.8								
30	14.1	2.0	−12.0								

A = 1981–1985; B = 1985–2000.

at least through the mid-1990s. Given this background, an 18% annual growth rate for investment in electronics and communication equipment does not seem to be extremely high. If so, 28% of the Japanese total investment in the year 2000 would be shared by this sector.

[2] But this figure, based on the period 1977 to 1985, was characterized by a massive effort to redesign and downsize US automobiles, which required an unusually large investment in new plants and equipment. This, period of extraordinary investment ended in 1982. (It also resulted in a sharp cutback by US machine tool manufactures, which left many of them financially weak.)

[3] They also cover a period when the US dollar was probably significantly overvalued.

[4] This probably reflects the lower Japanese propensity to import capital goods.

Appendix I

Uniform input–output (UIO) classification (53 Sectors) for Japan, the USA, and the FRG.

UIO	<Intermediate Sectors>		
1	Agriculture	39	Other manufacturing
2	Livestock	40	New construction
3	Agricultural services	41	Construction repairs
4	Forestry & fishery	42	Electricity, gas, & water
5	Metal ore mining	43	Trade
6	Nonmetal ore mining	44	Financial services
7	Coal mining,	45	Real estate
8	Crude petroleum	46	Transportation & warehousing
9	Food & beverage	47	Communication
10	Tobacco	48	Radio & TV
11	Weaving & knitting	49	Public administration
12	Apparel & other textile	50	Health, education, social service
13	Lumber & wood products	51	Business service
14	Furniture & fixture	52	Other personal service
15	Pulp & paper	53	n.e.c.
16	Paper products	54	Intermediate input, total
17	Printing & publishing	UIO	<Value-Added Sectors>
18	Basic chemical products	55	Business consumption
19	Plastics & syn. materials	56	Wages & salary
20	Other chemicals	57	Property income
21	Petroleum refinery	58	Indirect taxes less subs.
22	Rubber & plastic products	59	Double deflation discount
23	Leather products	60	Gross value-added
24	Glass products	61	Gross Outputs (Row)
25	Stone & clay products	UIO	<Final Demand Sectors>
26	Primary iron & steel	55	Business consumption
27	Primary nonferrous metals	56	Private consumption
28	Fabricated metal products	57	Government consumption
29	General industrial machinery	58	Fixed capital formation (private)
30	Special industrial machinery	31	Office & service machinery
32	Household electrical machinery	33	Electronics & communication equipment
34	Heavy electric machinery	35	Other electrical machinery
36	Automobile & repairs	37	Other transport equipment
38	Scientific equipment	59	Changes in inventory
60	Exports	61	Final demand (73~79)
62	Total demand (72+80)	63	Imports at cif
64	Import duties	65	Import sales taxes
66	Import, total (82~84)	67	Final demand – imports (80–85)
68	Gross Outputs (Column)		

Appendix II

Input–output sector classification for the FRG and the common sector classification (UIO53). Common Sector Classification (UIO53).

I–O#	Description	I–O#	Description
	< Intermediate Sectors (1–59) >		< Intermediate Sectors (1–54) >
1	Agriculture	1	Agriculture [includes (2),(3)]
2	Forestry & fishery	4	Forestry & fishery
3	Electricity, steam, & hot water	42	Electricity, gas, & water
4	Gas	42	Electricity, gas, & water
5	Water	42	Electricity, gas, & water
6	Coal mining	7	Coal mining
7	Other mining	5	Other mining [includes (6)]
8	Crude petroleum & natural gas	8	Crude petroleum & natural gas
9	Chemical products	18	Chemicals
10	Petroleum refinery	21	Petroleum refinery
11	Plastic materials	19	Plastic & synthetic resin
12	Rubber products	22	Rubber & plastic products
13	Stone & construction materials	25	Stone & clay products
14	Fine ceramic products	25	Stone & clay products
15	Glass & glass products	24	Glass & glass products
16	Iron & steel	26	Basic iron & steel
17	Basic nonferrous metals	27	Basic nonferrous metals
18	Casting	26	Basic iron & steel
19	Rolling mill	26	Basic iron & steel
20	Structural metal products	28	Metal products
21	Industrial machinery	29	Industrial machinery [includes (30)]
22	Office & computing machine	31	Office & service machinery
23	Land-transport equipment	36	Automobiles
24	Ship building	37	Other transport equipment
25	Airplane	37	Other transport eqpt.
26	Electric machinery	32	Electrical machinery [includes (33)(34)(35)]
27	Scientific equipment	38	Scientific equipment
28	Fabricated metal products	28	Metal products
29	Music equipment, sport goods, etc.	39	Other manufacturing
30	Lumber	13	Lumber & wood products
31	Wood products	13	Lumber & wood products
32	Pulp & paper	15	Pulp & paper
33	Paper products	16	Paper products
34	Printing & publishing	17	Printing & publishing
35	Leather products	23	Leather products
36	Spinning, weaving, & knitting	11	Spinning, weaving & knitting
37	Clothing	12	Apparel & textile goods
38	Foods	9	Foods & beverages
39	Beverages	9	Foods & beverages
40	Tobacco processing	10	Tobacco processing

Input–output sector classification for the FRG and the common sector classification (UIO53). Common Sector Classification (UIO53): Continued.

I–O#	Description	I–O#	Description
	< Intermediate Sectors (1–59) >		< Intermediate Sectors (1–54) >
41	New construction	40	New construction
42	Construction repairs	41	Construction repairs
43	Wholesale trade	43	Trade
44	Retail trade	43	Trade
45	Railway transportation	46	Transportation
46	Water transportation	46	Transportation
47	Postal & telecommunication	47	Communication service
48	Services related to transport	46	Transportation
49	Banking	44	Financial services
50	Insurance	44	Financial services
51	Real estate	45	Real estate
52	Hotel	52	Other personal services
53	Social, educational, & cultural services	50	Health, educational, & social services
54	Medical services	50	
55	Other personal services	52	Other personal service [includes (51)]
56	Government	49	Public administration
57	Social insurance services	50	Social insurance services
58	Non profit organization	52	Other personal service ([51])
59	Intermediate input, total	54	Intermediate input, total
	< Value-Added Sectors (61–72)>		< Value-Added Sectors (55–61) >
61	Value-added tax not reinversab	57	Property income
63	Depreciation	57	Property income
64	Net indirect tax	58	Net indirect taxes
65	Wages & salary	56	Wages & salaries
66	Operating surplus	57	Property income
67	Gross value-added at market price	60	Gross value-added
68	Gross output	61	Gross output
	< Final Demand Sectors (60–67) >		< Final Demand Sectors (55–68) <
60	Private consumption expenditure	56	Private consumption expenditures
61	Government consumption expenditures	57	Government consumption expenditures
62	Capital formation to equipment	58	Fixed capital formation
63	Capital formation to construction	58	Fixed capital formation
64	Changes in inventory	59	Changes in inventory
65	Exports	60	Exports
66	Final demand, total	61	Final demand, total
67	Total demand	62	Total demand
68	Imports, total	66	Imports, total
69	Gross output	68	Gross output

References

Ayres, R.U., 1989, *Technological Transformations and Long Waves* RR-89-1, IIASA, Laxenburg, Austria.

Department of Commerce, 1984, 1985, Versions of Survey of Current Business, Washington, DC.

MITI, (Ministry of International Trade and Industry), 1977, 1982, Japan, various versions of Updated Input-Output Table for Japan, Tokyo, Japan.

Statistisches Bundesamt, Volkswirtschaftliche Gesamtrechnung, Fachserie 18, Reihe S.12, Ergebnisse der Input-Output-Rechnung, 1970 bis 1986.

Stone, R. *et al.*, 1962, *A Programme for Growth*, No.2: *A Social Account Matrix for 1960*, Chapman and Hall, London, UK.

Uno, K., 1989, *Measurement of Services in an Input–Output Framework*, Elsevier Science Publishers, Amsterdam, Netherlands.

Chapter 10

An Econometric Analysis of Technological Change in Japan, the USA, and the FRG

Mitsuo Yamada

The introduction of CIM is expected to achieve greater flexibility (in the sense of quick response to market signals), improve quality control, and increase benefits to individual firms. Furthermore, this technological change would have many other impacts on national economies in general, both domestically and internationally.

In the long run, the industrial structure is changing. In particular, the agricultural and manufacturing sectors are shrinking in most countries following the great postwar expansion in the share of employment in manufacturing. The tertiary sector (services) is growing in most industrialized countries. Given declining birth rates, restricted immigration, and rapid growth of both domestic and foreign markets, labor shortages in the manufacturing sector become an acute problem (especially in Japan), and the accelerated introduction of CIM in Japan seems to be a strategic response to this problem.[1]

Of course, economic patterns differ among countries; each has a different history and different problems. How will industrial structures change in the future? What new problems will be created in the labor market? Low-priced and high-quality products, which will be possible with the new technology, will not only strengthen the competitiveness of the producing firms but also have an influence on other sectors through changes in demand and relative prices. Such questions must be discussed in an economywide context.

This chapter tries to evaluate some of these economic impacts and to consider the impacts of CIM. The discussion focuses on the three countries (Japan, the USA, and the FRG) that are currently leaders in CIM diffusion. Their economies are described using multisectoral econometric models.

Generally speaking, analyses of econometric models offer the following advantages. First, one can check whether an econometric model maintains internal consistency among the model variables from the point of view of economic theory. Second, the model can be tested to explain the actual changes of the variables within the sample periods. Third, for conditional forecasts, we can show explicitly what is assumed and what is kept invariant in the model. Of course, different assumptions yield different forecasts. But, it is important to be able to declare what is assumed.

Section 10.1 explains the framework of our model. The next section describes actual changes in input–output structures resulting from the technological changes. Section 10.3 gives the assumptions for the simulation. The simulation results and what they may mean are discussed in Section 10.4.

10.1 Framework of the Model

The common framework of our three econometric models for Japan, the USA, and the FRG are described in this section.[2] They are the so-called Keynesian-Leontief type models, *viz.*, large multisectoral models including input–output information in them, that simultaneously explain both the input–output relationships and Keynesian macroeconomic behavior.

In this model, the domestic economy in each country is disaggregated into 21 sectors. These are listed in *Table 10.1*. Among them, 18 sectors are related to manufacturing. So-called tertiary activities – mainly services – are aggregated into one sector to simplify the analysis.

Table 10.1. Disaggregation of the sectors.

1	AG	Agriculture, forestry, and fishing
2	MI	Mining
3	FD	Food, beverage, and tobacco
4	TX	Textiles
5	AP	Apparel
6	LT	Leather products and footwear
7	WD	Wooden products and furniture
8	PP	Pulp, paper, printing, and publishing
9	RB	Rubber and plastic products
10	CH	Chemicals
11	PC	Petroleum and coal products
12	NM	Nonmetallic mineral products
13	IS	Iron and steel products
14	NF	Nonferrous metals
15	MT	Fabricated metal products
16	MC	Machinery, except electrical
17	EM	Electrical machinery
18	TE	Transport equipment
19	PI	Precision instruments
20	MM	Miscellaneous manufacturing products
21	SV	Construction and tertiary industry

In the multisectoral model, the input–output structure is fundamental. In the input–output framework, the supply and demand equilibrium in the commodity market is expressed as

$$X + M = AX + F^d + E , \tag{10.1}$$

where X is the output vector, M is the import vector, A is the input coefficient matrix, F^d is the domestic final-demand vector, and E is the export vector. Supposing that the import vector M is proportional to the level of domestic demand, we can express

$$M = \overline{M}(AX + F^d) , \tag{10.2}$$

where $\overline{M}$ is an import coefficient matrix. The diagonal elements of $\overline{M}$ are import coefficients for each sector.

Given some level of final demand, domestic and export, we can determine the output that satisfied market equilibrium, *viz.*,

$$X = [I - (I - M)A]^{-1}[(I - \overline{M})F^d + E] . \tag{10.3}$$

Based on the output level determined by equation (10.3), we can then determine labor demand, assuming the stability of the labor coefficients for each sector. The basic equation is

$$\mathbf{L} = \overline{\mathbf{L}}\mathbf{X} , \tag{10.4}$$

where $\overline{\mathbf{L}}$ is a labor coefficient matrix. The diagonal elements of $\overline{\mathbf{L}}$ are labor coefficients for each sector. These elements can be regarded as a labor coefficient vector, l.

Now, using the duality relationships of the input–output framework, we can introduce a price determination equation. From the cost relation in input–output structures, one can write

$$\mathbf{P} = \mathbf{A}'(\mathbf{I} - \overline{\mathbf{M}})'\mathbf{P} + \mathbf{A}'\overline{\mathbf{M}}'\mathbf{P}^{\mathbf{m}} + \mathbf{w}\mathbf{l} + \mathbf{R}^{\mathbf{s}} , \tag{10.5}$$

where $\mathbf{P}$ is the output-price vector, $\mathbf{A}'$ is the transpose of the input coefficient matrix, $\mathbf{M}'$ is the transpose of the import coefficient matrix, $\mathbf{P}^{\mathbf{m}}$ is the import-price vector, $\mathbf{w}$ is the diagonal matrix of wage rate, l is the labor coefficient vector noted above, and finally $\mathbf{R}^{\mathbf{s}}$ means "other costs" (mainly capital costs).

In our multisectoral model, these fundamental relations should be included. In practice they are modified a bit. This is partly because we use only single-year information with respect to the input–output relations (1980 for Japan, 1977 for the USA, and 1980 for the FRG) and partly because we subsequently neglect some factors that seem to be of secondary importance for our analytical purposes. In spite of the simplifications, the model has some useful features. In particular, the main final-demand levels are endogenously determined, and price variables and quantity variables are mutually interdependent.

10.1.1 Architecture of the I–O model

The fundamental structure of the model and some modifications in its input–output structure will be discussed briefly in this section. Our multisectoral model includes the following elements of any economywide model: (1) final-demand block; (2) input–output and output block; (3) factor-demand block; (4) income generation block; (5) price-cost block; (6) wage and employment block; and (7) export and import block. *Figure 10.1* shows the basic structure of the model. Looking at this diagram, one can see schematically how each block relates to the others and to the model as a whole.

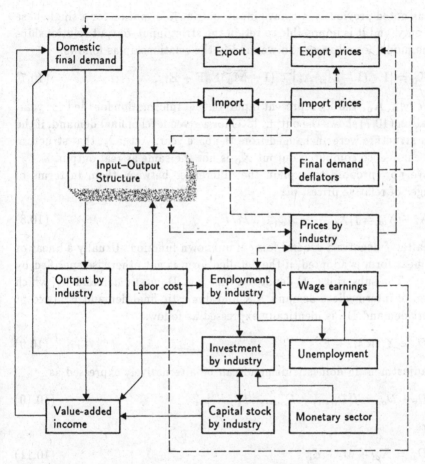

Figure 10.1. Structure of a multisectoral model.

Next, we consider the detailed structure of our model, and the modifications noted earlier. We decompose the final domestic demand vector into the following two elements:

$$\mathbf{F}^d = \overline{\mathbf{F}}\mathbf{F} , \tag{10.6}$$

where $\overline{\mathbf{F}}$ is the final-demand coefficient matrix and $\mathbf{F}$ is a vector of total final demand. Given a final demand $\mathbf{F}$, and the export vector $\mathbf{E}$, one can explain the output by sector from equation (10.3)

$$\mathbf{X} = [\mathbf{I} - (\mathbf{I} - \overline{\mathbf{M}})\mathbf{A}]^{-1}[(\mathbf{I} - \overline{\mathbf{M}})\overline{\mathbf{F}}\mathbf{F} + \mathbf{E}] . \tag{10.3'}$$

But as noted previously, we have input–output information for a single base year only, and it is impossible to retain the strict input–output relationships in the model for all years. So we modify the relationship as follows:

$$\mathbf{X_o} = [\mathbf{I} - (\mathbf{I} - \overline{\mathbf{M}}_o)\mathbf{A}_o]^{-1}[(\mathbf{I} - \overline{\mathbf{M}}_o)\overline{\mathbf{F}}_o\mathbf{F} + \mathbf{E}] \ , \tag{10.7}$$

where $\overline{\mathbf{M}}_o$, $\mathbf{A}_o$, and $\overline{\mathbf{F}}_o$ represent input–output information for the base year. Equation (10.7) shows the output level for a given level of final demand, if the input structure were unchanged from the base year. Actually, that structure does change, so computed output $\mathbf{X_o}$ is not the same as real output $\mathbf{X}$.

We now propose to explain the differences between them in terms of changes of relative prices, *viz.*,

$$X_i - X_{oi} = f[X_{i(-1)} - X_{oi(-1)}, P_i/P] \ . \tag{10.8}$$

Hereafter f denotes a general form of unknown function. Usually a linear or log-linear form is adopted, if the detailed form is not otherwise specified by theory. The subscript (-1) denotes a time lag. Domestic demand D_i, which is sum of intermediate demand $\mathbf{AX}$ and domestic final demand $\mathbf{F^d}$ (except export demand $\mathbf{E}$), is identically expressed as follows:

$$D_i = X_i + M_i - E_i \ . \tag{10.9}$$

An adjustment in domestic demand can be alternatively expressed as

$$D_i - D_{oi} = f[D_{i(-1)} - D_{oi(-1)}, P_i^d/P^d] \ , \tag{10.10}$$

where

$$D_{oi} = X_{oi} + M_i - E_i \ , \tag{10.11}$$

whence

$$X_i = D_i + E_i - M_i \ . \tag{10.12}$$

Here, P_i^d and P^d denote the price deflators of sectoral and average domestic demand, respectively.

Given an output level, employment by sector can be derived as either a labor demand function or the inverse of a production function. The latter case is written as follows:

$$L_i = f[X_i, K_{i(-1)}, T] \ , \tag{10.13}$$

where T is an index of technical progress.

Imports by sector are also explained as a demand function of the usual sort:

$$M_i = f[D_i, P_i^m/P_i, M_{i(-1)}] \ . \tag{10.14}$$

Sectoral output prices can be explained in terms of intermediate input costs and unit labor costs,

$$P_j = f(P_{oj}^r, W_j L_j/X_j) \ . \tag{10.15}$$

Here P_{oj}^r denotes the input price deflator by sector, computed using base-year input–output information, *viz*.,

$$\mathbf{P_o^r} = [P_{oj}^r] = \mathbf{A_o}'(\mathbf{I} - \overline{\mathbf{M}}_\mathbf{o})'\mathbf{P} + \mathbf{A_o}'\overline{\mathbf{M}}_\mathbf{o}'\mathbf{P^m} \ . \tag{10.16}$$

Of course, (10.16) is a modification of the following equation, which is part of equation (10.5),

$$\mathbf{P^r} = [P_j^r] = \mathbf{A}'(\mathbf{I} - \overline{\mathbf{M}})'\mathbf{P} + \mathbf{A}'\overline{\mathbf{M}}'\mathbf{P^m} \ . \tag{10.17}$$

Wage rates are determined in two steps. First, the changing rate of average wages is explained by a modified Phillips curve relation, namely:

$$\dot{W} = f(\dot{P}^c, 1/U^R, \dot{X}/L) \ , \tag{10.18}$$

where P^c is the consumption deflator, U^R is the unemployment rate, and X/L is labor productivity. The dot in the equation indicates the time derivative of its variable. Wage rates by industry are assumed to be determined by the average wage rate, given above, and the relative productivity of each industry. Thus,

$$W_j = f[W, X_j/L_j/(X/L)] \ . \tag{10.19}$$

The deflators of final expenditures are connected to the output deflators by sector through a final-demand coefficient matrix $\overline{\mathbf{F}}$:

$$\mathbf{P^{fd}} = \overline{\mathbf{F}}'\mathbf{P} \ . \tag{10.20}$$

But here again we have only base-year information, so equation (10.20) does not hold strictly in the model. In the same way as before, we modify it as follows:

$$P_j^{fd} = f(P_{oj}^{fd}) \ , \tag{10.21}$$

where

$$\mathbf{P_o^{fd}} = [P_{oj}^{fd}] = \overline{\mathbf{F}_o}'P \ . \tag{10.22}$$

The export deflator by sector is based on the output deflator, but that relation is modified by the condition of competition in the world market:

$$P_j^e = f(P_j, \varepsilon P_j^w) \ , \tag{10.23}$$

where ε denotes the exchange rate and P_j^w is the price of commodity j in the world market.

In the input–output framework, Y_j^g, the gross value-added by sector j, is determined as follows:

$$Y_j^g = P_j X_j - P_j^r R_j X_j \ , \tag{10.24}$$

where P_j^r and R_j are input price and input ratio through strict input–output relations, *viz.*,

$$P_j^r = \Sigma_i P_i a_{ij} / \Sigma_i a_{ij} \tag{10.25}$$

$$R_j = \Sigma_i a_{ij} \ . \tag{10.26}$$

But, as we cannot hold to such a strict relation in every year, we modify it as follows,

$$Y_j^g = f[(P_j - P_{oj}^r R_{oj}) X_j] \ , \tag{10.27}$$

where P_{oj}^r is the variable which is determined in equation (10.16) and R_{oj} is input ratio of sector j in the base period. Depreciation by industry D_j^p is assumed to be proportional to the value of capital stock, *viz.*,

$$D_j^p = f[P^{ip} K_{j(-1)}] \ , \tag{10.28}$$

where P^{ip} is the investment deflator and K_j is real capital stock.

The compensation of employees Y_j^w is defined as the product of the wage rate and employees by sector, *viz.*,

$$Y_j^w = W_j L_j \ . \tag{10.29}$$

We can treat indirect tax minus subsidies, T_j^x, as exogenous variables because these variables are determined by institutional relations and tax policy. Then net value-added by industry Y_j is defined by

$$Y_j = Y_j^g - D_j^p - T_j^x \ . \tag{10.30}$$

Investment by sector I_j is mainly explained as profitability or expected demand:

$$I_j = f[Y_j^g/P^{ip}, X_j, K_{j(-1)}, r - \dot{P}^{ip}] \qquad (10.31)$$

$$K_j = (1-d)K_{j(-1)} + I_j , \qquad (10.32)$$

where $r - \dot{P}^{ip}$ is the real interest rate, P^{ip} is the investment deflator, d is the depreciation rate, and K_j is the capital stock.

Finally, all these variables are aggregated by summing over to connect them to the macro variables. Major macro behaviors like the consumption function and the housing investment function are included in the model to close the causal chains. These functions are denoted as follows:

$$C = f[Y^d/P^c, P^c, C_{(-1)}] \qquad (10.33)$$

$$I^h = f[Y^d/P^{ih}, K^h(-1), r - P^{ih}] , \qquad (10.34)$$

where C is consumption expenditure, Y^d is disposable income, I^h is housing investment, P^{ih} is the deflator of housing investment, K^h is housing capital stock, and $r - \dot{P}^{ih}$ is the real interest rate.

10.1.2 CIM introduction within the model

The following causal chain can be considered to evaluate the economic impacts introducing CIM in manufacturing sectors. First, introducing CIM would reduce production costs and induce a price cut for manufactured products. That would increase demand in the market, (due to nonzero price elasticity of demand). This, in turn, induces firms to increase capacity in the industry or company (Salter cycle). As the result of the economies of scale, production costs would fall, which leads to increased profitability in the sectors. And that would make room for further reductions in prices.

It is not so clear, however, how effective economies-of-scale improvements are from the viewpoint of economywide relations, though such causality should be considered. Our model reflects the main parts of the causality

chain discussed earlier, though the model is basically demand-oriented and
the supply side or production function is not treated explicitly.

In the multisectoral model (*Figure 10.1*), the introduction of new technologies like CIM changes the input–output and final-demand structures,
labor demand, etc. Labor-saving technology reduces labor costs and then
output price and export price.

Changes in relative prices affect not only domestic demand but also international trade. This, in turn, will influence the total output level by sector.
Through the interdependence of input–output relationships, an increase of
demand in any sector is reflected indirectly in demand for products of other
sectors.

Expansion in production requires increased investment, which is one component of final demand. Also, increased capital stock has an indirect affect
on output through decreased labor demand. However, labor reduction in
one sector leads to increased unemployment in that sector. Whether unemployment in the whole economy increases or not depends on how much
labor demand will be created in other sectors. If aggregate unemployment
is increased, the rate of increase in wage rates is likely to fall and the same
will be true for prices. But if unemployment falls, the wage rate in turn will
be increased, partially offsetting the cost savings from the new technology.
These points will be elaborated later in the discussion of model results.

10.2   Recent Changes in Input–Output Structure

There are many possible impacts of the introduction of CIM, not all of which
can be examined in one model. Any model has its own restrictions. Here we
consider three factors.

First, the direct impact on the economy is to increase each output level
that will be induced by the changes in final demands. This is a rather short-
run effect. An increase in the future level of CIM investment or future share
of CIM within total investment of each industry would be required.

Second, in the long run, capital accumulation will accelerate the substitution of capital for labor. That is, there would be a labor-saving effect. This
is one of the most important factors acting to decrease production costs.

Third, we can expect changes in input coefficients associated with CIM.
For example, in the production process, an increase in the yield – output
per unit of physical input – would reduce the demand for some intermediate

inputs. This would reduce related production costs. CIM should have a beneficial effect on energy consumption in manufacturing, in particular.

There are several possible strategies to evaluate these economic impacts within the model:

(1) The first approach is to obtain information from previous studies at a more micro level or from survey research (Mori, 1987; Tani, 1987, 1989a, 1989b; Maly and Zaruba, 1989; Tchijov, 1989). For example, Kinoshita and Yamada (1989b) depend partly on the information from survey research on the use of robotics in Japan.
(2) Another approach is to extrapolate from statistics, such as the secular changes in input–output coefficients.
(3) A third approach is to rely on expert judgment based on engineering and microeconomic analysis. Dobrinsky (Chapter 14) relies on this sort of information, provided by IIASA.

But the integration of information from micro-level analysis to economy-wide purposes is not easy. This is because new technologies like CIM are mixtures of hardware (e.g., NC machine tools) and software systems, which depend on widespread use of computers. In our analysis, we start by observing what is changing and what is stable in input–output statistics, to extract trends in technology and tastes.

10.2.1 Data sources and aggregation

We have several comparable input–output tables of Japan, the USA, and the FRG; based on these tables we can observe recent changes in the input–output structures. For Japan, input–output tables exist for 1970, 1975, 1980, and 1985. The tables for the first three years are linked input–output tables with 71 sectoral classifications for 1970–1975–1980. These were compiled by the Japanese Administrative Management Agency in 1985. The latest table (1985), with 85 classifications, was made by the same Agency.

For the USA, we use the 85-sector input–output data for 1977 and 1982.[3] These are so-called Stone-type input–output tables. They consist of two parts: (1) the use of commodities by industries and (2) output of commodity by industries. We then construct traditional input–output tables from them, assuming the standard industrial technology hypothesis; an industry is assumed to have the same input structure whatever its products mix.

For the FRG, there are five input–output tables, one for every two years from 1978 to 1986, each having 58 classifications. These tables were compiled by the Bureau of Statistics (FRG).

For comparison, we have compiled these tables and aggregated the industries into 21 sectors, which correspond to the classifications in the econometric models. We emphasize the need to be careful in comparison of these tables, since they are all nominal, and changes from year to year stem partially from price changes.

10.2.2   Observable changes in the input structures

Figures 10.2 and *10.3* reveal the input coefficients of the manufacturing sector and the metalworking manufacturing sector in Japan as averages, from the year 1970 to 1985. We can readily see the changing input structures.

It is clear that inputs of iron and steel; agriculture, forestry, and fishing; and textiles are continuously declining in the manufacturing sector. On the other hand, inputs of chemical products, transport equipment, miscellaneous manufacturing, and the tertiary sector, including services, are gradually increasing.

The decline of iron and steel inputs corresponds to the tendency that manufacturing products are becoming "lighter, smaller, shorter, and thinner" from "heavier, larger, longer, and thicker." As a recent tendency, it is observable that inputs of general machinery products and electrical machinery are decreasing, and those of nonferrous metal and transport equipments are increasing rapidly.

In the metalworking manufacturing sector, one can observe similar tendencies. A decline in iron and steel inputs and an increase in service inputs are critical. But reductions in inputs of general machinery and electrical machinery mainly stem from reductions in own use, which is interpreted partially as the result of increased capital productivity within the sector. In transport equipment, the increased intrasectoral use of products is observed, but it is not easy to interpret.

Looking at the figures that show the input coefficient changes in the USA, (*Figures 10.4* and *10.5*) and the FRG (*Figures 10.6* and *10.7*), it is clear that declines of iron and steel inputs and increases in service inputs and in electrical machinery inputs (i.e., computers and electronics) have occurred. Of course, the magnitudes of these tendencies are variable, and there are some exceptions. In general, the change in input coefficients in the USA and the FRG seems more stable than that in Japan.

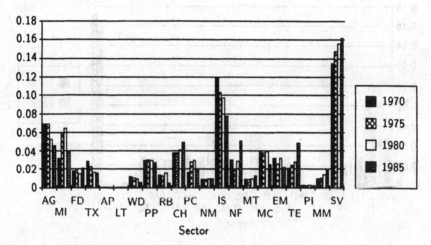

Figure 10.2. Input coefficients of manufacturing in Japan. See *Table 10.1* for disaggregation of the sectors.

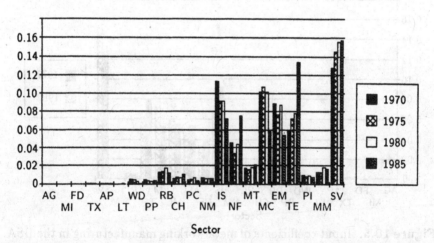

Figure 10.3. Input coefficients of metalworking manufacturing in Japan.

Figures 10.8 to *10.12* show the direction of recent change in input structure for the three countries on an annual basis.[4] In fabricated metal (*Figure 10.8*), the reduction in inputs of iron and steel is a common feature for all three countries, but the change in service inputs differs between Japan, the

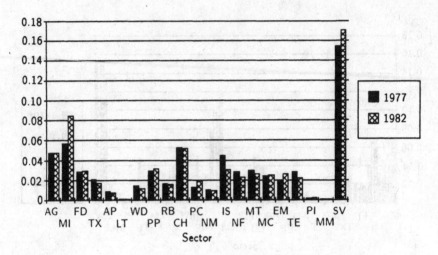

Figure 10.4. Input coefficients of manufacturing in the USA.

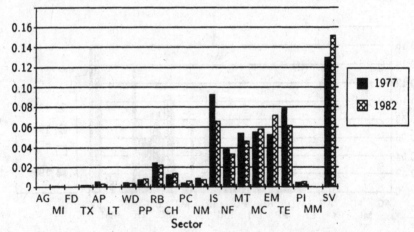

Figure 10.5. Input coefficients of metalworking manufacturing in the USA.

USA, and the FRG. The increase of inputs in nonferrous metal and fabricated metal and general machinery products is also distinctive in Japan.

In general machinery (*Figure 10.9*), Japan has reduced its own inputs and also that of electrical machinery inputs a great deal, but the other countries have not done so. Japan has also reduced its inputs of electrical

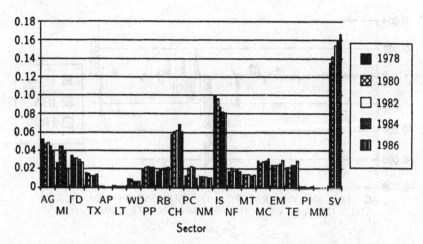

Figure 10.6. Input coefficients of manufacturing in the FRG.

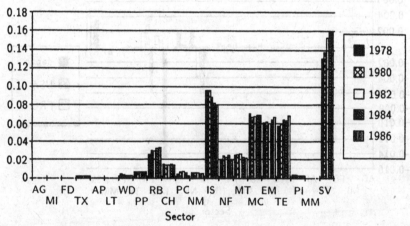

Figure 10.7. Input coefficients of metalworking manufacturing in the FRG.

machinery (*Figure 10.10*) and increased the inputs of nonferrous metal products. The USA augments the use of own products, but the FRG is relatively stable except in the use of iron and steel and services. In Japan, continued "electronization" is apparently going on in this sector, but the saving of own products seems dominant.

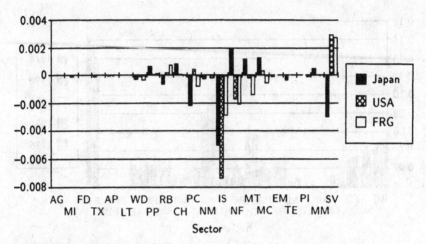

Figure 10.8. Recent changes in input coefficients: fabricated metal.

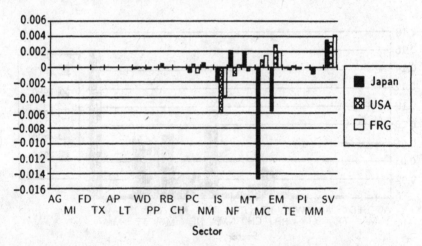

Figure 10.9. Recent changes in input coefficients: general machinery.

In transport equipment (*Figure 10.11*), a decrease in general machinery inputs and an increase in own products are observed as special features in Japan.

In precision instruments (*Figure 10.12*), the Japanese input pattern has changed more than that of the USA and the FRG. Japan has decreased inputs of iron and steel, nonferrous metals, and general machinery, but has

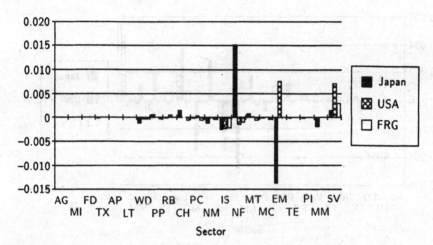

Figure 10.10. Recent changes in input coefficients: electrical machinery.

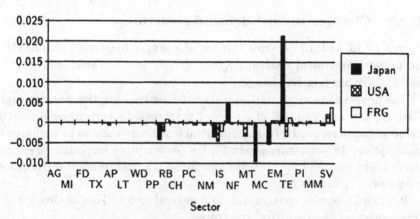

Figure 10.11. Recent changes in input coefficients: transport equipment.

increased inputs of fabricated metal products and machinery (except general machinery). It is especially interesting to note that Japan has increased its electrical machinery shares compared with the other countries. This reflects the electronization of the sector.

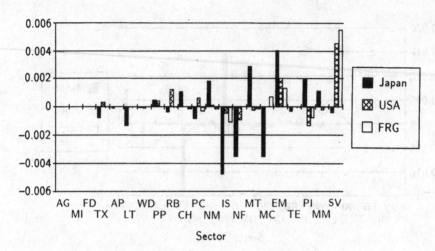

Figure 10.12. Recent changes in input coefficients: precision instruments.

10.2.3 Changes in final-demand structure

Figures 10.13 and *10.14* show the recent changes in private consumption and private investment coefficients, respectively for the three countries. Of course, these are relative changes.

The figures show two common features: food consumption has decreased relatively and consumption of services has increased. Agriculture products, textile and apparel products, and electrical machinery seem to be declining in consumption. In petroleum products, US consumption increases very little, which might partially reflect price increases. US consumption in transport equipment is decreasing.

Increase in service consumption is a general trend. This tendency has a large impact on the industrial structure.

Within the components of business investment, general machinery and electrical machinery are increasing. This tendency can be seen among all three countries, but it is strongest in Japan.

10.3 Technological Change Assumptions

To investigate the implications of a new technology like CIM, it is necessary to extend the input–output tables beyond the last year for which there is official data. For each country, we have input–output tables for more than

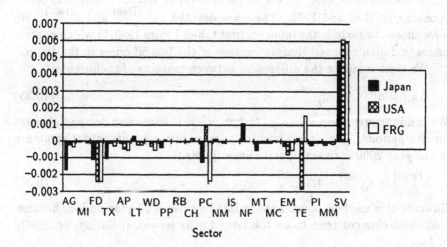

Figure 10.13. Recent changes in consumption coefficients.

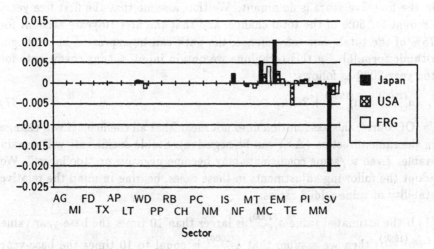

Figure 10.14. Recent changes in investment coefficients.

two points in time. We use this information to project recent trends in I–O coefficients and final-demand structure. Such trends are thought to be due to technological changes.

In the Japanese case, we use two sets of input–output coefficients corresponding to 1980 and 1985. These are denoted as $a_{ij}^{(1980)}$ and $a_{ij}^{(1985)}$. We have chosen to exclude the input–output tables before 1980, in which relative prices in Japan changed sharply because of the two oil crises in the 1970s.

We then compute the differences between two sets of coefficients:

$$\Delta a_{ij} = a_{ij}^{(1985)} - a_{ij}^{(1980)} \ . \tag{10.35}$$

We can interpret these values as technological changes and demand changes for five periods. One simple way to estimate future input–output structure in the year 2000 is to extrapolate linearly, *viz.*,

$$a_{ij}^{(2000)} = a_{ij}^{(1980)} + 4\Delta a_{ij} \ . \tag{10.36}$$

However this method is not appropriate to cover a 20-year period, because coefficient changes tend to be too large for linear extrapolation, especially in Japan.

Thus we must introduce some adjustments. For purposes of this analysis, we assume that the technological change is nonlinear, and that the change in the first five years is dominant. We thus assume that the first five years account for 50% of the total change, and that the first 10 years account for 75% of the total. For convenience, the path can be expressed in semilogarithmic form.[5] By this technique we obtain input–output coefficients for the year 2000 as follows:

$$a_{ij}^{(2000)} = a_{ij}^{(1980)} + 2\Delta a_{ij} \ . \tag{10.37}$$

Of course, this assumption does not mean that all coefficients will change in the same direction (Δa_{ij} can be negative). Stable coefficients will remain stable. Even so, some coefficients may become negative or "too large." We adopt the following adjustments in these cases, bearing in mind the relative stability of value-added ratios:

(1) If the estimated value $a_{ij}^{(2000)}$ is larger than 10 times the base-year value $a_{ij}^{(1980)}$, then we assume that $a_{ij}^{(2000)}$ is equal to 10 times the base-year value, *viz.*,

$$a_{ij}^{(2000)} = 10a_{ij}^{(1980)} \ . \tag{10.38}$$

(2) If the estimated value $a_{ij}^{(2000)}$ is smaller than one-tenth of the base-year value $a_{ij}^{(1980)}$, then $a_{ij}^{(2000)}$ is assumed to be equal to one-tenth of the base-year value, *viz.*,

$$a_{ij}^{(2000)} = 0.1 a_{ij}^{(1980)} \ . \tag{10.39}$$

(3) Finally, we exclude the changes in the industry's use of its own product. ("Own" use is represented by the diagonal elements in the original tables.) Change in the use of own product is one of the cost components, but it is not clear that such a change will be reflected in reduction of posted prices directly. Cost savings might be used for new investment or R&D, for example.

In the case of the FRG, we use the difference between the input–output tables for 1980 and 1986. We then obtain input–output coefficients for the year 2000 in the same manner as above. For the USA, we also obtain differences for the first five years, but the base year is 1977 and the comparison year is 1982. In spite of this, we assume that technological change starts in 1980 to facilitate comparisons among the three countries.

As to change of final-demand structure, we adopt the same method; however, we consider the change in only two main components, namely, private consumption and private business investment.

Unfortunately we cannot compile and analyze empirical information about the changing pattern in labor-demand structure for the three countries. Here we adopt an alternative method for assessing the labor-saving effects, which is the one examined by Dobrinsky (Chapter 14).[6]

Assuming several types of technologies are currently installed in each sector, we can express labor productivity of each sector as follows:

$$y_i = \Sigma_k S_{ik} y_{ik} \ , \tag{10.40}$$

where y_{ik} is labor productivity of technology k in sector i and S_{ik} is labor share for technology k in sector i ($\Sigma_k S_{ik} = 1$). In Dobrinsky's analysis, the following four types of production technology corresponding to four levels of successively more advanced automation are specified for the metalworking sectors:

Level–0: Traditional, nonautomated, stand-alone machines.
Level–1: Stand-alone NC and CNC machine.
Level–2: FMS and FMC.
Level–3: CIM.

Based on scenarios prepared by IIASA, Dobrinsky adopted the following relative magnitude of labor productivity for each level of automation:

Table 10.2. Production shares by technology, in percent.

Country	Sector	Level–0 1984	Level–0 1999	Level–1 1984	Level–1 1999	Level–2 1984	Level–2 1999	Level–3 1984	Level–3 1999
Japan	MT	87	60	12	31	1	8	0	1
	MC	78	30	18	48	4	20	0	2
	EM	81	40	17	45	2	12	0	3
	TE	82	20	14	58	4	20	0	2
	PI	86	40	13	45	1	12	0	3
USA	MT	93	65	6	28	1	6	0	1
	MC	84	35	13	47	3	16	0	2
	EM	88	45	10	42	2	10	0	3
	TE	84	25	13	57	3	16	0	2
	PI	90	45	9	42	1	10	0	3
FRG	MT	95	62	4	31	1	6	0	1
	MC	83	32	14	48	3	18	0	2
	EM	90	42	8	45	2	10	0	3
	TE	86	22	11	58	3	18	0	2
	PI	90	42	9	45	1	10	0	3

From Chapter 14.

Level–0: $y_0 = 1$.
Level–1: $y_1 = 2y_0$.
Level–2: $y_2 = 3y_1 = 6y_0$.
Level–3: $y_3 = 5y_2 = 30y_0$.

These relations were applied to the five metalworking sectors. Dobrinsky assumes the shares of production by technology in each sector as shown in *Table 10.2*. Adopting the relative magnitude of labor productivity by each technology that are mentioned above, employment shares are induced (*Table 10.3*).

Then, based on this formulation, one can induce the labor productivity of each sector as compared with traditional technology. The results are shown in *Table 10.4*, which we apply to the simulation hereafter.

It is clear that the difference of each productivity heavily depends on the difference of labor shares by technology. This corresponds to the state of CIM technology diffusion in each sector. In *Table 10.4*, we find that the difference in the changing rate of productivity among the sectors is more important than that among countries. Our model is based on the idea that increases in labor productivity induce a downward shift in the labor-demand function.

Table 10.3. Employment shares by technology, in percent.

Country	Sector	Level–0 1984	Level–0 1999	Level–1 1984	Level–1 1999	Level–2 1984	Level–2 1999	Level–3 1984	Level–3 1999
Japan	MT	93.38	78.06	6.440	20.16	0.179	1.735	0.0	0.043
	MC	88.97	52.26	10.270	41.81	0.760	5.807	0.0	0.116
	EM	90.17	61.92	9.462	34.83	0.371	3.096	0.0	0.155
	TE	91.45	38.17	7.807	55.34	0.743	6.361	0.0	0.127
	PI	92.81	61.92	7.014	34.83	0.180	3.096	0.0	0.155
USA	MT	96.71	81.22	3.120	17.49	0.173	1.249	0.0	0.042
	MC	92.31	57.16	7.143	38.38	0.549	4.355	0.0	0.109
	EM	94.29	66.40	5.357	30.99	0.357	2.459	0.0	0.148
	TE	92.31	44.46	7.143	50.68	0.549	4.742	0.0	0.119
	PI	95.07	66.40	4.754	30.99	0.176	2.459	0.0	0.148
FRG	MT	97.77	78.95	2.058	19.74	0.172	1.273	0.0	0.042
	MC	91.71	54.18	7.735	40.63	0.552	5.079	0.0	0.113
	EM	95.41	63.38	4.240	33.95	0.353	2.515	0.0	0.151
	TE	93.48	40.69	5.978	53.64	0.543	5.549	0.0	0.123
	PI	95.07	63.38	4.754	33.95	0.176	2.515	0.0	0.151

Table 10.4. Change in labor productivity in each sector.

Country	Sector	1984	1999	Ratio	Annual rate
Japan	MT	107.3	130.1	1.212	0.012
	MC	114.1	174.2	1.527	0.027
	EM	111.3	154.8	1.391	0.021
	TE	111.5	190.8	1.711	0.034
	PI	107.9	154.8	1.434	0.023
USA	MT	104.0	124.9	1.202	0.012
	MC	109.9	163.3	1.486	0.025
	EM	107.1	147.6	1.377	0.020
	TE	109.9	177.8	1.618	0.031
	PI	105.6	147.6	1.397	0.021
FRG	MT	102.9	127.3	1.237	0.013
	MC	110.5	169.3	1.532	0.027
	EM	106.0	150.9	1.424	0.022
	TE	108.7	185.0	1.702	0.034
	PI	105.6	150.9	1.429	0.023

10.4 Simulations

10.4.1 Boundary conditions

Simulations require explicit assumptions regarding many exogenous economic variables. In our model such variables relate to world trade, government activities, and other factors.

Table 10.5. The assumption in the main exogenous variables for projection.

World trade	Annual growth rate
World trade, total, real	5.5%
Price of world trade, total	4.5%
World trade, mining, real	3.0%

Originally the model for each country was connected to a commodity-based international trade-flow model, which links the import demands in one country to the export supply in the others and links export price in one country to import price in the others in a consistent way (see note [2]). For simulation purposes, however, we treat each country model separately. To do this it is necessary to provide some common assumptions with respect to world trade and its connection to each model.

First, we assume plausible growth rates for both total world trade and world trade of natural resources (products of the mining sector) and for corresponding world prices. These are listed in *Table 10.5*. We then assume relationships between total world trade and world trade of each sector as follows:

$$\ln W_i^{T\$} = a + b \ln W^{T\$} + c \ln(P_i^{WT\$}/P^{WT\$}) , \tag{10.41}$$

where $W_i^{T\$}$ means real world trade of sector i, $W^{T\$}$ is real total world trade, while $P_i^{WT\$}$ and $P^{WT\$}$ denote their prices, respectively. Of course, some adjustment is necessary to maintain the standard accounting identities in order that the sum of each sector's imports must be equal to total.[7]

We also assume a relationship between the prices of total world trade and the prices for each sector as follows:

$$\ln P_i^{WT\$} = a + b \ln P^{WT\$} . \tag{10.42}$$

Parameters in these equations are estimated from the world trade data of the data base. Extrapolation of this simple world trade model is shown in *Table 10.6* in terms of annual growth rates. These values are applied to the simulation for each model.

Other main exogenous variables, the numerical values of which we must provide from outside the model, are listed in *Table 10.7*. Some values are determined by considering the recent trend or movement of each variable, and some are based on the projections by other researchers and organizations.

Table 10.6. Projections in the simple world model: annual growth rate in 1990–2000, in percent.

Sector		Name	Real world trade	Export price
1	AG	Agriculture, forestry, fishing	2.63	3.21
2	MI	Mining	3.00	8.09
3	FD	Food, beverage, tobacco	4.01	2.67
4	TX	Textiles	2.16	3.09
5	AP	Apparel	11.23	3.73
6	LT	Leather products, footwear	5.64	4.22
7	WD	Wood products, furniture	5.49	3.37
8	PP	Paper, pulp, printing	3.19	3.84
9	RB	Rubber and plastics	4.52	4.05
10	CH	Chemicals	5.40	4.18
11	PC	Petroleum and coal products	2.56	8.87
12	NM	Nonmetallic mineral products	3.56	4.21
13	IS	Iron and steel products	2.57	3.48
14	NF	Nonferrous metals	4.18	3.80
15	MT	Fabricated metals	5.77	3.94
16	MC	Machinery except electrical	5.96	3.62
17	EM	Electrical machinery	8.04	2.90
18	TE	Transport equipment	4.29	3.69
19	PI	Precision instruments	11.81	3.31
20	MM	Miscellaneous manufacturing	7.54	3.97

10.4.2 The standard (control) scenario

Using the assumed values for the exogenous variables, we conducted a simulation for the period from 1980 to 2000. The computed values of the main economic variables in each model are listed in *Table 10.8* in terms of annual growth rates.

We note that the projected values for real GNP are slightly lower than most growth scenarios (e.g., the OECD). In our model, labor supply is fixed in relation to population, and it does not react to any economic variable. As a consequence, any increase in labor demand, induced by the expansion of production, will reduce unemployment, and then increase wage and price levels. This adds to inflation and reduces real GNP growth.

In fact, if we introduce a more flexible labor-supply function in the same model, real GNP or GDP will grow at the rate of 5.66% for Japan, 1.78% for the USA, and 2.17% for the FRG in the second decade of our simulation

Table 10.7. The assumption in the main exogenous variables for projection.

Model	Annual growth rate
Japan	
Exchange rate	120 yen/$ in 2000
Population, over 15 years old[a]	107.60 million in 2000
Discount rate of Bank of Japan	4% in 1990–2000
Real government consumption	3.5%
Real public investment	2.0%
Gross investment, mining	0.0%
Real output, mining	0.0%
Real export, agriculture, forestry, and fishery	0.0%
Real export, mining	0.0%
Employee rate to total employment	3.0%
US	
Population, over 16 years old[b]	208.185 million in 2000
Real output, mining	0.5%
Real export, mining	3.6%
Real government expenditures	1.5%
Money supply	10.0%
Prime rate	10% in 1990–2000
FRG[c]	
Exchange rate	1.8 DM/$ in 2000
Population[d]	58.952 million in 2000
Discount rate of Central Bank	4.7% in 1990–2000
Real public consumptions	2.8%
Real government investment	2.0%
Real export, mining	0.0%

[a]Population over 15 years old in Japan model is the value estimated by the Institute of Population Problem, Japan in 1986.
[b]Population over 16 years old in the US model is the value estimated by US Bureau, as projection of total population: 1990 to 2000.
[c]Estimates made prior to October 1990 and do not account for German unification.
[d]Population in FRG model is the value in *World Population Projections*, 1987–1988 Edition, K.C. Zachariah and My T.Vu, Johns Hopkins University Press.

period. But in this case, increased labor demand would induce more labor supply than seems compatible with the future level of the population.

This suggests that labor supply might be a real restriction for future possible economic growth in each country. Actually in Japan, the possibility that tight labor supply may restrict growth in the near future is often noted. On the other hand, there may be social changes, such as increased

Table 10.8. Control solution of the main macro variables: annual growth rate in 1980–2000, in percent.

Model	1980–1990	1990–2000
Japan		
Real GNP	3.13	2.64
Nominal GNP	3.28	9.17
Consumption	2.98	1.65
Housing investment	2.61	1.36
Business investment	3.38	1.35
Export	6.07	7.16
Import	4.50	4.45
Employment	1.06	0.80
Rate of unemployment	4.16	−7.16
GNP deflator	0.15	6.36
Wage rate	1.04	9.24
USA		
Real GNP	1.18	1.39
Nominal GNP	7.58	6.14
Consumption	1.74	0.27
Housing investment	0.44	−4.65
Business investment	2.15	4.19
Export	2.18	6.75
Import	6.14	4.05
Employment	1.07	1.10
Rate of unemployment	5.04	−1.65
GNP deflator	6.33	4.68
Wage rate	6.59	5.17
FRG[a]		
Real GNP	1.89	2.06
Nominal GNP	4.75	4.95
Consumption	1.80	1.40
Housing investment	0.29	0.36
Business investment	2.23	2.96
Export	3.57	4.19
Import	3.27	4.34
Employment	0.15	0.53
Rate of unemployment	3.36	−8.26
GNP deflator	2.81	2.83
Wage rate	4.17	4.67

[a]Calculations made prior to October 1990 and do not account for German unification.

Table 10.9. Possible factors considered in three cases in each country.

Factor	Case-1	Case-2	Case-3
Input–output coefficient			
Metalworking manufacturing	yes	yes	yes
Other sectors	yes	yes	no
Final-demand coefficient			
Private consumption	yes	yes	no
Private investment			
Metalworking manufacturing	yes	yes	yes
Other sectors	yes	yes	no
Labor productivity			
Metalworking manufacturing	no	yes	yes
Other sectors	no	no	no

employment of women or later retirement, that could increase the labor force without increasing population. Tight labor supply may also be a factor in accelerating the introduction of CIM technology, especially in the USA, where unemployment was quite severe in the late 1970s and early 1980s.

10.4.3 Variant cases

Table 10.9 considers three cases. In the table yes means that the factor noted on the left side is considered in the simulation; no means that it is not.

In Case-1, we examine the effects of the changing input–output coefficients and the final-demand structure on the whole economy. In other words, we explore the consequences of structural change in production technology and the revealed preferences. In Case-2, in addition to the factors considered in Case-1, we also study induced labor-saving effects in the five metalworking manufacturing sectors. In Case-3, we consider the effects of changes in metalworking manufacturing sectors only. The difference between Case-1 and Case-2 is attributable to labor-saving technological change (CIM) in metalworking manufacturing per se in the *presence* of overall structural changes, while the difference between Case-2 and Case-3 reflects the impacts of labor-saving technology (CIM) in the *absence* of overall structural changes.

Table 10.10 shows the effects on macroeconomic variables in the three countries in terms of the difference in annual growth rates between the standard (control) path and each case.

Table 10.10. Effects on macroeconomic variables: annual growth rates, in percent.

Variable	Case–1 1980–1990	Case–1 1991–2000	Case–2 1980–1990	Case–2 1991–2000	Case–3 1980–1990	Case–3 1991–2000
Japan						
Real GNP	0.22	0.29	1.03	0.68	0.66	0.56
Nominal GNP	2.19	1.04	1.79	0.83	-0.46	-0.23
Consumption	0.32	0.44	0.72	0.56	0.26	0.23
Housing investment	0.07	1.73	1.06	1.28	0.66	0.06
Business investment	0.43	1.52	2.34	2.41	1.65	1.50
Export	-1.34	0.05	-0.20	0.64	1.00	0.69
Import	-1.25	0.97	-0.93	1.19	0.25	0.40
Employment	-0.10	0.13	-0.03	0.07	0.00	-0.00
Rate of unemployment	2.69	-5.00	0.73	-2.46	-0.29	0.57
GNP deflator	1.90	0.71	0.74	0.11	-1.08	-0.80
Wage rate	2.02	1.05	1.28	0.70	-0.93	-0.54
USA						
Real GNP	-0.33	-0.29	-0.19	-0.23	0.18	0.11
Nominal GNP	0.25	0.62	-0.02	0.41	-0.26	-0.18
Consumption	-0.40	-0.38	-0.28	-0.43	0.12	-0.04
Housing investment	-1.01	-0.95	-0.75	-1.15	0.24	-0.15
Business investment	-0.53	-0.37	-0.47	-0.48	0.08	-0.13
Export	-0.25	-0.25	0.06	-0.07	0.31	0.21
Import	-0.56	-0.14	-0.58	-0.41	-0.19	-0.43
Employment	0.16	-0.06	0.16	-0.04	-0.01	0.03
Rate of unemployment	-1.56	0.41	-1.52	0.15	0.05	-0.32
GNP deflator	0.60	0.91	0.17	0.65	-0.44	-0.29
Wage rate	0.84	1.17	0.50	1.10	-0.34	-0.09
FRG						
Real GNP	0.22	-0.26	0.40	0.35	0.19	0.45
Nominal GNP	0.40	0.17	0.22	-0.29	-0.14	-0.41
Consumption	0.03	0.02	-0.02	-0.16	-0.02	-0.15
Housing investment	0.06	0.08	-0.02	-0.04	-0.05	-0.10
Business investment	0.72	-0.35	1.39	1.33	0.83	1.36
Export	-0.03	-0.09	0.18	0.43	0.20	0.44
Import	-0.39	0.59	-0.52	0.12	-0.03	-0.29
Employment	0.11	-0.11	-0.27	-0.05	-0.33	-0.04
Rate of unemployment	-1.86	1.89	3.59	4.05	4.25	4.23
GNP deflator	0.17	0.43	-0.19	-0.63	-0.32	-0.86
Wage rate	0.11	0.39	-0.16	-0.97	-0.21	-1.05

Case-1

Based on the assumptions in *Table 10.10*, possible future change in technology and demand structure affects GNP growth positively in Japan, but negatively in the US economy. In our Case-1 simulation, annual growth rates of real GNP in Japan increase by 0.22% and 0.29% for the 1980s and the 1990s, respectively, as compared with the base case. In the USA, annual growth rates of real GNP decreases by 0.33% and 0.29% for the two decades, respectively. For the FRG, real GNP growth at first increases by 0.22% in the first decade but falls by 0.26% in the second decade.

Why do such differences appear among the three countries? In the case of the USA, we note a strong increase in employment in the first decade. This reduces the unemployment rate by 1.56%, while increasing the wage rate by 0.84%. The GNP deflator increases by 0.60%. Higher domestic prices affect international trade, decreasing exports by 0.25%, due to deteriorating price competitiveness. Because of reductions in output in the US economy, import demand decreases by 0.56%. Consumption and housing investment are affected negatively by the increased price level. Export reduction decreases production in manufacturing sectors, which induces reduction in employment and investment. These changes, except reduction in imports, influence GNP growth negatively.

In Japan, by contrast, employment decreases at least in the first decade by 0.10%, and the unemployment rate increases by 2.69%. In spite of this change, the wage rate increases by 2.02%, mainly because of increased productivity. Average output prices increase by 1.76%, which is far larger than in the USA. Increases in prices cause decreased exports, but import demand moves in the opposite direction, as would be expected. This is because of internal structural changes, which work to reduce the volume of imports by sector. These changes affect GNP growth positively. In Japan, increased prices do not significantly reduce private consumption. Labor demand increases in the second decade, inducing a more severe labor shortage in Japan.

In the FRG, the situation seems similar to that of Japan in the first decade. Employment increases by 0.11%, and unemployment decreases by 1.86%. This would induce an increase in wage rate by 0.11% and prices by 0.17%. Increased business investment is one consequence, but import demand decreases by 0.39%. In the second decade, however, labor demand drops by 0.11% and the unemployment increases by 1.89%. Because of price increases, investment slows down and imports increase. These things affect GNP growth negatively.

Recent change in input–output coefficients and final-demand structure affect the three countries in different ways. Based on the model assumptions, the growth rate of real GNP in Japan and the FRG in the first period (1980–1990) would be accelerated. On the other hand, for the USA and the FRG in the second period (1990–2000), GNP growth would apparently be slightly inhibited.

Case-2

Here we examine the combination of overall structural changes and CIM technology in the metalworking sectors, subject to the model assumptions. Again we find that the effect on real GNP growth is positive in Japan and the FRG, but negative in the USA. In Japan, the growth rate of GNP is apparently increased by 1.03% and 0.68%, respectively, for the two decades. In the FRG, the enhancement is 0.4% and 0.35%, respectively. On the other hand, the growth rate of US GNP is inhibited by 0.19% and 0.23%, for the two decades, respectively.

In the USA, expanded labor demand in the tertiary sector raises wage rates and price levels through reduction in unemployment in the first decade. But a slowdown in the economy requires less employment in the latter period. In spite of this, wage rates and price levels grow at an accelerated rate, causing further GNP growth inhibition.

In the FRG, labor-saving CIM technology decreases labor demand in the metalworking manufacturing sectors, while structural change creates new labor demands in other sectors. But (because the metalworking sector is so large) total employment is somewhat reduced, and unemployment increases throughout the whole period. This tendency holds back price and wage inflation, which, in turn, reduces private consumption and housing investment through reduction in private income.

In the case of Japan, effects on wage rates and price levels are not so marked as those in Case-1. Labor-saving CIM technology works to contain the increase in output prices. Final-demand components are increased.

In the model calculation, the US economy is adversely affected through the changing input–output structure and final-demand structure. But this does not mean that new CIM technology adopted in the metalworking manufacturing sectors is meaningless from the macroeconomic viewpoint. Actually comparing the effects on real GNP in Case-1 and Case-2, we note that the introduction of labor-saving CIM technology in these sectors affects real GNP positively in each country. This point is examined in Case-3.

Case-3

This case shows the "pure" effects of the changes in production technology (CIM) in the metalworking manufacturing sectors per se.

Reduction in labor demand saves labor costs, other factors remaining constant. This results in a decrease in prices. Price reductions stimulate demand, and production is correspondingly increased. We can evaluate this causality chain (the Salter cycle) in our model.

In Japan, price reductions as measured by the GNP deflator are 1.08% and 0.80% for the two decades, respectively; real GNP growth rate is increased by 0.66% and 0.56%. In the USA, prices are reduced by 0.42% and 0.24%, respectively; real GNP growth increases by 0.18% and 0.11%. In the FRG, the price reduction is 0.27% and 0.71%; real GNP growth increases by 0.19% and 0.45%, respectively. The effects on GNP for each country are comparable to those suggested by Dobrinsky (Chapter 14) in scale and order of magnitude.

Why is the apparent effect on the US economy the smallest among the three countries? First, the share of the metalworking manufacturing industries in the US economy is relatively small compared with both Japan and the FRG. Thus the direct effect is correspondingly smaller. Second, the tertiary (service) sector is by far the largest in the USA. Relatively large spillover effects apparently induce large labor demand in the tertiary sector, which would weaken the reduction in wage rates and prices. It must be recognized, however, that the link between productivity in the metalworking sectors and labor demand in the tertiary sectors is questionable. There is no theoretical basis for such a link, and the empirical data were taken from the 1970s and early 1980s, which might not be typical, for various reasons.

Recent changes in metalworking production technology will hold back price increases and expand demand. These effects are favorable for the growth of GNP, but in the USA this effect is too small (based on model assumptions) to offset contrary effects elsewhere in the economy.

Of course, in our calculation, the labor-saving effects of CIM in other sectors are not considered. Nor are other impacts of CIM (capital saving, quality improvement). In particular, the model may exaggerate the impact of price changes on demand. If we were able to consider all of these effects simultaneously, the results of the analysis might be significantly different.

It must also be pointed out that the model results for the FRG may be largely irrelevant, due to major macroeconomic changes resulting from

German unification (which was not considered as a significant possibility when this analysis was begun).

10.4.4 Sectoral impacts

Figure 10.15 shows the ratio of output, employment, and investment by sector compared with the values of the control path in the year 2000 in Japan. This figure refers to Case-2. From this figure, we note that production in the five metalworking manufacturing sectors would be expected to increase, due to price reductions induced by the adoption of CIM technology. Recent changes in input–output structure and final-demand structure imply reduced output in such sectors as agriculture, fishery, and forestry; wooden products and furniture; rubber and plastics; petroleum and coal products; and iron and steel. Labor demand and investment demand in these industries would also be correspondingly reduced.

On the other hand, other sectors, such as chemicals and pharmaceuticals, appear to be increasing their output. Especially important is the growth in services (tertiary sector).

Figure 10.16 shows structural changes in the USA. Almost all industries are declining, at least in relative terms. Energy sectors, electrical machinery, precision instruments, and the tertiary sector are the major exceptions (energy demand is no longer growing). The tertiary sector is the biggest, and its output increases because of structural changes in technology and final demand.

Structural changes in the FRG is shown in *Figure 10.17*. As in Japan, price reductions in the metalworking sectors induce increased demand, leading to growth in production and investment. Again we note the reduced importance of production in agriculture, forestry, and fishery; mining; and petroleum and coal industries. Compared with Japan, induced demand in the other sectors is relatively small, especially in the tertiary sector.

Figure 10.18 shows the effects on exports and imports in Japan for the year 2000. The figure indicates the differences between Case-2 and the control path. There are some similarities in trade patterns between Japan and the FRG (see *Figure 10.20*). In both countries, exports from the metalworking manufacturing sectors increase, and imports of agricultural products and energy drop. Imports of electrical machinery rise, though imports of transport equipment and precision instruments decline in the FRG.

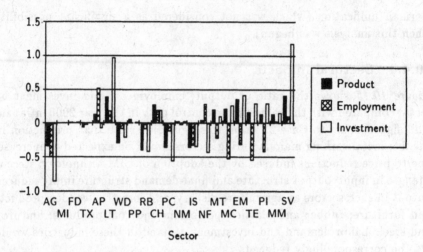

Figure 10.15. Effects on sectoral activities in Japan.

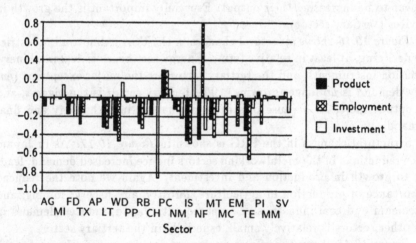

Figure 10.16. Effects on sectoral activities in the USA.

The USA has a different pattern (see *Figure 10.19*). We observe here an increase in imports of raw materials (mining product) and decreases in imports of general machinery and transportation equipment. However, exports of general machinery and transport equipment rise.

Changing patterns in international trade depend on the relative prices of commodity, at least in the long run. Success in cost reduction in Japanese

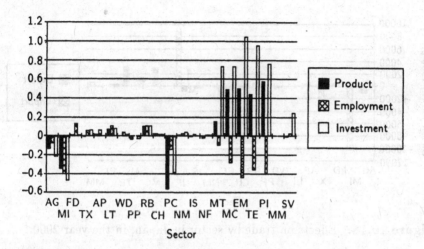

Figure 10.17. Effects on sectoral activities in the FRG.

and German metalworking manufacturing sectors seems to accelerate the increase of their exports to the world. And, based on the past, it would be presumed that US imports from Japan and the FRG would grow. This might offset the US import reduction. On the other hand, international financial conditions (exchange rates, interest rates) also have a strong influence. Such interdependence should be evaluated through a complete international trade linkage model, which falls outside the scope of this chapter.

10.4.5   Impact on labor demand

Tables 10.11, 10.12, and *10.13* compare Case-2 with the control scenario. They display the effects on employment in each country. From these tables, one can compare the differences in the patterns of labor demand among various sectors.

In Japan (*Table 10.11*), the total decrease in labor demand by five metalworking manufacturing sectors (Case-2) is about 4 million workers in the year 2000 from the control scenario. In our model, the employment in these sectors nearly doubles from about 5.9 million workers in 1980 to 9.8 million in the year 2000. In our simulation, which covers the period from 1980 to 2000, we assume that overall labor productivity increases by 1.67 in the year 2000 as compared with the case without structural changes. This would

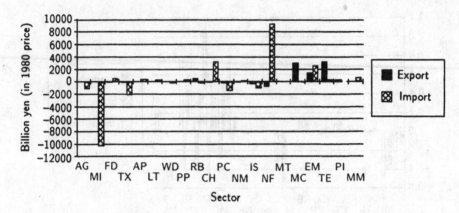

Figure 10.18. Effects on trade by sector in Japan, in the year 2000.

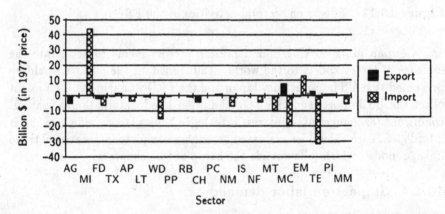

Figure 10.19. Effects on trade by sector in the USA, in the year 2000.

result in a 40% reduction in labor demand in the year 2000, which corresponds to the 4 million workers.[8] In this case, there is no increase in labor required, despite increased output.

The employment differential between the two cases in the USA (*Table 10.12*) is 4.1 million workers, which is almost the same as that of Japan. In the FRG (*Table 10.13*), it is 1.5 million, which is less than half of the reduction in the two other countries.

Projected effects on total employment are not the same among the three countries. The change in total employment is negligible in Japan, but in

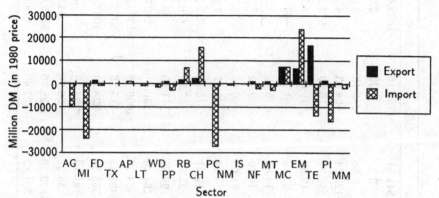

Figure 10.20. Effects on trade by sector in the FRG, in the year 2000.

the USA total employment increases by about 1.4 million workers. This contributes to reducing overall unemployment, of course. On the other hand, in the FRG case, total employment would fall by about 840,000 workers.

These apparent differences can be traced to the creation of new jobs in other sectors, especially in services. In Japan, the tertiary sector grows by about 5.4 million workers, whereas 1.2 million jobs are lost in other industrial sectors except the metalworking sectors. But the tertiary sector in the USA creates 7 million new jobs, while 1.6 million jobs are lost in other industrial sectors (except metalworking). As noted above, 4.1 million jobs are lost in metalworking. As a result, total labor demand in the US economy will increase by 1.4 million. But in the FRG, the projected increase in labor demand by the tertiary sector is minor. Only 600,000 jobs are created in that sector. This could have been 2 million, if the German economy had a structure similar to that of Japan or the USA.

Increased labor demand in the tertiary sector depends on increased aggregate output (and income). The extent to which output will be increased in the tertiary sector is related to the interdependence of the economy. This is attributable to the input–output structure and the changing final demand structure.[9]

Compared with the Japanese case, total employment in the USA will probably increase more. This is because the service share is already relatively large, and a continued shift to that sector is expected. On the other hand, relative lack of structural change in the FRG results in failure to create labor demand in the tertiary sector necessary to offset the projected labor saving in manufacturing sectors.

Table 10.11. Effects on employment in Japan: Case–2, in thousands.

Sector	Name	1980	1985	1990	1995	2000
AG	Agriculture, forestry, fishing	6.73	-1025.56	-1641.91	-1751.16	-1825.43
MI	Mining	0.00	0.00	0.00	0.00	0.00
FD	Food, beverage, tobacco	0.50	6.70	-5.66	-5.13	-0.97
TX	Textiles	2.25	-55.70	-125.44	-64.53	-14.27
AP	Apparel	0.86	127.89	232.97	416.69	570.24
LT	Leather production, footwear	0.02	15.34	19.62	19.68	15.37
WD	Wooden production, furniture	0.20	-0.31	-37.78	-51.22	-63.03
PP	Paper, pulp, printing	0.21	48.59	14.83	15.69	15.33
RB	Rubber and plastics	0.20	-6.88	-16.40	-19.22	-18.03
CH	Chemicals	0.55	61.31	71.63	90.86	96.77
PC	Petroleum and coal production	0.01	-1.93	-4.88	-6.52	-7.95
NF	Nonmetallic mineral production	0.94	75.01	17.91	5.85	-12.57
IS	Iron and steel production	0.28	3.23	-35.59	-67.58	-79.16
NF	Nonferrous metals	0.27	68.41	99.93	111.95	53.48
MT	Fabricated metal production	-26.08	-18.73	-164.35	-230.01	-269.17
MC	Machinery except electrical	-40.76	-148.24	-584.50	-1008.78	-1383.78
EM	Electrical machinery	-36.77	41.12	-407.35	-715.36	-1078.10
TE	Transport equipment	1.27	-177.45	-553.58	-828.15	-1057.92
PI	Precision instruments	0.20	-16.57	-91.15	-151.10	-213.36
MM	Miscellaneous manufacturing	0.21	55.48	36.32	36.70	34.97
SV	Tertiary industry	1.27	1534.18	2873.50	4328.76	5355.60
TO	Total	-87.65	585.89	-301.89	127.40	118.03

Table 10.12. Effects on employment in the USA: Case-2, in thousands.

Sector	Name	1980	1985	1990	1995	2000
AG	Agriculture, forestry, fishing	0.00	-2.99	-5.13	-7.57	-1.84
MI	Mining	0.00	0.04	-0.70	-7.07	-25.97
FD	Food, beverage, tobacco	-0.00	-15.34	-38.50	-63.11	-85.84
TX	Textiles	-0.00	-20.05	-46.47	-73.39	-100.87
AP	Apparel	-0.00	-14.81	-17.59	-11.20	-1.46
LT	Leather production, footwear	-0.00	-24.95	-40.42	-46.57	-47.44
WD	Wooden production, furniture	-0.03	-249.70	-377.54	-451.28	-495.94
PP	Paper, pulp, printing	-0.03	11.59	-5.88	-35.17	-60.24
RB	Rubber and plastics	-0.03	-99.60	-163.54	-220.36	-266.82
CH	Chemicals	-0.00	-8.99	-19.65	-26.64	-19.31
PC	Petroleum and coal production	-0.00	9.85	23.68	35.08	41.62
NF	Nonmetallic mineral production	-0.01	-60.83	-79.64	-69.16	-42.94
IS	Iron and steel production	-0.01	-133.17	-177.19	-232.47	-280.07
NF	Nonferrous metals	-0.01	-92.87	-144.23	-183.30	-214.41
MT	Fabricated metal production	-0.03	-258.75	-429.23	-547.82	-635.80
MC	Machinery except electrical	-0.03	-312.08	-567.32	-803.06	-1035.02
EM	Electrical machinery	-0.01	-179.48	-359.85	-576.82	-818.18
TE	Transport equipment	-0.03	-454.69	-797.24	-1065.14	-1279.82
PI	Precision instruments	-0.01	-26.01	-100.45	-209.58	-325.18
MM	Miscellaneous manufacturing	-0.00	-5.80	-4.53	2.44	9.58
SV	Tertiary industry	-0.22	2777.38	5076.66	6310.10	7127.70
TO	Total	-0.48	838.76	1725.24	1717.92	1441.74

Table 10.13. Effects on employment in the FRG: Case–2, in thousands.

Sector	Name	1980	1985	1990	1995	2000
AG	Agriculture, forestry, fishing	–0.00	–103.54	–117.48	–109.87	–76.06
MI	Mining	0.00	–40.07	–69.84	–88.94	–96.00
FD	Food, beverage, tobacco	–0.00	–14.11	26.47	65.37	105.74
TX	Textiles	–0.00	–1.30	4.33	7.52	12.19
AP	Apparel	0.00	1.51	7.81	11.29	17.85
LT	Leather production, footwear	–0.00	–1.95	0.05	3.03	6.79
WD	Wooden production, furniture	–0.00	–8.83	–4.25	1.78	13.14
PP	Paper, pulp, printing	–0.00	–12.47	–18.33	–15.66	5.31
RB	Rubber and plastics	0.00	3.10	12.83	27.88	51.56
CH	Chemicals	0.00	2.34	4.05	6.95	13.81
PC	Petroleum and coal production	–0.00	–1.80	–2.62	–3.42	–3.62
NF	Nonmetallic mineral production	–0.00	–15.45	–13.55	–3.67	9.33
IS	Iron and steel production	–0.00	–9.66	–10.31	–3.58	16.66
NF	Nonferrous metals	–0.00	–1.65	–2.26	–1.44	1.09
MT	Fabricated metal production	–0.00	–42.45	–59.20	–58.69	–44.49
MC	Machinery except electrial	–0.00	–126.42	–229.56	–296.91	–293.78
EM	Electrical machinery	–0.00	–111.33	–294.00	–467.11	–573.00
TE	Transport equipment	–0.00	–89.79	–228.59	–379.44	–522.22
PI	Precision instruments	–0.00	–32.10	–64.43	–86.08	–88.16
MM	Miscellaneous manufacturing	–0.00	–3.35	–4.06	–3.51	0.67
SV	Tertiary industry	–0.01	231.62	379.62	500.33	602.38
TO	Total	–0.01	–377.70	–683.32	–894.18	–840.85

10.5 Concluding Remarks

This chapter describes a set of large-scale multisectoral econometric models for Japan, the USA, and the FRG to examine the effects of changes in input structure and final-demand structure on production and employment. The changing patterns are extracted in part from the actual statistical data on input–output tables, with some modifications, and in part from "expert judgment." New technologies introduced to the manufacturing sector (CIM) are reflected in these changes. In considering the effects of CIM, labor saving is important. Therefore we focus especially on the metalworking sectors, which are, and continue to be, the main users of CIM.

What will be the consequences of such structural changes? What are the differences between countries? What is the role of CIM? What happens to the labor market? These are the questions which need to be explored. Based on our model projections, we can say the following.

Changes in input structure and final-demand structure apparently tend to accelerate Japan's rate of GNP growth from 1980 to 2000. By contrast, the effect on GNP growth for the USA appears to be slightly depressed over the whole period. For the FRG, GNP growth appears to be accelerated at first but decelerated in the second decade (after 1990). These differences are mainly due to differences in induced labor demand in the tertiary sectors for the three countries. Because the elasticities of labor demand are based largely on data from the 1970s and early 1980s, which may not be representative of conditions in the 1990s, these conclusions remain tentative.

Labor-saving efforts in metalworking manufacturing per se appear to affect GNP growth positively for all three countries, though the magnitude is not so large as to offset the indirect effects in the case of the USA. New technology adopted in metalworking manufacturing clearly reduces the price level of metal products and increases the demand, and hence the output. Here again, the results are based on standard assumptions regarding the relationship between demand and prices. These assumptions may not be as robust as one would like.

Industrial structures are changing. One probable impact of CIM is that metalworking manufacturing industries (as well as tertiary industries) will increase in relative shares. On the other hand, agriculture, energy, and heavy industry like iron and steel will almost certainly continue to decline.

As a result of structural shifts, we can expect many additional jobs to be created, especially in services. These more than offset the reduction of labor demand in the metalworking manufacturing sector, except in the case of the FRG. In the USA, especially, labor demand in the tertiary sector will apparently increase much more than the decrease in other sectors. On the contrary, in the FRG, labor demand in the tertiary sector is not expected to grow fast enough to compensate for the reduction in metalworking, because of more moderate structural changes.[10] Japan seems to fall in between these extremes.

These results depend on econometric models and many underlying assumptions. We have neglected many possible relationships for the sake of analytical simplicity, and concentrated on the effects of structural changes and changes in technology.

For example, all currency exchange rates are assumed to be determined exogenously for the Japanese and the FRG models. Future movements are assumed on the basis of recent tendencies. They do not appear in the US model. Interest rates are fixed at recent levels for the model runs. Labor participation rates are also based on recent tendencies. All of these assumptions are questionable.

Clearly, future devaluation in the US dollar might increase exports and decrease imports. Macroeconomic policy in the USA is neglected in the model, but is important in the real world. Renewed inflation could drive up US interest rates. But, by the same token, a "peace dividend" might reduce the persistent federal deficit. Lower interest rates would increase investment. Labor shortages would increase wage rates, and higher wages would induce more labor-saving investment (as in Japan and the FRG).

We neglect the foreign investment problem. But labor shortages in Japan and large trade surpluses in both Japan and the FRG force increased foreign investment. Eventually new capacity will be built abroad, much of it in the USA. This is a very interesting point especially in considering future economic policy for Japan. Manufacturing strategy is to replace the plants of low value-added products, and to save costs in domestic factories by adopting advanced technology. But, recent foreign investment is attempting to substitute overseas production for Japanese exports, to thwart protectionism. All of these points should be addressed in a worldwide trade model, which links individual country models.

Notes

[1] In its long-run forecast of the labor market, the Ministry of Labor in Japan projects a labor shortage of more than 1 million workers by the year 2000,

considering recent Japanese economic growth and shifts in the labor market (*Nihon Keizai Shimbun*, European Edition, 14 June 1989).

[2] Kinoshita, Yamada, and others have been developing a multicountry and multisectoral econometric model since 1979. The first version of the model was reported in Kinoshita *et al.* (1982). The model consists of three components: (1) several large-scale multisectoral I–O country models, (2) small regional models, and (3) trade linkage submodels by commodity. A revision of the model was started in 1986. Some results of this second version were reported as the international and domestic impacts of robotization (Kinoshita and Yamada, 1989b). Recent changes in the international trade structure based on this data base are discussed in Kinoshita and Yamada (1989a).

The Japanese and the US models are those adapted from second version of the multicountry and multisectoral model. The data base, which covers the years from 1965 to 1984, is the same, but some modifications are added to the present version. Several equations are reestimated with minor changes in specification. Labor supply is changed to be exogenously determined.

The FRG model is newly developed for the IIASA Project. Its data base covers the years from 1965 to 1986. Each model consists of about 400 equations.

[3] Published in *The Survey of Current Business*, US Department of Commerce (1984 and 1988).

[4] These figures are calculated as

$$\Delta a_{ij} = [a_{ij}^{(t1)} - a_{ij}^{(t2)}]/(t_2 - t_1 + 1) \ ,$$

where $a_{ij}^{(t)}$ means the input coefficient of commodity i in sector j at time t.

[5] The functional form we adopt is

$$s = 0.455 \ln(t + 2.5) - 0.417 \ ,$$

Where s is an index of changing pattern. In *Figure 10.A*, this pattern is compared to the case of linear extrapolation. This assumption is arbitrary in the sense that we have no data to justify it. But we adopt it as one expression of a plausible possible future pattern that we can analyze.

[6] In Chapter 14, the five three-digit subsectors of ISIC sector 38, fabricated metal products, are selected to analyze the diffusion of CIM technologies. This classification is consistent with the ones used in this chapter.

[7] To keep the consistency between the total world export $W^{T\$}$ and the summation of world export by commodity $W_i^{T\$}$, we give the following restriction to get the values of $W_i^{T\$}$:

$$\Sigma_i W_i^{T\$} = W^{T\$} \ .$$

[8] According to a report of the Japan Industrial Robot Association (JIRA), the future robot population in Japan is estimated 451,900 in 1993 and 715,600 in 1998. For the five metalworking manufacturing sectors, the population is 317,400 and 510,900, respectively. The projected reduction of labor demand

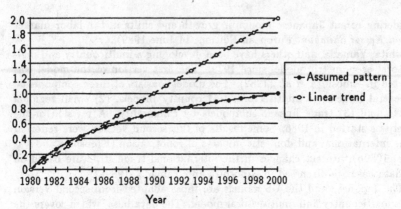

Figure 10.A. Possible patterns of input–output coefficient.

Table 10.A. Effect of a unit increase in metalworking sector production on the production in the tertiary.

Country	15:MT 34	16:MC 35	17:EM 36	18:TE 37	19:PI 38	Sum
Japan	0.357	0.366	0.360	0.344	0.329	1.756
USA	0.351	0.317	0.344	0.328	0.318	1.658
FRG	0.339	0.317	0.276	0.326	0.259	1.517

is seven to eight times larger than these values. Of course, the application of robots is only one component of new technology considered here.

Saito (NIRA, 1988) estimated the impacts of robotics on labor demand, using a multisectoral econometric model. According to his analysis, 71,000 jobs will be lost due to the introduction of 155,000 robots for the period from 1985 to 1990, which is a direct effect. Increased investment and price reduction induce 48,000 jobs, which is an indirect effect.

[9] The Leontief inverse matrix of each country's input–output table for the base year shows that the effect of a unit increase in the metalworking sector production on demand in the tertiary sector is largest in Japan, intermediate in the USA, and smallest in the FRG. Calculations by sector are shown in *Table 10.A*. The labor-demand elasticity with respect to production in the tertiary sector is another important factor. Estimated elasticities are 0.37 (Japan), 0.33 (USA), and 0.26 (FRG). These estimates are based on historical data.

[10] The reasons for slow growth in the tertiary sector in the FRG may be attributable to legislated restrictions on retail trade (e.g., shopping hours) and new business formation.

Appendix

Table 10A.1. Variable used in the model.

Code	Name	Note
A	input coefficient matrix	$A = [a_{ij}]$
A_o	input coefficient matrix, base year	$A = [a_{oij}]$
E	export vector	$E = [E_i]$
F	final-demand vector	$F = F^d + E$
F^d	domestic final-demand vector	$F^d = \bar{F}^d\, F$
$\bar{F}_o$	final-demand coefficient matrix, base year	
$\bar{F}$	final-demand coefficient matrix	
l	labor-coefficient vector	
L	labor-demand vector	$L = \bar{L}\, X$
$\bar{L}$	diagonal matrix of labor coefficients	
M	import-demand vector	$M = [M_i]$
$\bar{M}$	diagonal matrix of import coefficients	
$\bar{M}_o$	diagonal matrix of import coefficients, base year	
p_o^{fd}	final-demand deflator vector	$p_o^{fd} = [P_{oi}^{fd}]$
p^{fd}	final-demand deflator vector	$p^{fd} = [P_i^{fd}]$
p^m	import-price vector	$p^m = [P_i^m]$
p^r	intermediate input-price vector	$p^r = [P_j^r]$
p_o^r	intermediate input-price vector, base year	$p_o^r = [P_{oj}^r]$
P	output-price vector	$P = [P_i]$
R^s	"other costs" vector (mainly capital costs)	
w	diagonal matrix of wage rate	$w = [W_i]$
X_o	output vector, base year	$X_o = [X_{oi}]$
X	output vector	$X = [X_i]$

Table 10A.1. Continued.

Code	Name
ε	exchange rate
a_{ij}	input coefficient
C	consumption expenditure
D_j^p	depreciation by sector
D_{oi}	domestic demand by sector, computed using base-year input–output information
D_i	domestic demand by sector
d	depreciation rate
E_i	export by sector
I^h	housing investment
I_j	business investment by sector
K^h	housing capital stock
K_j	real capital stock by sector
L_j	employment by sector
L	employment, total
M_i	import by sector
P_i	output deflator by sector
P	output deflator, average
P^c	consumption deflator
P_i^d	domestic-demand deflator, sectoral
P^d	domestic-demand deflator, average
P_i^e	export deflator by sector
P_j^{fd}	final-demand deflator
P_{oj}^{fd}	final-demand deflator by sector, computed using base-year input–output information

Table 10A.1. Continued.

Code	Name
P^{ih}	housing-investment deflator
P^{ip}	business-investment deflator
P_i^m	import deflator by sector
P_j^r	intermediate input deflator by sector
P_{oj}^r	intermediate input deflator by sector, computed using base-year
	input–output information
P_j^w	price of commodity j in the world market
R_j	input ratio by sector
R_{oj}	input ratio by sector j, base year
r	interset rate
T_j^x	indirect tax minus subsidies by sector
T	index of technical progress
U^R	unemployment rate
W_j	wage rate by sector
W	wage rate, average
X_i	output by sector
X_{oi}	output by sector, computed using base-year input–output information
X	output, total
Y^d	disposable income
Y_j^g	gross value-added by sector
Y_j^w	compensation of employees by sector
Y_j	net value-added by sector

Table 10A.2. Control solution of Japanese model: annual growth rate in 1980–2000, in percent.

Sector	Name	Production		Price	
		1980–1990	1990–2000	1980–1990	1990–2000
1 AG	Agriculture, forestry, fishing	−1.71	−3.17	0.52	7.88
2 MI	Mining	−0.63	−0.00	−0.17	6.43
3 FD	Food, beverage, tobacco	1.30	0.51	0.06	5.90
4 TX	Textiles	−1.31	−2.21	−0.96	5.29
5 AP	Apparel	2.77	1.40	−0.50	6.39
6 LT	Leather production, footwear	0.66	−0.80	−0.46	7.59
7 WD	Wooden production, furniture	−0.15	−1.44	0.11	9.53
8 PP	Paper, pulp, printing	1.82	−0.09	−0.23	8.62
9 RB	Rubber and plastics	6.48	6.63	−1.74	4.24
10 CH	Chemicals	2.99	1.28	−1.92	6.23
11 PC	Petroleum and coal production	2.75	1.46	−0.61	9.13
12 NM	Nonmetallic mineral production	.38	−1.48	−0.71	8.19
13 IS	Iron and steel production	0.97	−3.66	−0.79	7.83
14 NF	Nonferrous metals	2.76	1.26	−2.78	6.64
15 MT	Fabricated metal production	2.81	0.58	−0.73	6.66
16 MC	Machinery except electrial	6.29	5.96	−0.89	5.52
17 EM	Electrical machinery	6.70	9.42	0.60	1.71
18 TE	Transport equipment	6.22	3.86	−1.27	5.35
19 PI	Precision instruments	3.21	6.73	−0.66	3.82
20 MM	Miscellaneous manufacturing	5.56	2.19	−1.97	7.84
21 SV	Tertiary industry	3.38	2.50	0.03	6.69
00 TO	Total	3.34	2.77	−0.29	6.19

Table 10A.2. Continued.

Sector	Name	Export 1980–1990	Export 1990–2000	Import 1980–1990	Import 1990–2000
1 AG	Agriculture, forestry, fishing	-4.00	0.00	-2.57	-3.75
2 MI	Mining	-1.27	0.00	2.06	1.33
3 FD	Food, beverage, tobacco	-3.88	-1.05	5.97	2.79
4 TX	Textiles	-1.37	-0.62	12.25	5.62
5 AP	Apparel	2.96	1.03	7.39	4.91
6 LT	Leather production, footwear	0.41	-0.79	6.03	3.07
7 WD	Wooden production, furniture	3.01	5.23	-0.11	-2.88
8 PP	Paper, pulp, printing	0.28	0.97	7.62	6.81
9 RB	Rubber and plastics	11.53	11.46	4.01	4.47
10 CH	Chemicals	2.93	0.27	5.62	7.98
11 PC	Petroleum and coal production	2.76	-1.11	2.41	1.53
12 NM	Nonmetallic mineral production	1.68	-3.93	8.80	13.22
13 IS	Iron and steel production	-0.58	-3.95	10.94	7.93
14 NF	Nonferrous metals	13.07	5.78	8.52	6.80
15 MT	Fabricated metal production	1.90	1.13	4.74	2.42
16 MC	Machinery except electrical	8.19	9.82	4.69	4.92
17 EM	Electrical machinery	5.43	11.81	11.22	10.86
18 TE	Transport equipment	8.53	5.39	8.77	6.88
19 PI	Precision instruments	7.80	9.61	2.97	7.56
20 MM	Miscellaneous manufacturing	3.65	3.66	11.23	5.80
21 SV	Tertiary industry	5.40	6.06	5.07	4.16
00 TO	Total	5.90	7.06	3.85	3.78

Table 10A.2. Continued.

Sector	Name	Investment 1980–1990	Investment 1990–2000	Employment 1980–1990	Employment 1990–2000
1 AG	Agriculture, forestry, fishing	-2.03	- 4.53	-2.96	-3.71
2 MI	Mining	-0.66	0.00	-1.77	-0.80
3 FD	Food, beverage, tobacco	-1.53	- 1.33	0.13	-0.13
4 TX	Textiles	-7.45	-12.50	-2.72	-3.17
5 AP	Apparel	1.89	0.65	2.50	1.43
6 LT	Leather production, footwear	-1.27	- 3.45	0.13	-0.25
7 WD	Wooden production, furniture	-3.03	5.13	-0.42	-0.23
8 PP	Paper, pulp, printing	-1.09	- 0.08	0.64	-0.05
9 RB	Rubber and plastics	2.78	6.94	0.90	-0.06
10 CH	Chemicals	0.35	- 0.72	-0.52	0.44
11 PC	Petroleum and coal production	3.18	- 1.58	0.66	0.61
12 NM	Nonmetallic mineral production	2.76	- 3.08	0.58	-1.06
13 IS	Iron and steel production	-7.40	- 8.41	-1.74	-2.95
14 NF	Nonferrous metals	3.35	0.46	-0.77	0.84
15 MT	Fabricated metal production	1.42	0.79	1.02	-0.05
16 MC	Machinery except electrical	5.04	5.51	3.29	2.79
17 EM	Electrical machinery	4.19	10.18	5.10	2.89
18 TE	Transport equipment	5.88	1.48	2.38	0.99
19 PI	Precision instruments	8.76	7.70	1.62	2.24
20 MM	Miscellaneous manufacturing	6.11	1.91	1.71	0.60
21 SV	Tertiary industry	4.98	0.15	1.48	1.15
00 TO	Total	3.24	0.90	0.99	0.75

Table 10A.3. Control solution of Japanese model: employment, in thousands.

Sector	Name	1980	1985	1990	1995	2000
1 AG	Agriculture, forestry, fishing	7454.03	6217.25	5520.35	4620.56	3782.57
2 MI	Mining	146.00	131.17	122.07	116.32	112.61
3 FD	Food, beverage, tobacco	1397.07	1400.02	1414.99	1402.89	1397.38
4 TX	Textiles	1353.02	1254.55	1026.45	851.88	743.43
5 AP	Apparel	726.71	806.72	929.84	1065.71	1072.23
6 LT	Leather production, footwear	80.25	80.88	81.29	81.47	79.26
7 WD	Wooden production, furniture	844.84	821.94	810.19	809.54	791.66
8 PP	Paper, pulp, printing	975.57	1002.59	1039.39	1040.68	1033.80
9 RB	Rubber and plastics	199.00	206.33	217.75	213.54	216.50
10 CH	Chemicals	478.30	458.83	454.00	471.03	474.18
11 PC	Petroleum and coal production	65.60	64.93	70.07	73.04	74.43
12 NM	Nonmetallic mineral production	677.82	662.72	717.87	713.67	645.01
13 IS	Iron and steel production	391.71	374.14	328.59	298.50	243.59
14 NF	Nonferrous metals	178.72	181.65	165.50	188.27	179.96
15 MT	Fabricated metal production	1139.08	1220.98	1260.30	1290.32	1253.80
16 MC	Machinery except electrical	1466.02	1653.30	2027.06	2403.89	2669.71
17 EM	Electrical machinery	1574.25	2183.14	2588.41	2975.13	3443.02
18 TE	Transport equipment	1400.51	1568.88	1772.41	1874.94	1955.89
19 PI	Precision instruments	356.29	384.13	418.42	470.74	522.04
20 MM	Miscellaneous manufacturing	786.80	851.40	932.39	965.22	989.67
21 SV	Tertiary industry	37081.57	40016.32	42934.51	45981.21	48149.50
00 TO	Total	58773.15	61541.85	64831.85	67908.56	69830.23

Table 10A.4. Control solution of US model: annual growth rate in 1980–2000, in percent.

Sector	Name	Production		Price	
		1980–1990	1990–2000	1980–1990	1990–2000
1 AG	Agriculture, forestry, fishing	0.34	1.10	3.91	2.30
2 MI	Mining	0.48	0.50	7.92	9.27
3 FD	Food, beverage, tobacco	0.31	0.38	4.58	3.67
4 TX	Textiles	−0.76	0.02	5.70	4.61
5 AP	Apparel	−0.47	−0.19	5.51	4.32
6 LT	Leather production, footwear	−0.05	0.69	5.72	3.97
7 WD	Wooden production, furniture	−0.56	−0.17	5.29	4.57
8 PP	Paper, pulp, printing	0.67	1.36	6.67	4.49
9 RB	Rubber and plastics	1.17	1.44	6.23	5.44
10 CH	Chemicals	−0.49	0.83	7.60	5.95
11 PC	Petroleum and coal production	0.51	0.08	7.35	8.18
12 NM	Nonmetallic mineral production	−1.80	−4.40	6.94	5.18
13 IS	Iron and steel production	−0.44	−0.48	5.04	5.40
14 NF	Nonferrous metals	−0.05	1.35	2.82	4.20
15 MT	Fabricated metal production	0.13	0.94	6.17	4.70
16 MC	Machinery except electrical	0.87	3.04	4.62	3.65
17 EM	Electrical machinery	1.70	1.57	5.59	4.50
18 TE	Transport equipment	1.61	1.13	6.04	5.21
19 PI	Precision instruments	1.96	3.98	6.60	4.46
20 MM	Miscellaneous manufacturing	0.28	−0.22	4.91	4.25
21 SV	Tertiary industry	1.68	1.66	6.66	5.29
00 TO	Total	1.20	1.42	6.36	5.27

Table 10A.4. Continued.

Sector	Name	Export		Import	
		1980–1990	1990–2000	1980–1990	1990–2000
1 AG	Agriculture, forestry, fishing	− 3.34	3.85	0.40	0.54
2 MI	Mining	− 1.66	3.60	0.25	1.61
3 FD	Food, beverage, tobacco	− 0.23	3.40	6.72	1.99
4 TX	Textiles	− 9.39	−0.97	13.85	4.11
5 AP	Apparel	− 4.72	1.36	6.16	2.41
6 LT	Leather products, footwear	− 1.58	5.67	5.16	−0.08
7 WD	Wooden products, furniture	− 3.57	1.16	14.11	5.67
8 PP	Paper, pulp, printing	− 2.38	2.49	4.52	3.15
9 RB	Rubber and plastics	3.75	4.00	3.09	2.77
10 CH	Chemicals	− 1.42	3.99	9.68	4.13
11 PC	Petroleum and coal products	5.83	0.26	1.31	−0.43
12 NM	Nonmetallic mineral products	− 1.79	1.82	17.35	9.02
13 IS	Iron and steel products	−10.05	−3.17	9.06	3.63
14 NF	Nonferrous metals	− 4.09	3.50	1.62	1.94
15 MT	Fabricated metal products	− 3.54	3.05	10.90	6.41
16 MC	Machinery except electrical	1.31	5.74	10.50	6.53
17 EM	Electrical machinery	2.66	5.12	11.97	6.74
18 TE	Transport equipment	− 3.69	−0.20	7.69	3.34
19 PI	Precision instruments	3.10	8.00	6.17	6.30
20 MM	Miscellaneous manufacturing	− 5.30	1.97	6.17	3.85
21 SV	Tertiary industry	6.41	8.90	6.09	4.13
00 TO	Total	− 1.10	4.18	6.91	4.31

Table 10A.4. Continued.

Sector	Name	Investment 1980–1990	Investment 1990–2000	Employment 1980–1990	Employment 1990–2000
1 AG	Agriculture, forestry, fishing	1.48	−0.87	−0.14	0.24
2 MI	Mining	6.51	5.50	2.91	4.24
3 FD	Food, beverage, tobacco	1.88	4.58	−0.92	−0.81
4 TX	Textiles	−2.13	1.84	−2.25	−1.04
5 AP	Apparel	0.21	−1.22	−1.30	−0.54
6 LT	Leather production, footwear	5.18	12.99	−2.50	−2.71
7 WD	Wooden production, furniture	−4.47	−0.16	−0.59	−0.37
8 PP	Paper, pulp, printing	3.07	8.46	−0.31	−0.79
9 RB	Rubber and plastics	0.79	5.14	−0.14	0.87
10 CH	Chemicals	−2.90	11.32	−0.98	−0.63
11 PC	Petroleum and coal production	6.70	8.10	−0.91	−2.35
12 NM	Nonmetallic mineral production	0.63	−0.41	−2.20	−5.82
13 IS	Iron and steel production	1.27	2.97	−1.72	−1.05
14 NF	Nonferrous metals	−4.90	4.69	−1.44	0.79
15 MT	Fabricated metal production	1.56	2.25	−0.62	0.02
16 MC	Machinery except electrical	5.41	7.80	−0.78	1.06
17 EM	Electrical machinery	4.65	1.36	1.53	0.85
18 TE	Transport equipment	−0.39	1.78	−0.29	0.74
19 PI	Precision instruments	−3.64	12.74	2.32	2.76
20 MM	Miscellaneous manufacturing	3.46	4.80	−1.91	−1.93
21 SV	Tertiary industry	1.37	2.25	1.49	1.30
00 TO	Total	2.18	4.22	1.07	1.10

Table 10A.5. Control solution of US model: employment, in thousands.

Sector	Name	1980	1985	1990	1995	2000
1 AG	Agriculture, forestry, fishing	3362.17	3324.49	3315.38	3331.77	3397.21
2 MI	Mining	1012.12	1043.00	1348.38	1683.33	2042.84
3 FD	Food, beverage, tobacco	1768.14	1668.42	1611.81	1552.42	1486.13
4 TX	Textiles	870.82	743.92	693.58	656.19	624.76
5 AP	Apparel	1293.72	1191.02	1135.14	1101.61	1074.83
6 LT	Leather production, footwear	243.11	197.33	188.74	166.35	143.44
7 WD	Wooden production, furniture	1176.30	1166.71	1108.48	1099.59	1067.96
8 PP	Paper, pulp, printing	1968.92	1956.74	1907.88	1858.98	1762.34
9 RB	Rubber and plastics	757.52	787.98	746.70	789.40	814.43
10 CH	Chemicals	1113.37	1034.28	1008.61	985.31	946.52
11 PC	Petroleum and coal production	197.23	187.37	179.92	163.38	141.81
12 NM	Nonmetallic mineral production	667.77	622.18	534.53	421.89	293.52
13 IS	Iron and steel production	756.11	557.32	635.40	601.21	571.68
14 NF	Nonferrous metals	435.96	409.90	376.94	392.78	407.70
15 MT	Fabricated metal production	1605.03	1538.17	1507.89	1510.60	1510.45
16 MC	Machinery except electrical	2546.84	2288.64	2354.48	2477.66	2615.64
17 EM	Electrical machinery	2173.54	2386.44	2528.84	2644.66	2752.86
18 TE	Transport equipment	1881.14	1833.86	1827.25	1899.88	1967.95
19 PI	Precision instruments	714.58	741.04	899.17	1064.45	1180.59
20 MM	Miscellaneous manufacturing	427.40	387.70	352.27	322.36	289.95
21 SV	Tertiary industry	75,264.01	82,249.06	87,233.50	92,492.85	99,271.22
00 TO	Total	100,235.79	106,315.59	111,494.92	117,216.68	124,363.82

Table 10A.6. Control solution of the FRG model: annual growth rate in 1980–2000, in percent.

Sector	Name	Production		Price	
		1980–1990	1990–2000	1980–1990	1990–2000
1 AG	Agriculture, forestry, fishing	1.46	1.43	0.57	1.36
2 MI	Mining	−0.22	0.45	4.86	5.71
3 FD	Food, beverage, tobacco	1.00	0.84	1.75	2.07
4 TX	Textiles	0.23	0.71	2.33	1.85
5 AP	Apparel	−0.97	2.45	2.67	1.62
6 LT	Leather production, footwear	−0.33	0.15	2.44	2.09
7 WD	Wooden production, furniture	1.06	0.19	1.85	2.67
8 PP	Paper, pulp, printing	1.87	1.53	2.76	3.76
9 RB	Rubber and plastics	2.64	2.24	2.17	2.52
10 CH	Chemicals	2.56	2.85	2.20	2.74
11 PC	Petroleum and coal production	0.81	1.64	3.60	5.10
12 NM	Nonmetallic mineral production	−0.41	−0.26	2.71	3.15
13 IS	Iron and steel production	0.67	0.74	2.14	3.25
14 NF	Nonferrous metals	3.15	2.74	1.93	2.87
15 MT	Fabricated metal production	0.23	0.90	3.60	2.64
16 MC	Machinery except electrical	0.43	1.13	3.05	2.53
17 EM	Electrical machinery	2.47	3.19	1.85	1.67
18 TE	Transport equipment	4.01	3.60	3.15	2.36
19 PI	Precision instruments	1.12	1.16	3.28	2.54
20 MM	Miscellaneous manufacturing	−0.99	1.41	3.63	3.74
21 SV	Tertiary industry	2.47	2.79	2.96	3.11
00 TO	Total	2.02	2.37	2.75	2.96

Table 10A.6. Continued.

Sector	Name	Export		Import	
		1980–1990	1990–2000	1980–1990	1990–2000
1 AG	Agriculture, forestry, fishing	6.24	5.90	1.53	1.78
2 MI	Mining	-3.55	0.00	-1.39	0.16
3 FD	Food, beverage, tobacco	5.89	5.85	3.77	2.96
4 TX	Textiles	2.22	2.39	1.60	1.12
5 AP	Apparel	5.45	6.96	1.85	-1.35
6 LT	Leather production, footwear	2.58	3.80	1.93	1.61
7 WD	Wooden production, furniture	3.83	4.47	0.39	2.83
8 PP	Paper, pulp, printing	4.53	4.29	3.24	3.14
9 RB	Rubber and plastics	2.15	3.69	6.26	6.40
10 CH	Chemicals	4.43	4.84	5.11	5.45
11 PC	Petroleum and coal production	2.56	2.71	3.35	1.87
12 NM	Nonmetallic mineral production	3.12	2.85	2.18	1.36
13 IS	Iron and steel production	1.91	1.88	0.36	1.64
14 NF	Nonferrous metals	3.30	4.03	2.03	1.59
15 MT	Fabricated metal production	2.00	2.94	5.01	4.99
16 MC	Machinery except electrical	1.12	1.48	6.10	6.40
17 EM	Electrical machinery	3.67	5.61	8.60	7.95
18 TE	Transport equipment	6.25	5.44	7.86	7.67
19 PI	Precision instruments	3.16	4.20	8.89	8.16
20 MM	Miscellaneous manufacturing	-2.12	2.18	-0.63	0.70
21 SV	Tertiary industry	3.52	3.89	2.27	4.92
00 TO	Total	3.57	4.19	3.27	4.34

Table 10A.6. Continued.

Sector	Name	Investment		Employment	
		1980–1990	1990–2000	1980–1990	1990–2000
1 AG	Agriculture, forestry, fishing	0.16	2.89	-1.95	-1.66
2 MI	Mining	1.92	3.66	-0.60	1.06
3 FD	Food, beverage, tobacco	-0.45	0.78	-0.83	-0.81
4 TX	Textiles	1.74	-0.82	-2.84	-2.32
5 AP	Apparel	-1.17	1.31	-2.70	-0.01
6 LT	Leather production, footwear	7.37	-0.92	-4.34	-2.44
7 WD	Wooden production, furniture	-1.94	-0.65	-1.71	-1.60
8 PP	Paper, pulp, printing	0.80	1.34	0.35	2.22
9 RB	Rubber and plastics	-0.02	1.12	1.02	0.64
10 CH	Chemicals	1.16	1.73	-0.52	-0.30
11 PC	Petroleum and coal production	5.25	0.32	0.36	0.28
12 NM	Nonmetallic mineral production	-1.54	-0.71	-1.79	-1.33
13 IS	Iron and steel production	3.05	-0.78	-1.12	-0.10
14 NF	Nonferrous metals	-0.41	0.25	-0.74	-0.13
15 MT	Fabricated metal production	3.40	3.50	-0.36	-0.65
16 MC	Machinery except electrical	2.31	3.23	-0.54	-0.65
17 EM	Electrical machinery	4.84	1.91	0.62	1.21
18 TE	Transport equipment	4.39	4.44	1.56	1.84
19 PI	Precision instruments	3.95	3.86	-0.02	-0.73
20 MM	Miscellaneous manufacturing	-1.74	2.17	-1.41	1.54
21 SV	Tertiary industry	2.39	3.36	0.61	0.82
00 TO	Total	2.23	2.96	0.15	0.52

Table 10A.7. Control solution of the FRG model: employment, in thousands.

Sector	Name	1980	1985	1990	1995	2000
1 AG	Agriculture, forestry, fishing	1433.91	1343.63	1177.26	1093.21	995.66
2 MI	Mining	229.83	223.69	216.46	232.07	240.43
3 FD	Food, beverage, tobacco	890.42	861.05	819.46	784.77	755.43
4 TX	Textiles	316.67	263.63	237.46	211.39	187.69
5 AP	Apparel	326.00	264.49	248.00	244.72	247.75
6 LT	Leather production, footwear	115.08	88.43	73.83	64.20	57.66
7 WD	Wooden production, furniture	413.70	378.73	348.01	322.54	296.27
8 PP	Paper, pulp, printing	394.06	395.36	408.09	455.63	508.46
9 RB	Rubber and plastics	379.00	420.93	419.49	437.69	447.09
10 CH	Chemicals	593.01	583.80	563.10	556.34	546.27
11 PC	Petroleum and coal production	24.43	27.10	25.33	25.88	26.04
12 NM	Nonmetallic mineral production	370.56	334.92	309.41	290.72	270.71
13 IS	Iron and steel production	721.88	647.18	644.74	651.41	638.60
14 NF	Nonferrous metals	72.72	73.95	67.54	67.70	66.65
15 MT	Fabricated metal production	534.53	549.92	515.39	496.66	482.87
16 MC	Machinery except electrical	1167.98	1155.58	1106.72	1094.18	1036.45
17 EM	Electrical machinery	1087.58	1075.28	1157.03	1265.28	1304.31
18 TE	Transport equipment	1036.43	1097.30	1209.39	1343.66	1451.57
19 PI	Precision instruments	234.29	233.71	233.75	233.75	217.33
20 MM	Miscellaneous manufacturing	99.84	94.03	86.64	94.64	100.91
21 SV	Tertiary industry	15504.07	15751.51	16468.20	17193.39	17863.56
00 TO	Total	25946.00	25864.21	26335.32	27159.85	27741.71

References

Ayres, R., 1991, *Computer Integrated Manufacturing*, Vol. I: *Revolution in Progress* Chapman and Hall, London, UK.

JIRA (Japan Industrial Robot Association), 1989, *The Report (Manufacturing) on the Long-term Forecast of the Demand for Industrial Robots*, Tokyo, Japan (in Japanese).

Kinoshita, S., and M. Yamada, 1989a, "An Econometric Analysis of the Changing World Trade Linkage Structure: 1970–84," in *Economic Research* No. 89, March (in Japanese).

Kinoshita, S., and M. Yamada, 1989b, "The Impacts of Robotization on Macro and Sectoral Economies within a World Econometric Model," *Technological Forecasting and Social Change* 35(2–3)April:211–230.

Kinoshita, S., *et al.*, 1982, *The Development and Application of the World-Trade Model for the Analysis of the International Industry-Trade Structure around Japan*, Economic Planning Agency, Tokyo, Japan (in Japanese).

Krelle, W., ed., 1989, *The Future of the World Economy: Economic Growth and Structural Change*, Springer-Verlag, Berlin, Heidelberg, New York.

Maly, M., and P. Zaruba, 1989, "Prognostic Model for Industrial Robot Penetration in Centrally Planned Economies," in J. Ranta, ed., *Trends and Impacts of Computer Integrated Manufacturing*, WP-89-1, IIASA, Laxenburg, Austria.

Mori, S., 1987, *Social Benefits on CIM: Labor and Capital Augmentation by Industrial Robots and NC Machine Tools in the Japanese Manufacturing Industry*, WP-87-40, IIASA, Laxenburg, Austria.

NIRA (National Institute for Research Advancement), 1988, *The Impacts of Robotics Related Technology on the Industry and Economy and Corresponding Strategy*, Tokyo, Japan (in Japanese).

Tani, A., 1987, *Future Penetration of Advanced Industrial Robots in the Japanese Manufacturing Industry: An Econometric Forecasting Model*, WP-87-95, IIASA, Laxenburg, Austria.

Tani, A., 1989a, "International Comparison of Industrial Robot Penetration," in J. Ranta, ed., *Trends and Impacts of Computer Integrated Manufacturing*, WP-89-1, IIASA, Laxenburg, Austria.

Tani, A., 1989b, "Saturation Level of NC Machine-Tool Diffusion," in J. Ranta, ed., *Trends and Impacts of Computer Integrated Manufacturing*, WP-89-1, IIASA, Laxenburg, Austria.

Tchijov, I., 1989, *FMS World Data Bank*, WP-89-33, IIASA, Laxenburg, Austria.

Chapter 11

Impact of CIM on the Economy: Simulations Based on Macromodels

Lucja Tomaszewicz, Grazyna Juszczak, Czeslaw Lipinski, Witold Orlowski, and Mariusz Plich

The discussion devoted to the CIM technologies comprises two main topics:

(1) Forecasts and penetrations of CIM systems.
(2) Analyses of the economic and social impact.

Empirical investigations of these topics demand proper statistical data at the level of CIM which includes many elements of factory automation, as well as data on measures of technological progress in CIM technologies, and the definitions of the measures themselves. The methodology of empirical research is another important problem. These problems were solved to a great extent in the framework of IIASA's Computer Integrated Manufacturing Project.

In our simulations, we used the characteristics of CIM described in IIASA working papers, especially those by Tchijov (see, for example, Tchijov, 1991a, 1991b). Theoretical foundations, historical surveys (Ayres, 1989a,

1989b), and analyses also helped us to understand the complexity of problems connected with CIM technologies.

However, we obtained direct advantages for our investigations from empirical researches done at IIASA in which significant formal methods of analysis of the two main topics were applied (see, Ayres and Bodda, Chapter 3; Mori, Chapter 4; Tani, 1991; Maly and Zaruba, 1989).

Our investigations analyzed concerns of specific countries and particular aspects and elements of CIM penetrations and benefits.

Our attempts are oriented to the complex analysis of CIM introduction; so the investigations are carried out at the macro level, and a macromodel is our main tool.

Before we describe the model in detail we would like to underline the essence of such a tool as the macromodel for economic simulations and forecasts.

By defining the economic categories at the macro level and using macromodels, it is possible to understand the specification and construction of the equations used in the analysis to study the relationships and feedbacks of the economic processes considered.

According to Ayres and Bodda (Chapter 3) the main benefits of CIM are labor saving; capacity augmenting; capital sharing; product quality improvement; and acceleration of product performance. The first three short-run effects increase profit and motivate CIM usage.

Considering these benefits at the macro level, e.g., investigating particular branches of the economy, long-term effects can be measured by analyzing the relationships between the output and the particular elements of capital equipment (especially CIM elements and others as their effectiveness is totally different). Special considerations are needed for the relationship between output and such equipment as robots, which can be regarded as additions to the labor force (with the cost of this capital treated as wages).

There is another important problem. CIM equipment and robots are direct substitutes for human labor. From this point of view the medium- and long-term employment effects of CIM implementation should be measured. The model should allow one to simulate the labor-saving effects and answer the question: What impacts do higher productivity and higher income (resulting in higher demand), as indirect effects of CIM applications, have on employment?

We begin by analyzing labor shifts and changes in the occupational structure. Adoption of more efficient capital equipment changes demand for fixed capital investments, thus the dynamic (accelerator) equations characterizing the relationships between demand for output and fixed assets, and then for investments of particular equipment, should be the main elements of the macromodel.

These last two effects of CIM applications are the long-term indirect benefits which are passed on to the consumers through lower prices and higher quality. To catch these effects another loop should be introduced to the model, for example, productivity, costs of production (wages), and prices.

Various methodological (formal) approaches can be applied to construct the macromodel, fulfilling the above properties. We concentrate on input–output models with endogenous final demand and related categories, described by econometric equations (see, for instance, Almon *et al.*, 1974; Barker and Peterson, 1989).

Although it was not the subject of our studies, the input–output approach makes it possible to analyze the changes of material costs (through the changes of input–output coefficients caused by CIM implementation) and, which is probably the main advantage of this approach, allows one to study employment effects (levels and occupational categories) in all sectors of the economy (see, for example, Leontief and Duchin, 1986; Ayres and Mori, Chapter 6).

The application of input–output models integrated with behavioral equations for simulations of effects of efficient technology adoption was first treated on the basis of the INFORUM-type model for the Polish economy (see Juszczak *et al.*, 1989). Models for Japan and the USA have also been constructed based on the data from the IIASA CIM Project (Juszczak *et al.*, 1990).

11.1 The Models

11.1.1 General information

The main theoretical concept of the models follows the Keynesian approach, hence the models are demand-driven systems of equations. They integrate input–output and conversion identities with stochastic behavioral description of economic agents activities in particular markets.

Such integrated models have input–output relationships as their central part. These relationships are the set of commodity (industry) balance equations which underline the original specification of the open input–output model:

$$(I - A)X = f \; , \tag{11.1}$$

where X and f denote gross output and final-demand vectors and A is a square matrix of the input–output coefficients.

The elements of final demand (as well as the gross output vector) are disaggregated in equation (11.1) by industries (commodities). The distribution of gross output for final needs is, however, connected with the economic functions of final demand (to fulfill demand for consumption, investments, exports, and so on). So, the disaggregation of final demand by economic functions should be considered:

$$(I - A)X = f = Gg \; , \tag{11.2}$$

where g is a vector of final demand disaggregated by economic functions and G is a non-square conversion matrix characterizing the industry structure of final-demand categories.

To close model (11.2), the elements of g are included as the outcome of behavioral equations which seek to describe the actions of economic agents. So, on one hand, disaggregated final-demand equations are used to generate, through conversion and input–output identities, demand for gross output and imports by industries. On the other hand, the equations concerning final demand through their explanatory variables need the links with those describing the financial side of the economy (wages, prices, taxes). In turn, there are the interactions of output demand with the production factors (fixed capital and labor) as well as links between the goods market and labor and financial assets markets. The most important exogenous variables are economic development abroad and elements of government policy.

As input–output identities are the central part of the model, an input–output framework is used to organize the statistical data and check the model solution. The A and conversion matrices were calculated from input–output balances for 1980 in the Japanese model and for 1977 in the US model. The estimation of behavioral equations is based on time-series data mostly in 1980 (Japan) and 1977 (USA) constant prices. These data as well as special data concerning CIM equipment were derived from the IIASA CIM Project.

The behavioral equations are estimated by the OLS method with the use of the "G" package. Models are installed in the *SYMPHONY* package

(*SYMPHONY* is a product of the Lotus Development Corporation). Special macroinstructions were created to solve the problems.

Some characteristics of disaggregation are the following: number of industries, 21; personal consumption expenditures, 3; investments, exports, imports, and prices by industries, value-added, 3; elements by branches, inventory increase, and government expenditures, 1.

Generally, models are designed to analyze and forecast changes in the economic growth and structure, with special attention paid to the impact of technological changes performed by the computer industrial manufacturing assessment.

This section on methodology and data background is followed by a description of the relationships in the model. A more detailed description can be found in Juszczak *et al.* (1990). Sections 11.1.2 and 11.1.3 give the standard specification of the equations. The changes connected with the CIM block implementation are described in Section 11.2. Finally, the results of implementation in the form of long-term simulation experiments (until 2010) are presented.

11.1.2 Final-demand elements

Personal Consumption Expenditures

In both models the consumers' demand for goods and services is composed of demand for durable, nondurable, and other goods. For their generation dynamic functions of the Houthakker–Taylor type are used, which include lagged personal consumption expenditures, personal disposable income, and interest rate in the case of durables. The standard specification of this equation is

$$C_k = \alpha \, C_{k,-1} + \beta \, YDISP/PC_k + \gamma \, , \qquad (11.3)$$

where C_k is consumption of group k at constant prices; subscript -1 means the lagged value; $YDISP/PC$ is disposable income in real terms; and α, β, γ are parameters.

Fixed Investments

The linear accelerator model was used for the description of investment behavior. However, we have attempted to generalize this framework by also testing relationships between investment and profitability and interest rate. Empirical results show that the inclusion of additional explanatory variables is effective in a few branches only. The basic equation of fixed investments can be written as

$$GI_{jt} = f(K_{jt-1}, XX_{jt}, \Pi_{jt}, RLB_t) \ , \tag{11.4}$$

where GI_{jt} are fixed investments in the j-th branch in period t; K_{jt-1} are fixed assets value in the j-th branch in period $t - 1$; XX_{jt} is gross output (real) in period t; Π_{jt} are profits in period t; and RLB is the interest rate.

Inventory Changes

To describe the stock-building behavior, we have investigated for the USA and Japan the fluctuation of the ratio KJX:

$$KJX = \frac{KJ_t}{XX_t} \ , \tag{11.5}$$

where KJ_t is the total inventory stock (real) in period t and XX_t is the total output (real) in period t.

In both cases there exists a certain saturation level. Therefore, hyperbolic trend functions were constructed for this ratio. With the use of the ratio we find the stock level demanded, and eventually the inventory increase.

Exports and Imports

Both the export and import equations (disaggregated to 21 commodity groups) explain the observed value by the demand indicator and the relative price. The export equation is

$$E_i = \gamma (EW_i)^\alpha (PE_i/PWE_i)^\beta \ , \tag{11.6}$$

where E_i is export of commodity i; EW_i is the world export of commodity i; PE_i is the export price of commodity i; PWE_i is the world market price of commodity i; and α, β, γ are parameters.

In the export equation both the PE and PWE prices are expressed in dollars (so the prices in domestic currency in the Japanese model must be adjusted according to the exchange rate).

The import equation has the form

$$M_i = \gamma (DD_i)^\alpha (PM_i/PX_i)^\beta \ , \tag{11.7}$$

where M_i is import of commodity i; DD_i is the domestic demand for commodity i; PM_i is the import price of commodity i; PX_i is the domestic price of commodity i; and α, β, γ are parameters. In this case both prices must be expressed in domestic currency.

11.1.3 Other blocks of the models

Fixed Assets

The stochastic equations approximate identities equalizing the fixed assets to their level from the previous year plus investments put into operation minus scrapping. The length of the investment lags was constrained up to two years. The difficulties with obtaining proper signs of estimates made us introduce some *a priori* assumptions.

Industrial Employment

In the Keynesian approach the labor market is treated as one in which the level of employment is demand-determined. Industrial employment can be seen as expressing the implicit relationship between the gross output and the labor input necessary for production subject to the given technology of the production process. Therefore, these equations are determined directly from the production function solution with the given level of fixed assets. This relationship is expressed

$$X_{jt} = \gamma K_{jt}^{\alpha} L E_{jt}^{\beta} , \tag{11.8}$$

where LE_{jt} is the level of employment in the j-th branch in period t; K_{jt} is fixed capital; X_{jt} is net output; and α, β, γ are parameters.

The solution of (11.8) subject to LE_{jt} allows one to determine

$$LE_{jt} = \gamma^{-1/\beta} X_{jt}^{1/\beta} K_{jt}^{-\alpha/\beta} . \tag{11.9}$$

Wages and Incomes

The wage equations include both the determination of an average wage and the differences of wages among industries. The equation of the average wage is based on the Phillips curve hypothesis. The specification allows for the changes in earnings affected by the change in unemployment rate. Additional factors were included to catch the effects of the wage-bargaining process caused by inflation and the long-term influence of the labor productivity factor. The form of the wage equation is the following:

$$(\Delta W/W_{-1}) = \alpha + \beta UR + \gamma(\Delta PC/PC_{-1}) + \delta(\Delta XL/XL_{-1}) , \tag{11.10}$$

where W is average wage; UR is unemployment rate; PC is consumer price index; XL is labor productivity in the economy; and $\alpha, \beta, \gamma, \delta$ are parameters.

The parameter α in the equation (11.10) is connected with the "natural" rate of unemployment (when unemployment is given at this level and there is no growth of prices or productivity, the rate of growth of wage should be equal to zero).

The specification of the wage equation (together with the price block equations) allows for a good explanation of the inflationary process in the economy and includes some elements connected with long-run properties of the INFORUM-type model (see Almon, 1983). The next element explained in the model is the ratio of wages by industries to the average wage. The standard specification of the equation is as follows:

$$W_i/W = \alpha(XL_i/XL) + \beta(\text{time}) , \tag{11.11}$$

where XL_i is productivity of labor in sector i and α and β are parameters.

The wages by industries multiplied by the employment rate give the wage income by industries (YW), which constitutes one of the most important elements of the personal income.

The second element, transfers to households (YTR), is considered to be a function of the unemployment level and the price level. The dividend income $(YDIV)$ is generated as the linear function of the gross profits in the economy. The payments from households to the social security system $(CSIP)$ are assumed to be proportional to the personal income. Finally, personal income is generated as follows:

$$YP = \alpha(YW + YTR + YDIV + CSIP) + \beta , \tag{11.12}$$

where α and β are parameters.

Domestic Industrial and Final User Prices

To close the income-expenditure loop in the models, several prices are modeled – namely, industrial prices and final user prices. Industrial prices are determined by using the input–output relationship, e.g., by costs per unit of output (material costs, labor costs, and other elements of value-added; profits are treated as a markup of global costs).

The conversion matrix is applied to derive final user prices, excluding export prices. The prices of exports are calculated in a different way. They are assumed to adjust both to the domestic costs of production increase and to the world market prices (prices of the main trade competitors). The following specification was adopted:

$$PE_i = \alpha PX_i + \beta PEC_i , \tag{11.13}$$

where PE_i is export price of commodity i; PX_i is producer price of commodity i; PEC_i is the main trade competitor export price of commodity i; and α and β are parameters.

The export price and the price of trade competitor (USA for Japan, Japan for USA) are expressed in the domestic currency. The parameters α and β can be considered as the weights of the "importance" of internal and external factors in the creation of export prices.

11.2 Labor Relationships and CIM Implementation

11.2.1 CIM in the US and Japanese models

The implementation of the CIM technological factor into the input–output model forced several changes in the traditional structure including the following:

(1) Different effectiveness of the traditional tools and CIM machinery.
(2) Different demand effect caused by the investment in CIM (the technology and cost of production of CIM equipment).
(3) The effects of substituting CIM equipment for labor.

The solution we have chosen is in some way similar to the one used by Mori (Chapter 4). It is based on the change of the description given in the model to the production process and the explicit distinction of the traditional fixed capital and new technologies (however, both types of machinery are assumed to be substitutes). Some clear assumptions are made about the effectiveness of different machinery types, substitutability of labor and the technological process. These presuppositions, as well as the forecasts of the CIM diffusion in the USA and Japan, are open to discussion and are presented in this chapter.

Tchijov (1991b) distinguished four levels of manufacturing technologies: (1) TR, traditional machinery; (2) NC, numerical control machines; (3) FMC/FMS, flexible manufacturing cells and systems; and (4) IR, industrial robots. The first assumption was connected with the CIM diffusion: only five industrial branches are considered as CIM users (branches 15–19, that is, metal products and machinery including electrical, transport equipment, and precision instruments).

From the comparisons of the level of the CIM diffusion in the sample period (until 1984) and the forecasts, it is to be assumed that the relatively insignificant role played by CIM in the 1960s and 1970s does not allow the correct estimation of the CIM machinery effectiveness parameters. Instead of estimation we have changed the production function for the forecast period by using exogenous parameters. In the case of NC and FMC/FMS, it was the average price ratio (with respect to traditional tools) that was assumed to reflect the relative productivity. Thus, the homogeneous fixed capital in the j-th branch (K_j) (with the average effectiveness of the traditional one) is given by the expression

$$K = RP_{nc} \times KNC_j + RP_{fms} \times KFMS_j + KTR_j , \qquad (11.14)$$

where $K...$ is the number of machinery of a certain type and $RP...$ is the relative price with respect to traditional tools.

The relative prices are given by Tchijov (1991b): for NC the price ratio is at level 3–4; for FMS with respect to NC equipment, level 4–6. In our research we used $RP_{nc} = 3.5$ and $RP_{fms} = 18$.

Industrial robots are included in the production functions in a different way. As the robots are assumed to be perfect substitutes for labor (the rate of substitution given by Tchijov is 1 robot for 1.3 persons), they are incorporated with the same production elasticity parameter. Finally, the production functions in branches 15–19 are written as follows:

$$
\begin{aligned}
X_{jt} &= \gamma (3.5 \times KNC_{jt} + 18 \times KFMS_{jt} + KTR_{jt})^{\alpha} \\
&\quad LE_{jt}^{\beta} (1.3 \times KIR_{jt})^{\beta} ,
\end{aligned}
\qquad (11.15)
$$

where X is net output; LE is labor; KIR is number of robots; and α, β, γ are parameters. The functions are, then, solved to find the short-run demand for labor by branches.

The investment-demand functions must be changed to capture the effect of the CIM implementation. The demand for homogeneous fixed capital (generated by the model equations) is met partly by the exogenously given NC and FMS machinery increases and partly by the traditional equipment (the check was installed in the model to assure the minimal 10% ratio of traditional tools in total investment in selected branches). The demand for the traditional investment is given by the expression

$$InvTR_{jt} = f1(...) - 3.5 \times KNC_{jt+1} - KNC_{jt})$$
$$-18 \times KFMS_{jt+1} - KFMS_{jt}) , \qquad (11.16)$$

where $f1(...)$ is the investment demand function (demand for homogeneous capital).

At the same time, the investment demand for CIM machinery is given by the expression

$$InvCIM_{jt} = 3.5 \times (KNC_{jt+1} - KNC_{jt}) \qquad (11.17)$$
$$+18 \times KFMS_{jt+1} - KFMS_{jt})$$
$$+1.3 \times (KIR_{jt+1} - KIR_{jt}) ,$$

where $K...$ are taken from the forecast of Tchijov (1991b).

Two clear assumptions are made: the projected numbers of CIM equipment are the *demanded* ones, the investment lag is one period, and there is no significant capital depreciation in the CIM equipment.

The last significant change of the model is connected with CIM equipment production. The assumption is made that this machinery is produced by three branches of industry: 16 (machinery), 17 (electrical machinery), and 21 (construction and tertiary industry). The elements of the bridge vector for CIM investment are 0.325, 0.45, and 0.225 for branches 16, 17, and 21, respectively.

11.2.2 Short- and long-run effects of CIM diffusion

The incorporation of the CIM technologies diffusion into the macroeconomic model leads to the appearance of various short- and long-run effects connected first of all with the substantial increase of labor productivity in the branches absorbing the technological progress.

In the short-run effect, one can expect results that are rather discouraging for the model user (*Figure 11.1*).

This short-run effect can be slightly diminished by the existence of a social security system. As the social security payments in the model depend on the unemployment level, their increase is a reaction to labor-demand decreases.

The second element that plays a role in the short-run effect is the wage-bargaining process reflected in the model by the incorporation into the wage

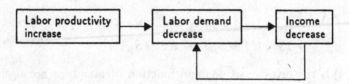

Figure 11.1. Short-run effect of CIM diffusion.

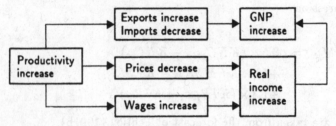

Figure 11.2. Long-run effect of CIM diffusion.

equation (based generally on the Phillips curve) of the share of employees in the increase of the labor productivity. However, as this share is not very big, it can hardly compensate for the fall of real income caused by the labor demand decrease.

The situation changes in the long-run effect. It is a common observation that not only, as Almon says, "the market economies do a remarkable job of providing work" for the growing labor force (see Almon, 1983), but they are able to absorb at the same time the remarkable technological progress. This feature observed in the past in both the US and Japanese economies must be incorporated in the model. The factors that allow this phenomenon are analyzed by Kendrick (1976); special attention is paid to the role of CIM technologies in this process by Ayres and Bodda in Chapter 3. The most important long-run effect of the implementation of CIM technologies is connected with the cost per unit reduction. This leads to the price-level decrease and certain effects connected with the international competitiveness. Finally, the wage-bargaining effect gives the labor part of the productivity growth. The most important factors are exhibited in *Figure 11.2*.

Incorporation of this mechanism into the macroeconomic model allows one to investigate the technological changes in the economy and the simulation experiments connected with its character and diffusion speed. In the specification of equations of both the US and Japanese input–output models, the price–wages loop, which generates the inflation level, plays an active role in the creation of an effective demand for production and (through the relative prices) in the international competitiveness of the economy.

11.3 CIM Diffusion in the USA and Japan: Simulations

11.3.1 The assumptions

In both the US and Japanese cases the CIM simulation was made in two steps. In the first one, the base solution of the model was found (until 2010). Then the CIM equations were introduced, and the models were solved once again. All the assumptions concerning the exogenous variables, however, were the same in the base run and simulation experiment.

The assumed diffusion of CIM technologies was based on Tchijov (1991b). As Tchijov gives only the levels for years 1990, 2000, and 2010, the interpolation was made to find values for other years. Tchijov also gives the relative prices of CIM and traditional equipment and substitutability parameter for industrial robots and labor.

11.3.2 Simulation results for the USA

The simulation of the implementation of CIM equipment in the USA shows the general positive impact of new technologies at the production and welfare level. The control solution results and the CIM experiment results are given in *Tables 11.1* to *11.5*.

The main components of final-demand consumption expenditures, are higher for all the simulations; in 2010 consumptions (PCEs) increase almost 5% (see *Table 11.1*) in comparison with the base solution (0.19% rate of growth increase in 1990–2000, 0.16% later on; see *Table 11.2*). Such a result is caused first of all by the price-level reduction, which exceeds the fall of the nominal wage and income. Price-level decrease can be observed first of all in five CIM-using branches, from a fall of 7 points in branch 15 to 36 points in branch 19 – precision instruments (see *Table 11.3*). [Branches 15 to 19 are subdivided in the tables by gross output deflators (PXX), exports (EXP), and employment (LAB).] The spillover effects cause price reductions in all the other branches: the consumption deflator falls by 8 points (*Table 11.1*) and the investment goods deflator by 10 points (*Table 11.1*). The average

Table 11.1. Selected CIM simulation results and base solution results for the USA.

Selected indicators	1990 Base (a)	CIM (b)	(c) b/a	2000 Base (d)	CIM (e)	(f) e/d	2010 Base (g)	CIM (h)	(i) h/g
Consumption (PCE)	1760.96	1781.08	101.1	2287.40	2356.77	103.0	2875.41	3009.60	104.7
Priv. investment	467.97	473.59	103.4	737.95	753.85	102.2	988.40	1012.02	102.4
Inventory investment	40.64	55.66	137.0	11.88	12.74	107.3	13.71	14.73	107.5
Exports	269.49	268.99	99.8	338.49	343.37	101.4	434.32	446.34	102.8
Imports	257.82	265.27	102.9	337.02	345.84	102.4	428.16	442.00	103.2
GNP	2785.24	2828.05	101.5	3634.42	3717.40	102.3	4575.96	4732.97	103.4
Gross output	5035.85	5174.96	102.8	6588.42	6765.01	102.7	8285.20	8616.06	104.0
Employment	109.10	110.69	101.5	117.82	117.87	100.0	126.19	126.47	100.2
Labor productivity	25.53	25.55	100.1	30.85	31.54	102.2	36.26	37.42	103.2
Unemployment rate	5.55	4.17	75.1	4.40	4.36	99.2	3.82	4.03	94.8
Average wage	31.29	31.25	99.9	75.04	73.46	97.9	196.32	184.52	94.0
Consumption deflator	204.10	203.00	99.5	418.70	404.10	96.5	969.0	891.40	92.0
Investment deflator	204.50	203.90	99.7	419.30	398.9	95.1	973.2	875.30	89.9
Disposable income	3856.74	3984.51	103.3	10,119.18	10,174.20	100.5	28,842.60	28,088.79	97.4

Table 11.2. Control solution and CIM simulation results for the USA.

Main macrovariable	Annual growth rate in base run		Effects of CIM: Growth rate increase	
	1990–2000	2000–2010	1990–2000	2000–2010
Consumption (PCE)	2.65	2.31	0.19	0.16
Business investment	4.89	2.97	−0.13	0.02
Export	2.31	2.52	0.17	0.13
Import	2.74	2.40	−0.05	0.09
Real GNP	2.70	2.33	0.08	0.11
Employment	0.77	0.69	−0.14	0.02
Labor productivity	1.91	1.63	0.22	0.10
Consumption deflator	7.45	8.75	−0.32	−0.52
Investment deflator	7.44	8.78	−0.50	−0.61

Table 11.3. Gross output deflators for CIM-related industries in the USA.

Industry branch	1990			2000			2010		
	Base (a)	CIM (b)	(c) b/a	Base (d)	CIM (e)	(f) e/d	Base (g)	CIM (h)	(i) h/g
PXX15	129.3	127.4	98.6	170.0	150.8	88.7	192.8	164.40	85.3
PXX16	133.6	130.8	98.0	171.7	149.3	87.0	187.8	156.61	83.4
PXX17	122.5	120.0	97.9	147.4	130.5	88.5	153.2	134.95	88.1
PXX18	129.4	127.1	98.2	170.1	149.1	87.6	192.0	161.17	83.9
PXX19	132.9	130.1	97.9	177.9	150.9	84.8	204.7	163.47	79.9

Table 11.4. Exports of CIM-related industries in the USA.

Industry branch	1990			2000			2010		
	Base (a)	CIM (b)	(c) b/a	Base (d)	CIM (e)	(f) e/d	Base (g)	CIM (h)	(i) h/g
EXP15	4.23	4.24	100.3	5.18	5.24	101.1	6.38	6.54	102.6
EXP16	31.07	31.09	100.1	41.24	41.55	100.7	54.77	55.50	101.3
EXP17	11.29	11.28	99.9	15.31	15.42	100.7	20.77	21.10	101.6
EXP18	16.82	16.01	95.2	13.86	15.67	113.0	11.84	14.56	123.0
EXP19	7.92	7.92	99.9	12.51	12.66	101.2	19.74	20.35	103.1

inflation rate is lower in the CIM experiment; the consumption deflator rises before year 2000: 0.3 points slower in 1990–2000 and 0.5 points slower in 2000–2010 than in the base run. The investment deflator rises even slower than the consumption deflator: 0.5 points in 1990–2000 and 0.6 points in 2000–2010 (*Table 11.2*).

At the same time the average wage falls by only 6 points (see *Table 11.1*) down in comparison with the base run (however, in five CIM industries it

Table 11.5. Employment in CIM-related industries in the USA.

Industry branch	1990 Base (a)	CIM (b)	(c) b/a	2000 Base (d)	CIM (e)	(f) e/d	2010 Base (g)	CIM (h)	(i) h/g
LAB15	1.79	1.84	102.9	2.06	1.92	93.1	2.29	2.08	91.1
LAB16	2.28	2.37	103.6	2.50	2.29	91.6	2.59	2.51	96.7
LAB17	2.10	2.33	110.9	2.40	2.05	85.3	2.70	1.93	71.4
LAB18	2.82	3.15	111.9	3.40	2.82	82.8	3.94	2.97	75.5
LAB19	0.63	0.67	106.0	0.73	0.53	72.1	0.80	0.39	48.8

is higher than in the base solution because of substantial labor productivity increases). The nominal disposable income is even higher (2.6 points down; see *Table 11.1*), as it partly depends on the profits.

The private investment level is 2.4% up in 2010 (see *Table 11.1*) in spite of the fact that the investment in CIM branches is substantially lower (in terms of quantity, but not value). At the same time, however, the trend of increasing investment can be observed in most other sectors of the economy. The overall effect obtained in the experiment is negative until the year 2000, then (as the forecast rate of diffusion of CIM technologies lowers) the effect changes to positive (0.02%; see *Table 11.2*). An increase can be seen in the inventory investment as well.

The international competitive position of the US economy does not change a lot in the CIM experiment. The exports improve substantially in CIM industries (up to 23% improvement in EXP 18 – transport equipment; see *Table 11.4*); they are higher in all branches. Finally, the growth rate of the exports is 0.17: 0.13 points higher than in the base run (see *Table 11.4*). At the same time, however, the economic growth causes the imports increase (after the year 2000, 0.09 point positive impact). Eventually, the trade balance remains unchanged.

Finally, the GNP level obtained in the experiment is 3.4 points higher than that in the base run (see *Table 11.1*). The rate of growth of the real GNP in the CIM experiment is 0.08–0.11 points higher than that in the base run (see *Table 11.2*). Employment structure changes: it is much smaller in the CIM industries [about 50% reduction in precision instruments industry (LAB 19), smaller in the others; see *Table 11.5*], but higher in the other branches. The productivity of labor growth (0.22 points higher rate of growth till year 2000, then 0.1 point higher; see *Table 11.2*) is fully absorbed by the economy: the unemployment rate remains almost unchanged (the growth of the demand for labor is lower in the CIM experiment till the year 2000, but then it becomes higher than the base solution).

11.3.3 Simulation results for Japan

The positive impact of the CIM diffusion can be also observed in the Japanese case (see *Tables 11.6* to *11.10*):

The PCEs level in the CIM experiment increases by 2.1 points (in 2010; see *Table 11.6*). The rate of growth of consumption is 0.08–0.09 points higher than in the base run (see *Table 11.7*). Again it is the result of the price-level reduction being much stronger than the nominal wage decrease.

The price levels in the CIM industries fall in comparison with the base run from 15 to 20 points. Again, the biggest reduction in the precision instruments industry (PXX 19 in *Table 11.8*). Eventually, the consumption deflator is reduced by 2 points (see *Table 11.6*), and the investment goods deflator by 5.4 points. For consumption deflator the rate of growth is lowered by 0.15 point until 2000, and 0.04 points later on; for investment, 0.38 and 0.13, respectively (*Table 11.6*).

The average wage reduction is much smaller (1 point; see *Table 11.6*). In the CIM industries the wage is higher than in the base run (from 11% to 25% increase in branch 15), in the others it is smaller or equal. One should keep in mind, however, that the Phillips curve decides the average wage; as there is no substantial change in the unemployment level, the wage does not change significantly. At the same time the nominal disposable income is 1.3 points higher than in the base solution (again the result of the profit increase; see *Table 11.6*).

The rate of growth of the business investment in the CIM experiment is lower than in the base run until 2000, but then substantially improves (0.82 points in the period 2000–2010) in spite of the fact of the investment reduction in the CIM industries (again, in quantity terms, not in value terms; see *Table 11.7*). The inventory investment remains almost at the same level in both solutions.

The biggest differences between the base run and the CIM experiment results can be found in the Japanese foreign trade sector. Exports grow by almost 15 points. This growth is caused by the international competitiveness improvement due to the price reduction (*Table 11.6*). The growth is drawn from the CIM industries exports increase: from 9 points in the branch 19 (precision instruments) to almost 30 points in 17 (electrical machinery). Therefore, we obtain more than 1 point of the additional annual growth of

Table 11.6. Selected CIM simulation results for Japan.

Selected indicators	1990			2000			2010		
	Base (a)	CIM (b)	(c) b/a	Base (d)	CIM (e)	(f) e/d	Base (g)	CIM (h)	(i) h/g
Consumption (PCE)	187,505.90	188,251.00	100.4	217,032.40	219,526.70	101.1	266,594.00	272,174.64	102.1
Priv. investment	58,210.50	56,687.90	97.4	67,849.80	65,977.00	97.2	95,853.10	100,865.84	105.2
Inventory investment	112.60	120.80	107.2	379.90	428.90	112.9	966.10	981.47	101.6
Exports	49,666.22	50,400.31	101.5	83,468.99	93,457.41	112.0	168,481.08	193,417.78	114.8
Imports	49,360.82	49,929.57	101.2	65,405.18	67,982.73	103.9	95,336.09	101,669.65	106.6
GNP	307,354.81	306,974.41	99.9	389,351.20	398,598.82	102.4	557,684.86	584,279.73	104.8
Gross output	670,574.37	669,438.62	99.8	860,017.85	883,026.77	102.7	1,262,872.50	1,328,508.56	105.2
Employment	62,574.98	62,439.75	99.8	66,680.54	66,760.29	100.1	70,832.96	71,028.58	100.3
Labor productivity	4.91	4.92	100.1	5.84	5.97	102.3	7.87	8.23	104.5
Unemployment rate	2.52	2.73	108.4	2.93	2.82	96.0	3.64	3.34	91.8
Average wage	451.40	447.40	99.1	716.50	703.00	98.1	1084.00	1070.80	98.8
Consumption deflator	133.00	132.70	99.8	183.60	180.60	98.4	219.60	215.22	98.0
Investment deflator	132.10	131.20	99.3	177.80	170.20	95.7	207.00	195.69	94.6
Disposable income	261,745.30	263,284.60	100.6	430,000.00	430,946.50	100.2	668,356.50	677,266.62	101.3

Table 11.7. Control solution and CIM simulation results for Japan.

Main macrovariable	Annual growth rate in base run		Effects of CIM: Growth rate increase	
	1990–2000	2000–2010	1990–2000	2000–2010
Consumption	1.47	2.08	0.08	0.09
Business investment	1.54	3.52	−0.02	0.82
Export	5.33	7.28	1.04	0.27
Import	2.85	3.84	0.28	0.27
Real GNP	2.39	3.66	0.25	0.24
Employment	0.64	0.61	0.03	0.02
Labor productivity	1.74	3.03	0.22	0.23
Consumption deflator	3.28	1.81	−0.15	−0.04
Investment deflator	3.02	1.53	−0.38	−0.13

Table 11.8. Gross output deflators for CIM-related industries in Japan.

Industry branch	1990			2000			2010		
	Base (a)	CIM (b)	(c) b/a	Base (d)	CIM (e)	(f) e/d	Base (g)	CIM (h)	(i) h/g
PXX15	201.7	200.4	99.4	422.2	410.4	97.2	991.2	930.2	93.8
PXX16	195.7	194.5	99.4	302.0	372.5	94.8	896.3	814.3	90.8
PXX17	205.0	206.4	100.7	398.5	377.8	94.8	879.9	785.6	89.3
PXX18	222.2	228.6	102.9	488.1	454.8	93.2	1197.5	1062.7	88.7
PXX19	228.0	230.6	101.1	553.3	460.1	83.2	1523.2	965.6	63.4

exports during the period 1990–2000, but the growth rate then reduces to 0.27 points (see *Table 11.7*). At the same time imports are up only by 6.6 points (0.28–0.27 rate of growth increase; see *Table 11.7*), and the trade surplus increases (one can argue if it is a positive phenomenon in the case of Japan!).

The total GNP growth in the CIM experiment is 4.8 points (0.25–0.24 points higher growth than in the base run; see *Table 11.6*). This is due to the 4.5% productivity growth (0.22–0.23 points higher annually; see *Table 11.6*). The labor demand is almost equal to this one from the base run (0.03–0.02 points in the growth rates; see *Table 11.7*). The 45% to 55% decrease in the employment level in the CIM industries is accompanied by the increase in the other industries (see *Table 11.10*). Again, the unemployment rate is even slightly lower than in the base run, and the economy is able to absorb the effects of the accelerated technological progress.

Table 11.9. Exports of CIM-related industries in Japan.

Industry branch	1990 Base (a)	CIM (b)	(c) b/a	2000 Base (d)	CIM (e)	(f) e/d	2010 Base (g)	CIM (h)	(i) h/g
EXP15	1392.69	1409.43	101.2	2199.21	2407.38	109.5	3741.98	4189.60	112.0
EXP16	6425.75	6561.45	102.1	12,192.02	13,956.61	114.5	26,352.89	31,254.50	118.6
EXP17	8503.70	8888.22	104.5	17,276.22	22,095.39	127.9	45,585.37	58,543.57	128.4
EXP18	13,195.40	13,409.28	101.6	24,206.37	27,133.97	112.1	49,792.10	57,698.90	115.9
EXP19	1689.28	1705.52	101.0	2883.59	3081.49	106.9	5119.73	5564.74	108.7

Table 11.10. Employment in CIM-related industries in Japan.

Industry branch	1990 Base (a)	CIM (b)	(c) b/a	2000 Base (d)	CIM (e)	(f) e/d	2010 Base (g)	CIM (h)	(i) h/g
LAB15	1274.3	1179.1	92.6	1392.3	841.1	60.4	1510.3	699.42	46.3
LAB16	1513.0	1432.3	94.7	1537.5	1076.3	70.0	1562.0	905.36	58.0
LAB17	1783.3	1656.5	92.9	1960.3	1229.5	62.7	2137.3	1059.89	49.6
LAB18	1517.2	1435.5	94.6	1664.2	1198.5	72.0	1811.2	1125.68	62.2
LAB19	384.5	362.7	94.3	421.5	295.7	70.2	458.5	273.22	59.6

11.3.4 Comparison of results

The CIM simulation results indicate some significant differences in the economic system response to the technological change in the case of the USA and Japan. This phenomenon was observed in other studies (Jorgenson, 1988; Jorgenson and Nishimizu, 1978). Generally speaking, the differences are connected with the distribution of the effects of the productivity growth and with the export behavior of firms.

A similar GNP growth (3.4% in the USA, 4.8% in Japan) leads to a much higher increase in consumption in the USA (4.7% compared with only 2.1% in Japan). At the same time the investment increase is much more rapid in Japan (5.2% compared with 2.4% for the USA). This is due to the larger real wage increase in the USA, connected with the influence of trade unions (more generally with the stronger position of labor in wage bargaining) and with more aggressive investment behavior of Japanese companies.

The difference in the inventory investment behavior should also be stressed. The stock increase is much higher in the USA (7.5%) than in Japan (1.6%).

Finally, the response of Japanese companies to the improvement of the relative prices in foreign trade is much stronger than the response of US companies. In spite of the smaller domestic prices reduction, Japanese exports increase much more rapidly (almost 15% compared with 2.8% for the USA).

Eventually, the CIM simulation results reflect the more aggressive and "outward-looking" type of growth of Japan and the more "inward-looking" US growth and response to the technological change.

Acknowledgments

We are grateful to Professors Furukawa, Yamada, Uno, and Tchijov for their consultations on the data. The research was a team effort. It is, however, obvious that certain specialization among the members of the team should exist. We are obliged to Professor Welfe whose experience in econometric modeling was very helpful.

References

Almon, C., 1983, "The Price-Income Block of the US Inforum Model," in M. Grassini and A. Smyshlyaev, eds., *Input-Output Modeling*, CP-83-502, IIASA, Laxenburg, Austria.

Almon, C., M.B. Buckler, L.M. Horwitz, and T.C. Reimbold, 1974, *Interindustry Forecasts of the American Economy*, Lexington Books, Lexington, MA.

Almon, C., M.B. Buckler, L.M. Horwitz, and T.C. Reimbold, 1974, *Interindustry Forecasts of the American Economy*, Lexington Books, Lexington, MA.

Ayres, R.U., 1989a, *Future Trends in Factory Automation*, RR-89-9, IIASA, Laxenburg, Austria.

Ayres, R.U., 1989b, *Complexity, Reliability, and Design: Manufacturing Implications*, RR-89-9, IIASA, Laxenburg, Austria.

Barker, T., and A. Peterson, 1989, *The Cambridge Multisectoral Dynamic Model of the British Economy*, Cambridge University Press, Cambridge, UK.

Jorgenson, D.W., 1988, "Productivity and Economic Growth in Japan and the United States," *American Economic Review* **78**(2):217–222.

Jorgenson, D.W., and M. Nishimizu, 1978, "US and Japanese Economic Growth, 1952–1974: An International Comparison," *Economic Journal* **88**(352):707–726.

Juszczak, G., C. Lipinski, W.M. Orlowski, M. Plich, and L. Tomaszewicz, 1989, "An Input–Output INFORUM–Type Model of the Polish Economy," Mimeo, University of Lodz, Lodz, Poland.

Juszczak, G., C. Lipinski, W.M. Orlowski, M. Plich, and L. Tomaszewicz, 1990, "Input–Output Models Development for CIM Application Analysis and Forecasting (the USA and Japanese models)," Mimeo, University of Lodz, Lodz, Poland.

Kendrick, J.W., 1976, "Productivity Trends and the Recent Slowdown: Historical Perspectives, Causal Factors, and Policy Options," in Fellner, ed., *Contemporary Economic Problems*, American Enterprise Institute, Washington, DC.

Leontief, V., and Duchin, F., 1986, *The Future Impact of Automation on Workers*, Oxford University Press, New York, NY.

Maly, M., and P. Zaruba, 1989, "Prognostic Model for Industrial Robot Penetration in Centrally Planned Economics," in J. Ranta, ed., *Trends and Impacts of Computer Integrated Manufacturing*, (WP-89-1), IIASA, Laxenburg, Austria.

Tani, A., 1991, "The Diffusion of Robots," in R. Ayres, W. Haywood, I. Tchijov, eds., *Computer Integrated Manufacturing*, Volume III: *Models, Case Studies, and Forecasts of Diffusion*, Chapman and Hall, London, UK.

Tchijov, I., 1991, "The Difusion of Flexible Manufacturing Systems," in R. Ayres, W. Haywood, I. Tchijov, eds., *Computer Integrated Manufacturing*, Volume III: *Models, Case Studies, and Forecasts of Diffusion*, Chapman and Hall, London, UK.

Tchijov, I., 1991, "Computer Integrated Manufacturing: International Diffusion Forecasts," in R. Ayres, W. Haywood, I. Tchijov, eds. *Computer Integrated Manufacturing*, Volume III: *Models, Case Studies, and Forecasts of Diffusion*, Chapman and Hall, London, UK.

Chapter 12

Occupation-by-Sector Matrices: Methodological Problems and Results

Hans-Ulrich Brautzsch

The rapid diffusion of CIM has unquestionably had an important influence on labor demand in the metalworking industries, which are the most important sectors for initial CIM application. There is also a delayed effect on labor demand in other economic sectors that are directly or indirectly linked with the metalworking industries, as suppliers or customers. Indeed, the estimation of structural changes in labor demand caused by CIM applications requires an assessment of input–output relations in the national economy. The input–output approach has been characterized by Brooks (1985) as the one that

> provides the most rigorous method for projecting employment effects of new technologies because it is capable of accommodating economy-wide effects arising out of the linkage among sectors and thus of tracing through the system-wide impacts of introduction of a particular technology.

An important precondition for using the input–output approach for such investigations is the elaboration of detailed labor matrices in which the single labor category makes it possible to address the labor impacts caused by CIM application. Occupation-by-sector matrices are available for many countries.[1] The main data sources for these matrices are national and local censuses or special surveys (as in the USA). In these matrices national classification systems are used for the occupational categories, as well as for the economic sectors. To achieve comparable occupation-by-sector matrices it is necessary to convert the national labor matrices into a common format. In particular, there are two requirements:

(1) Occupational categories should be classified according to the International Standard Classification of Occupations (ISCO, 1968).
(2) Economic sectors should be defined according to the International Standard Industrial Classification of all Economic Activities (ISIC, 1968).

A precondition for these computations is the availability of a conversion list that merges the national classification systems and the international classification systems (ISCO and ISIC, respectively). As such conversion lists are not available for some countries the detailed comparisons have to be restricted to Austria, Finland, the Federal Republic of Germany, the Netherlands, Sweden, and the USA. [Throughout this chapter the Federal Republic of Germany (FRG) refers to West Germany prior to German unification, October 1990. Likewise the German Democratic Republic (GDR) refers to the previously separate state of East Germany.]

For some countries the occupational structure of the metalworking sector and the logistic sector are available for two points of time. In 1969–1970 the OECD published a set of highly aggregated labor matrices for 53 countries. The most sophisticated study was carried out by Zymelman (1980); this study analyzed matrices with 120 occupations and 58 sectors for 26 countries during 1970–1971. Zymelman's work has not been updated. Nevertheless the study makes the investigation of changes in the occupational structure by sectors in some countries possible. In this last decade robots, flexible manufacturing systems, and other computer integrated production equipments have diffused rapidly. These observed structural changes could be a useful starting point for an investigation of the socioeconomic effects of CIM application.

This chapter focuses on the following problems:

(1) A detailed explanation of the structure of internationally comparable occupation-by-sector matrices.
(2) An elaboration of some methodological problems of investigating the employment impacts of CIM application with the help of labor matrices.

(3) A descriptive analysis of some observed changes in the occupational structure, especially in the metalworking industries and the logistic sector (transportation and distribution).

With regard to the analysis of changes in the occupational structure, it should be emphasized that this analysis cannot be considered as exhaustive. In fact, the existing data base allows further analysis concerning different aspects of the labor-input structure in the sectors.

12.1   Internationally Comparable Labor Matrices

12.1.1   The structure of occupation-by-sector matrices

The columns of the labor matrices correspond to the different economic sectors, whereas the rows denote the different occupational categories. The sectors are classified according to ISIC and the occupational categories are classified according to ISCO.

According to ISCO (1968), a *job* is a set of interrelated tasks and duties which "belong" together. Either they are regularly performed by one person or one person can, in principle, be trained to perform them. An occupation is a set of similar jobs, in terms of tasks and duties. "Type of work performed" is the basic selection principle used in the revised ISCO for defining and grouping occupations. The chosen code divides major and minor groups, unit groups, and occupational categories.

In practice, the basic principle of classification according to the type of work performed is easier to identify in ISCO at the level of occupations than at the unit group or minor group level. The major groups represent very broad fields of work rather than specific types of work performed. An "occupation" is the narrowest occupational category (i.e., the smallest segment of work) which is specially identified in ISCO (1968). This problem is important considering that the chosen occupational classification in the international labor matrices is based on the two-digit level.

The groups of the ISIC (1968) are defined in such a way that two conditions are satisfied in respect to the way in which activities are distributed among similar units:

(1) The production of the class of goods and services which characterizes a given group accounts for the bulk of the output of the units classified to the group.

(2) The group contains the units which produce most of the class of goods and services which characterizes it.

Hence, the main criteria employed in delineating the ISIC categories concern the characteristics of the activities of production units which are important in determining the degree of similarity in the structure and experience of the units and certain relationship in an economy. The major aspects of the activities considered are the character of the products and services produced; the purposes for which the goods and services are used; and the process, technology, and organization of production.

The classification system of occupations and sectors used in these matrices accords to the two-digit classification of ISCO and ISIC, respectively. The two-digit level was chosen for the computation of internationally comparable labor matrices for two reasons. First, at the three-digit level, many national classification systems do not correspond to the international standard classifications (e.g., the occupational classification system for Austria). Second, labor data at the three-digit level are collected less frequently than at the two-digit level in some countries (e.g., the Netherlands). The latter problem is especially important with regard to estimating historical data sets for labor matrices.

For some countries it was possible to subdivide the occupational groups and sectors (e.g., metalworking) that are most important for the investigation of labor impacts of CIM application. The classification of occupations and sectors in the labor matrix of the GDR differs considerably from the above-described internationally comparable matrices. With regard to the occupational classification there are some similarities to the ISCO (1968). These classification principles have some advantages from the point of view of investigating employment impacts of technological progress, as illustrated in the following description of the East German classification system.

The East German labor matrix consists of 19 columns which represent the productive and nonproductive sectors of the economy. The sector "industry" is subdivided into 10 branches. The rows represent different occupational categories. To explain the classification system it is necessary to describe some specific occupations.

Jobs are generally classified into relatively broad occupational groups. A job is a set of interrelated tasks and duties which belong together in the sense that they are regularly performed by one person. An occupation is a set of similar jobs. The East German census of 1981 counted approximately 20,700 different occupations which reflected different types of work performed, and these occupations are grouped into approximately 450 occupational groups.

The ability to carry out a certain type of work (to fulfill its requirements) will depend on the worker's possession of the appropriate skill. In the different occupational categories it is necessary to consider two aspects: the occupations as an indicator for the type of work performed, on the one hand, and the necessary skill level as an indicator for the ability to fulfill the work requirements, on the other.

The classification of occupations used in the East German labor matrix groups occupations according to 38 occupational units and five different skill levels (unskilled/semiskilled worker, skilled worker, supervisor, technical school graduate, and university graduate), so that each category indicates a certain occupation as well as a certain skill level.

Altogether there are 215 occupational entities in East Germany. It should be mentioned that in some occupational groups we found more than one category for one skill level. For example, the occupational group "electro-technical/electronic occupations" includes two different groups of skilled workers.

12.1.2   Problems with the data

The main problem is the quality of the primary data. For most labor matrices exact conversions from the national classification systems to the ISCO and ISIC systems were available. However, in some cases, this was not easy. For example, for the USA a very detailed labor matrix is available, but unfortunately there are difficulties in comparing the international classifications. For example:

(1) A list for converting the US classification of occupations to the ISCO classification was not available, so one had to be made (Zymelman, 1986; Norword, 1986). This required a direct comparison between approximately 1,500 occupational categories. However, the comparison of these occupational categories is less than perfect.

(2) The occupational classification of the US matrix appears to differ from international usage in some cases. Examples include the occupational groups "managers" and "supervisors."

(3) In the US matrix some occupations (e.g., "economists") are explicitly identified in the ISCO classification, but omitted from the original US labor matrix. Needless to say, employees who can be identified as "economists" are nevertheless working in all sectors of the US economy.

In other countries authorized conversion lists were not available. Thus, the elaboration of an internationally comparable labor matrices for a point of time in the 1970s was difficult. For example, for the Netherlands it was only possible to compute (tentatively) the occupational structure for the metalworking and the logistic sectors. For the USA we were able to estimate the shares of some occupational groups that may be affected by CIM, based on the total number of employees in the metalworking industries in 1970. With regard to the development of occupational structures in the US metalworking industries, it was possible to compute three alternatives using official projections for 1995.

For the FRG labor matrices were computed for 1950, 1961, 1970, and 1982. The matrices for 1950, 1961, and 1970 were estimated using the data compiled by Karr and Leupoldt (1976). For these years, complete conversion lists to ISIC and ISCO were not available. However, to identify changes in the occupational structure by sectors in the FRG a detailed study covering more than three decades was carried out by Hellwig (1988).

With regard to the quality of the primary data base, it should be mentioned that in some cases the quantity of labor was given in units of "100 employees." When proving the column sum or the row sum, differences in the given figures for the column or the row sum can be observed. This is mainly due to inconsistencies in the primary data base given by the statistical offices. In some cases this error is also reflected in the matrices entitled "labor input structure in the sectors" and "occupational structure of employees by sectors."

In summary, using the international labor matrices, a number of caveats should be considered. The most important are the following:

(1) The figures given in the primary labor matrices are usually extracted from surveys using subjective statements of the employees about their job or occupation.
(2) In some cases the data were based on a small sample (local census), which can lead to significant margins of errors.
(3) To compare the national classification systems with the ISIC and ISCO systems, supplementary computations were sometimes necessary. For example, to compare the labor matrix for the FRG, four different conversion lists have to be taken into account. Needless to say, this is a source of added uncertainty.

(4) In some cases important data about the occupational structure in different sectors were not given in the original national surveys. This leads unavoidably to distortions in the international matrices.

Unquestionably these shortcomings have to be considered when investigating changes in the occupational structures between different countries.

12.2 Technology-Induced Employment Effects: Methodological Problems Using Input–Output Models

By incorporating detailed labor matrices into input–output models insights into the following problems can be achieved.

(1) The labor replacement potential (by occupation) in the different sectors, due to technological progress.
(2) Estimates of additional labor demand (by occupations) in the different sectors, induced by technological progress.
(3) The possibility of moving displaced employees, without professional retraining, to other sectors of the economy.
(4) Estimates of needs for retraining and education.

However, it should be emphasized that the labor matrices do not, by any means, reflect all the important socioeconomic impacts induced by technological progress. Below, we note some of the methodological problems that arise in connection with investigations of the employment effects of technological progress.

In the literature such effects as changing work content and work environment, which are conditioned by the application of CIM, are emphasized. The less quantifiable effects are often more important than the labor replacement effects induced by CIM application. Ayres (1986) emphasized that the social importance of various issues may well be in inverse ratio to their quantifiability.

To estimate the influence of technological progress on the employment structure, it is necessary to classify the diverse working places into groups that are comparatively influenced by technological progress (Schäfer and Wahse, 1986) Similar tasks or occupations have been given special attention here.

Table 12.1. Task composition of different occupations in the FRG in 1981, in percent.

Task	Fitter	Mechanic	Toolmaker	Electrician	Assembler
Machine adjusting/ maintaining	18.4	12.0	18.9	22.7	21.2
Producing	45.2	22.2	64.4	34.6	68.3
Repairing	31.9	56.3	9.8	33.2	5.3
Other	4.5	9.5	6.9	9.5	5.2

Source: Statistisches Bundesamt Wiesbaden, 1982.

Tasks describe the actual work content of a job (Warnken, 1986). Nevertheless, the subdivision of the labor force by occupation has the advantage that it establishes a direct connection to educational planning. Hence, to estimate the influence of technological progress on the level and the structure of employment, and to achieve a link with educational planning, it would be very useful to have data on the occupational structure by sectors and by tasks, as well as the task composition by sectors (see Chapter 10 by Yamada). These statements are illustrated below for the FRG.

In 1982 the FRG machine-building industry employed 1,445,100 workers. Among these were 335,900 fitters; 143,800 mechanics; 59,400 toolmakers; 63,400 electricians; and 88,100 assemblers.[2] The tasks of the employees performed in these occupations differ considerably. *Table 12.1* shows the task composition of employees by different occupations in the FRG. The table shows that an employee with a given occupation may perform several distinct types of tasks.[3] As the different tasks are differentially influenced by technological progress, the occupation-by-task structure of the labor force is very important for purposes of analysis of the employment effects of CIM.

Furthermore, the mix of tasks that an employees with a given occupation (or education) performs in the different sectors can vary considerably. Thus, it is important to consider the occupation-by-task structure on a sectoral basis. All of the desired data are not available, but at least the task-by-sector structure of the labor force could be investigated. In the following, this structure is given for the total manufacturing sector of the FRG (*Table 12.2*).

In general, one can assume that tasks are more relevant to the labor impacts of progress than occupations. However, task-by-occupation matrices as well as task-by-sector matrices are available for only a few countries,

Table 12.2. The task composition of the manufacturing sector in the FRG in 1982, in percent.

Task	Manufacturing sector
Machine adjusting/maintaining	12.9
Producing	36.7
Repairing	10.3
Carry-on trade	5.8
Office work	13.6
Planning/research	6.5
Managing	5.0
Service	7.7
Securing	0.6
Education	0.7
Other	0.2

Source: Statistisches Bundesamt Wiesbaden, 1982.

whereas the occupation-by-sector matrices are available for many more countries. Furthermore, an international comparison of occupational categories is relatively practicable, whereas the comparison of task categories is more difficult.

The employment effects of CIM can only be addressed in the labor matrix in terms of the classifications used. This matrix can only reflect such occupations that are established and well documented in official statistical sources. But CIM – like any other basic innovation – will also create a demand for new occupations, such as NC-machine programmers, CAD designers, and robot mechanics. The basic feature of these new occupations is the use of computer control (Tchijov, 1988). One can easily understand that official data sources – especially census data as the main source for estimating labor matrices – cannot reflect the new qualification requirements for new occupations until they are revealed by the passage of time.

However, in this context we note that only a few innovations in the past stimulated the introduction of new occupations. Dörfer *et al.* (1977) analyzed 40 basic innovations which occurred during the period 1680–1970. Only 13 of these had created new occupations (*Figure 12.1*). They emphasized that the share of employees working in the new occupations is often overestimated. For example, only 20% of all employees in the USA are working in occupations that have been created in this century and only 3% of all current US occupations have been created since 1950. Dörfer *et al.* (1977) concluded that even with an accelerated pace of change in the year 2000 in the FRG only 10% to 12% of all employees will be working in occupations that were unknown at the end of the 1970s. The demand for employees with new occupations could be met by graduates from vocational schools and universities. Changes in the occupational structure of the labor force mainly caused by technological progress can only be partially investigated with the help of labor matrices, since only currently existing occupations are reflected. The estimation of demand for new skills and professions induced by CIM is a crucial problem. It cannot be solved using the labor matrices described earlier.

The occupational effects of CIM are sector-specific. But in each column an "average" technology of the corresponding (more or less aggregated) production process is reflected. An innovation like CIM causes exceptional effects that can hardly be reflected adequately in "average" technologies. This fact is illustrated by the following example, based on the US labor matrix. Unquestionably the communication-equipment industry is an innovative sector. In the labor matrix the communication-equipment industry is part of the sector manufacture of electrical machinery apparatus, appliances, and supplies (ISIC code 383). *Table 12.3* shows that the "average" labor-input structure of this sector differs considerably from the labor-input structure of the more aggregated communication-equipment sector.

In the US matrix employees are usually assigned according to their highest degree of professional education. The qualification an employee has gained in addition to his or her other level of formal education is not reflected in the occupational figures. This problem is important because an employee's qualification depends – in no small degree – on the qualification behind the so-called highest educational level. In this connection we should consider the important role of reeducation and retraining in innovations like CIM. Mastering these innovations could lead to a second (or third) qualification for many employees (e.g., in the field of programming).

To illustrate this problem, some analytical results concerning the so-called secondary educational level in East Germany are of interest. In the early 1980s nearly half of all East German employees possessing a university degree also had another educational degree. Approximately 28.2% of all employees with a university degree also had a technical qualification. Some 16% of all employees with a university degree were also qualified as skilled workers. Of all employees with a technical qualification, 43.9% were also qualified as skilled workers and 6% were qualified as supervisors. The share of employees with two or more qualifications increased from 11.2% in 1971 to 13.1% in 1981 (Lötsch, 1985; Brautzsch, 1990).

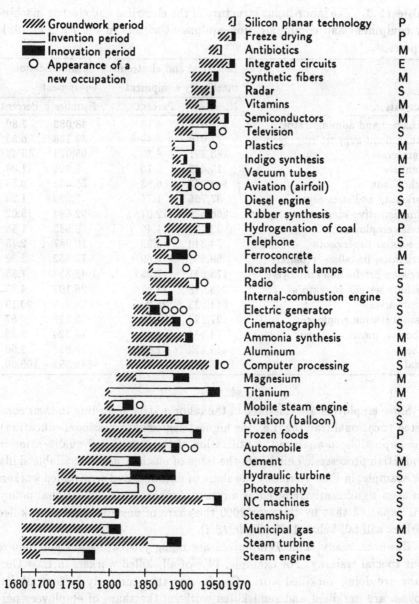

Figure 12.1. Innovations and new occupations: E, element invention; M, material invention; P, process invention; S, system invention. Source: Dörfer *et al.*, 1977.

Table 12.3. The labor-input structure of the electrical and electric machinery equipment and communication-equipment sectors in the FRG in 1981, in percent.

Occupation	Electrical and electronic machinery equipment		Communication equipment	
	Number	Percent	Number	Percent
Managers and administrators	135,313	6.13	48,083	7.80
Management-support employees	77,016	3.49	39,128	6.35
Engineers	183,291	8.30	95,074	15.42
Scientists	15,472	0.70	6,401	1.04
Technicians	144,263	6.53	58,802	9.54
Marketing and sales reps.	37,765	1.71	7,390	1.20
Administrative-support employees	266,538	12.07	92,584	15.02
Service employees	32,192	1.46	9,533	1.55
Blue-collar supervisors	74,840	3.39	16,987	2.75
Mechanics, installers, repairers	59,351	2.69	12,862	2.09
Precision production employees	175,541	7.95	45,334	7.35
Machine setters, operators	243,078	11.01	26,197	4.25
Handworkers	644,833	28.20	124,499	20.19
Transportation employees	27,190	1.23	3,512	0.57
Laborers, material movers	74,784	3.39	14,324	2.32
Other	16,822	1.75	15,888	2.56
Total	2,208,289	100.00	616,598	100.00

Since employees are classified in the labor matrix according to their completed vocational training (i.e., the highest degree of professional education), it is impossible to analyze the utilization of the employees' qualification in productive processes. This raises the issue of efficient use of available skills. For example, in East Germany the share of unskilled and semiskilled workers has been significantly reduced as a result of a consistent educational policy. It is expected that by the year 2000 the share of unskilled and semiskilled workers will fall below 11% (*Table 12.4*).

However, many skilled employees are doing jobs which do not require their special training. For example, 14% of all skilled workers in East Germany are doing unskilled work. Considering that currently 17% of all employees are unskilled and semiskilled workers, the share of employees performing unskilled work actually amounts to more than 30% (Schäfer and Wahse, 1981). Only full utilization of employees according to their qualification level makes it possible to use efficiently the enormous educational

Table 12.4. Qualification structure of employees in East Germany, in percent.

Year	Semi/ unskilled worker	Supervisor/ skilled worker	Technical school graduate	University graduate
1960	64.1	30.3	3.6	2.0
1970	41.3	47.3	7.2	4.2
1982	17.5	62.8	12.7	7.0
2000	10.9	63.0	15.0	11.1

Sources: Schäfer and Wahse, 1984; Lötsch, 1985.

Table 12.5. Changes in labor-input coefficients in East Germany between 1971 and 1981 (employees according to qualification per 1 million DM value-added), in percent.

Sector	Semi/ unskilled worker	Supervisor	Skilled worker	Technical school graduate	University graduate	Total
Industry	−11.13	−4.48	−1.20	0.18	0.42	−16.21
Construction	−8.42	−3.57	−0.78	0.75	0.69	−11.33
Agriculture	−16.29	4.40	−0.21	0.87	0.37	−10.86
Transport	−17.10	−1.73	−0.33	0.59	0.66	−17.91
Trade	−18.27	−4.84	−0.78	1.18	0.32	−22.39
Other	−8.54	−1.43	−0.38	−1.89	−0.04	−12.28
Total	−14.50	−4.42	−0.94	1.04	0.34	−18.48

Source: Personal computations using census data and the Statistical Yearbooks of the GDR.

expenditures. *Table 12.5* shows the distribution of the educational funds in East Germany from 1971 to 1981.

Important changes can also be seen in the labor-input structure over a period of 10 years. For instance, changes in the labor-input coefficients in East Germany between 1971 and 1981 are shown in *Table 12.5*. An evolutionary development in the labor-input structure of the sectors can be observed. In all sectors the total labor demand per unit of value-added decreased considerably. There is a significant decrease of specific labor demand for unskilled and semiskilled workers in all sectors. The specific demand for skilled workers shows a different picture. In the agricultural sector there was a demand for skilled workers. However, in the other sectors there is a more or less significant decrease of the labor-input coefficients. With the exception of the sectors labeled other, the demand for university and technical graduates per unit of value-added increased in all sectors.

Absolute changes in the sectoral labor force by occupation (*Table 12.6*) are mainly influenced by the development of labor-input coefficients. The influence of changes in the output structure was very low. It is evident that the changes in labor-input coefficients are not only the result of technological progress. Unquestionably, since the 1970s, microelectronic-based technologies have realized an important diffusion and have started to have a significant influence on the labor-input structure at the macroeconomic level.

But the labor-input structure of the different economic sectors is influenced by various factors in addition to technological change, such as organization, the educational structure (especially vocational training), national policies, and social objectives, (Knabe, 1982). Hence, the share of the changed occupational structure that is directly attributable to technological progress cannot be determined at this time.

12.3 Changes in the Occupational Structures of Different Sectors

The data base provides many possibilities of analyzing changes in the occupational structure of the countries investigated. This aims at identifying both similarities and differences in the dynamic changes of occupational structures of the metalworking industries in these countries. Hence, the analysis now concentrates on the occupational groups that seem to be important to the rapid diffusion of CIM. It should be emphasized that the analysis does not exhaust the broad analytical possibilities contained within the computed data base.

It is necessary to identify the indicators that seem to be suitable for investigating changes in occupational structures of the different sectors, on the one hand, and that are contained in the labor matrices, on the other hand. Such indicators include the following:

(1) Changes in the shares of the sectors within the total number of employees.
(2) Changes in the shares of different occupations within the total number of employees.
(3) Changes in the occupational structures of the different sectors.
(4) Changes in the sectoral distribution of employees by occupations.

Table 12.6. Changes in labor force by occupation in the FRG between 1950 and 1982, in percent.

ISCO	ISIC 37				ISIC 382				ISIC 383			
	1950	1961	1970	1982	1950	1961	1970	1982	1950	1961	1970	1982
7-21	10.30	10.02	9.82	3.84	0.11	0.24	0.18	0.31	0.53	0.35	0.40	0.09
7-24	13.06	15.08	11.47	7.10	2.98	6.93	5.08	3.52	3.19	7.06	6.83	1.91
7-26	0.98	1.99	1.78	0.67	0.00	0.68	0.34	3.55	3.56	1.42	1.94	0.25
8-31	11.35	5.32	3.32	1.97	1.80	1.06	0.74	1.30	1.06	0.27	0.28	0.03
8-32	0.68	1.19		3.07	1.79	2.37		2.51	3.48	2.64		2.00
8-33	0.03	0.60	1.65	0.71	0.33	0.51	2.17	0.92	0.71	0.94	2.48	1.09
8-34	5.37	6.46	0.47	4.65	10.32	9.29	0.47	4.53	7.72	4.61	0.82	1.53
8-35			6.48	1.83			7.81	1.14			4.33	0.57
8-39	1.82	2.34	1.20	0.68	0.81	0.52		0.68	5.61	3.93	4.49	0.65
8-41	23.29	15.47	14.11	14.45	25.03	18.53	15.27	14.26	7.08	5.00	5.24	3.17
8-42	0.86	1.06	1.35	3.08	12.43	11.61	12.32	6.93	10.50	6.75	4.94	6.88
8-51	0.08	0.17	0.28	0.24	0.17	0.32	0.52	0.65	3.10	4.88	5.68	5.36
8-53	0.02	0.06	0.20	0.10	0.34	0.32	0.94	0.11	4.41	7.03		4.34
8-55	1.50	1.93	2.34	1.83	1.58	1.78	2.23	2.66	7.74	7.05	8.53	11.35
8-71	0.48	0.52	0.84	0.71	2.52	3.58	4.82	3.84	0.95	0.65	0.63	0.25
8-72	1.61	1.85	2.35	2.51	2.74	3.13	3.26	2.89	2.21	2.45	2.54	1.95
9-7	7.71	6.36	6.12	3.33	5.15	3.97	3.31	1.88	7.62	4.03	3.02	1.99

The elements of these matrices reflect the number of employees required by occupation in the different sectors. The labor force (by occupation) required in the different sectors depends on many factors. These include the development of the output of the sectors and the development of input of labor (by occupation) per unit output of the sectors, i.e., the labor intensity. The main factor influencing the output structure of the economy as well as the labor intensity is technological progress.

To compare labor intensity matrices at the international level it would be necessary to have data concerning the output of the different sectors in terms of the ISIC classification. Moreover, the sectoral outputs in different countries and at different points in time have to be comparable with regard to prices and exchange rates. Such data were not available at the sectoral level. Hence, the labor intensity could not be computed in most cases.

However, to compare changes in the occupational structures in different countries the analysis has to be limited to the labor-input structures in the different sectors, and the sectoral distribution of the employees by occupations. The labor-input structure means the distribution of employees according to the occupations in a certain sector in percentage terms, i.e., the shares of the different occupational groups within the total number of employees of a certain sector. The distribution of employees by occupation in the sectors refers to the distribution of the employees performing a certain occupation across different sectors, i.e., the shares of the different sectors within the total number of employees characterized by a certain occupation.

Additionally, it should be noted that an economically feasible interpretation of the sectoral distribution of employees by occupations can be given using the concept of professional *flexibility* or adaptability. The term professional *flexibility* means the ability of the employees to perform different tasks. It depends on qualification level and professional experience, among other factors. The flexibility potential of employees differs considerably.

To estimate the flexibility potential of a given occupation it is possible to analyze the sectoral distribution of the employees in that occupation. If these employees are distributed relatively uniformly across sectors, then the flexibility potential is high. This is important because employees who are displaced as a consequence of technological progress, or other factors, can be allocated to other sectors without professional retraining. Occupations that are concentrated on some sectors indicate a high degree of specialization or a low flexibility potential. If these employees were replaced, the chance to shift them to other sectors without professional retraining would be low. Reeducation here seems to be unavoidable.

It should be emphasized that the possibility to reallocate people to other sectors does not only depend on professional flexibility. For example, the geographical, social, and family factors also have an important influence on this issue. However, investigation of the sectoral distribution of employees possessing the skills for a given occupation allows some conclusions concerning the vulnerability of the jobs caused by certain innovations.

Table 12.6 shows the sectoral distribution of employees in several occupations. A relatively homogeneous distribution across sectors suggests the ability to fulfill different tasks in different branches of the economy. The chance to move to other branches without professional reeducation is relatively high. By contrast, some occupations exist only in limited sectors. The demand for these employees depends mainly on the development of those sectors.

In some publications (see, for example, Schäfer *et al.*, 1982) it has been suggested that to quantify the professional flexibility potential, the use of concentration measures (for example, Gini coefficient and Lorenz curve) could be used. This could be very useful to investigate the advantages of such indicators. This refers especially to the ability of these indicators to reflect the vulnerability of jobs and occupations created by technological progress.

Changes in the occupational structures of the different sectors could also be given such an economic feasibility interpretation. Dostal (1982) studied four hypotheses concerning changes in the qualification requirements that are caused by technological progress. It is claimed that technological progress generally leads to higher qualification, de-qualification, polarization, and de-polarization (i.e., equalization).

Assuming that a proper one-dimensional indicator to measure the qualification is adequate, these different hypotheses are illustrated in *Figure 12.2*. To verify these hypotheses it would be useful to investigate changes in the labor-input structures of the different sectors. The changes in the shares of different occupations or skill levels within the total number of employees in a certain sector could be interpreted as an indicator of an increasing polarization, a de-qualification, etc., in the sector. Furthermore, it could be useful to prove if the tendencies of qualification requirements can be identified using concentration measures (e.g., Lorenz curve).

The description and analysis of changes in occupational structures lead to the question, How far can changes in the occupational structures be traced back to technological progress? The changes in the occupational structures

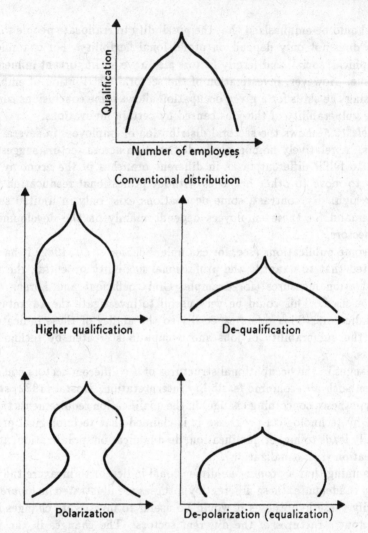

Figure 12.2. Hypotheses concerning the development of the qualification structure. Source: Dostal, 1982.

of the different sectors are caused by several factors. Unquestionably, technological progress is one of the most important, though it is impossible to trace back changes in the occupational structure to technological progress alone. Furthermore, it is hardly possible to isolate the influence of technological progress (or of a single technology) on the occupational structure of a given sector. However, the changes observed in the occupational structure are the result of the influence of many factors in the past. These observations can yield useful clues concerning the possible directions of changes in the occupational structure in the future. It has to be assumed that the same complex of factors will influence changes in the future.

The characteristics of the metalworking industries are given in *Table 12.7*. Up to now it has been very difficult to identify the occupational groups in the occupation-by-sector matrices that are most likely to be influenced by CIM. Currently, it is a reasonable assumption that engineers and technicians and some metalworking employees will be especially affected by this technology. Accordingly, the analysis of occupational structures in the metalworking industries is limited to these occupational groups, given in *Table 12.8*. The restriction seems justified because the shares of the employees in these categories typically amount to between 60% and 70% of the total number of employees of the metalworking industries.

In a cross-country comparative study of occupational structures and their dynamics it should be emphasized that there are some differences between the sectoral and occupational categories of the ISIC and ISCO, and the categories used in the estimated national labor matrices. A review of these differences is given in Brautzsch (1988).

This analysis of the metalworking industries begins with an investigation of the shares of the manufacturing industries within the total number of employees. The observed trends in the shares of the manufacturing, as well as the metalworking industries within the total number of employees, can be summarized as follows:[4]

(1) There is a widespread tendency for the number of employees in the manufacturing industries to decrease as a percentage of the total number of employees. In West Germany this share had increased significantly from 1950 (29.8%) to 1970 (33.5%).

(2) The shares of the manufacturing industries within the total number of employees differ considerably between countries. Whereas this share amounted to 33.5% in West Germany (1982), in the USA it was only 19.9% (1984), and in the Netherlands only 19.3% (1985). Differences in the output structure of the economies, as well as differences in the sectoral labor productivities, explain this spread.

(3) In Sweden and Austria the shares of the metalworking industries in the total number of employees increased. The opposite tendency is observed in West Germany. The shares of the metalworking industries within the

Table 12.7. Classification of basic metal industries and metal-processing industries according to ISIC (ISIC, 1968).

ISIC	Category	Description of activities
37	Basic metal industries	
371	Iron and steel basic industries	Manufacture of primary iron and steel products, consisting of all processes from smelting in blast furnaces to the semifinished stage in rolling mills and foundries
372	Nonferrous metal industries	Manufacture of primary nonferrous metal products, consisting of all processes: smelting, alloying and refining, rolling and drawing, and founding and casting
38	Manufacture of fabricated metal products, machinery, and equipments	
381	Manufacture of fabricated metal products, except machinery and equipment	Manufacture of cutlery, hand tools, and general hardware; furniture and fixtures (primarily of metal); structural metal products, fabricated metal products except machinery and equipment not elsewhere classified
382	Manufacture of machinery except electrical	Manufacture of engines and turbines; agricultural machinery and equipment; metalworking and woodworking machinery; special industrial machinery and equipment; and office computing and accounting machinery
383	Manufacture of electrical machinery	Manufacture of electrical industrial machinery and apparatus; radio, television, and communication equipment and apparatus; and electrical appliances and houseware
384	Manufacture of transport equipment	Ship building and manufacture of railroad equipment, motor vehicles, motorcycles, bicycles, and aircraft
385	Manufacture of professional and scientific, measuring and controlling equipment, photographic and optical	Manufacture of professional and scientific measuring and controlling equipment; photographic and optical goods; and watches and clocks

Table 12.8. Occupational groups considered in the analysis.

ISCO	Category
0-2/0-3	Architects, engineers, and related technicians
7-2	Metal processers
8-3	Blacksmiths, toolmakers, and machine tool operators
8-4	Machinery fitters, machine assemblers, and precision instrument makers
8-5	Electrical fitters and related electrical and electronics workers
8-7	Plumbers, welders, and sheet metal and structural metal preparers and erectors
9-7	Material-handling and related equipment operators, dockers, and freight handlers

total number of employees in the manufacturing sector also differ considerably between the investigated countries, (e.g., 39.2% in the Netherlands and 48.4% in the USA).

(4) In the 1970s, changes also took place with regard to the distribution of employees within the metalworking sector. Unfortunately, the metalworking industries could not be subdivided for all countries studied. However, figures for Austria, Sweden, and the Netherlands verify this shift. Whereas in Sweden and Austria the shares of the basic metal industries decreased, they increased in West Germany during the 1970s. The shares of the subsector manufacture of fabricated metal products, machinery, and equipment also show divergent tendencies (see Brautzsch, 1988, pp. 34–35).

Several changes can also be observed with regard to the shares of occupational groups that are likely to be affected by CIM (*Table 12.9*). The shares of engineers and technicians increased significantly in all the investigated countries; in 1980, in Sweden this share amounted to 6.86%, whereas in Austria it was 3.19% (1981), and in the USA 3.21% (1984). With regard to typical metalworking production workers, increased shares are observed as a percentage of the total number of employees in West Germany and Austria. Significant differences concerning the shares of these occupational groups between countries also exist. The figures range from 11.43% in the Netherlands (1985) and 11.55% in the USA (1984) to 18.69% in West Germany (1970).

The analysis of the shares of the manufacturing industries as well as of the metalworking industries within the occupational groups labeled architects,

Table 12.9. The shares of selected occupational groups within the total number of employees, in percent.

Country	Year	ISCO code 0-1, 0-2, 0-3	ISCO code 7-2, 8-3, 8-4, 8-5, 8-7, 9-7
Finland	1980	4.59	13.74
USA	1984	3.21	11.55
Netherlands	1985	4.14	11.43
FRG	1950	1.82	14.07
	1961	3.18	17.00
	1970	4.67	18.69
	1982	5.58	15.00
Sweden	1970	6.81	16.17
	1980	6.86	14.58
Austria	1971	2.77	14.63
	1981	3.19	15.54

Table 12.10. Subdivision of the occupational group architects, engineers, and related technicians.

ISCO	Category
0-21	Architects and town planners
0-22	Civil engineers
0-23	Electrical and electronics engineers
0-24	Mechanical engineers
0-25	Chemical engineers
0-26	Metallurgists
0-27	Mining engineers
0-28	Industrial engineers
0-29	Engineers not classified
0-31	Surveyors
0-32	Drafting engineers
0-33	Civil engineering technicians
0-34	Electrical and electronics engineering technicians
0-35	Mechanical engineering technicians
0-36	Chemical engineering technicians
0-37	Metallurgical technicians
0-38	Mining technicians
0-39	Engineering technicians not classified

engineers, and related technicians in *Table 12.10* leads to several conclusions (Brautzsch, 1988, pp. 38–41). It is no surprise that mechanical engineers, chemical engineers, metallurgists, and industrial engineers are concentrated in the manufacturing sector. These groups are especially concentrated in the metalworking industries. The sectoral distribution of the engineering specialties differs considerably among countries. For example, whereas the share of the metalworking industries within the total number of electrical and electronics engineers amounted to 29.4% in Finland (1980), 66.5% of electrical and electronics engineers were employed in these industries in the USA (1984).

An analysis of the shares of the metalworking industries within selected occupational groups of production and related workers, transport equipment operators, and laborers (*Table 12.11*) shows special features concerning the sectoral distribution of these occupational groups.

(1) The shares of the metalworking industries within the occupational group production and related workers and transport equipment operators and laborers are considerably higher than the shares of these sectors within the total number of engineers and technicians (Brautzsch, 1988, pp. 44–46). The *flexibility* of engineers and technicians is indisputably higher. As mentioned above, the concentration of production workers on a small number of sectors means that the demand on employees performing these occupations depends mainly on the growth of these sectors.

(2) In some cases the shares of the metalworking industries within the total number of production workers decreased considerably in the 1970s (Brautzsch, 1988, Table 35). Only in a small number of cases did the concentration of these occupational groups in the metalworking industries increase. This fact can also be verified using the West German data on the shares of the metalworking industries within the total number of production workers between 1950 and 1982.

(3) Significant differences are found concerning the shares of the different metalworking industries within the total number of production workers. For example, in the USA the share of the metalworking industries within the total number of metal processors amounted to 50.3% (1984), whereas this figure amounted to 93.0% in Austria (1981) and to 96.2% in the Netherlands (1985). The share of the metalworking industries within the total number of blacksmiths, toolmakers, and machine tool operators amounted to 20.0% in the USA, 68.3% in Austria, and 34.6% in the

Table 12.11. Subdivision of selected occupational groups.

ISCO	Category
	Metal processors
7–21	Metal smelting, converting, and refining furnace workers
7–22	Metal rolling-mill workers
7–23	Metal melters and reheaters
7–24	Metal casters
7–25	Metal molders and coremakers
7–26	Metal annealers, temperers, and case-hardeners
7–27	Metal drawers and extruders
7–28	Metal platers and coaters
7–29	Metal processers not classified
	Blacksmiths, toolmakers, and machine tool operators
8–31	Blacksmith, hammersmith, and forging-press operators
8–32	Toolmakers, metal patternmakers, and metal markers
8–33	Machine tool setter operators
8–34	Machine tool operators
8–35	Metal grinders, polishers, and tool sharpeners
8–39	Blacksmith, toolmakers, and machine tool operators not classified
	Machinery fitters, machine assemblers, and precision instrument makers (except electrical)
8–41	Machinery fitters and machine assemblers
8–42	Watch, clock, and precision instrument makers
8–43	Motor vehicle mechanics
8–44	Aircraft engine mechanics
8–49	Machinery fitters, machine assemblers, and precision instrument makers (except electrical) not classified
	Electrical Fitters and related electrical and electronics workers
8–51	Electrical fitters
8–52	Electronics fitters
8–53	Electrical and electronic equipment assemblers
8–54	Radio and television mechanics
8–55	Electrical installers
8–56	Telephone and telegraph installers
8–57	Electric cable jointers
8–59	Electrical fitters and related electronics workers not classified
	Plumbers, welders, and sheet metal and structural metal preparers
8–71	Plumbers and pipe fitters
8–72	Welders and flame-cutters
8–73	Sheet metal workers
8–74	Structural metal preparers and erectors

Netherlands. These differences can be traced to the different output structures or to differing labor productivity or to both factors.

(4) Significant differences can also be observed with regard to the shares of the different metalworking industries within the total number of production workers. The shares of the metalworking industries within the total number of production workers differ considerably between the countries investigated (Brautzsch, 1988, Part II).

These statements can be verified by studying the shares of production workers and transport equipment operators, laborers, and technicians within the total number of employees in the metalworking industries in the FRG.

Notes

[1] The terms "occupation-by-sector matrix" and "labor matrix" are used synonymously in this chapter.

[2] The figures refer to the sector and occupational classification used in the FRG.

[3] Only the aggregated task-by-sector matrix of the FRG was available for the author's investigations. *Table 12.1* shows that the task composition of the sectors differs considerably and can be verified also with aggregated data.

[4] The figures used in this comparison are estimated using the computed labor matrices. With regard to the sectoral distribution of the employees, certain differences with respect to published data cannot be excluded in some cases.

References

Ayres, R.U., 1986, *Socio-economic Impacts of Robotics*, Carnegie-Mellon University, Pittsburgh, PA.

Ayres, R.U., H.-U. Brautzsch, and S. Mori, 1987, *Computer Integrated Manufacturing and Employment: Methodological Problems of Estimating the Employment Effects of CIM Application at the Macroeconomic Level*, WP-87-19, IIASA, Laxenburg, Austria.

Brautzsch, H.-U., 1988, The Occupational Structure by Sectors in Selected Countries – Final Report, Unpublished study.

Brautzsch, H.-U., 1990, Personal estimates using census data.

Brooks, H., 1985, "Automation Technology and Employment," *ATAS Bulletin*, UNO, New York,, NY.

Dörfer, G., W. Dostal, K. Kostner, M. Lahner, and E. Ulrich, 1977, *Technik und Arbeitsmarkt: Auswirkungen technischer Änderungen auf Arbeitskräfte*, Report, Quint 6/77, Nuremberg, Germany.

Dostal, W., 1982, *Bildung und Beschäftigung im technischen Wandel*, Beitrag 65, Nuremberg, Germany.

Hellwig, K., 1988, Die Matrizen der Verteilung der Beschäftigten nach Berufen und Wirtschaftssektoren in der BRD und in Schweden, Methodische Probleme bei der Ermittlung international vergleichbarer Matrizen des Arbeitskräfteeinsatzes, Diplomarbeit, Hochschule für Ökonomie "Bruno Leuschner," Berlin, Germany.

ILO, 1986, The Revision of the International Classification of Occupations, Background of first draft proposals, Document prepared for the Round Table Discussion, 24–28 November, Geneva, Switzerland.

ISCO, 1968, *International Standard Classification of Occupations*, ILO, Geneva, Switzerland.

ISIC, 1968, *International Standard Industrial Classification of all Economic Activities*, UN Statistical Papers, Series M, No. 4, Rev. 2, New York, NY.

Jacob, K., B. Stiehler, 1985, Entwicklung und Nutzung der Disponibilität der Werktätigen – ein Beitrag zur Erschließung qualitativer Leistungsreserven der Ressource Arbeitskraft im Sozialismus, in Wissenschaftliche Zeitschrift der Hochschule für Ökonomie "Bruno Leuschner", Berlin, Vol. 2.

Kessler, C., 1988, Die Matrix der Verteilung der Beschäftigten nach Berufen und Wirtschaftssektoren in Finnland, Diplomarbeit, Hochschule für Ökonomie "Bruno Leuschner," Berlin, Germany.

Knabe, M., 1982, Grundlagen und Tendenzen der Berufsstrukturenentwicklung der Facharbeiter in der DDR, Habilitationsschrift, Hochschule für Ökonomie "Bruno Leuschner," Berlin, Germany.

Lötsch, I., 1985, Zur Entwicklung des Bildungs- und Qualifikationsniveaus in der DDR, in *Jahrbuch für Soziologie und Sozialpolitik 1985*, Akademie-Verlag, Berlin, Germany.

Norwood, J., 1986, US Department of Labor, Personal communication.

Schäfer, R., and J. Wahse, 1981, Zur qualitativen Entwicklung des gesellschaftlichen Arbeitsvermögens und seiner Nutzung, *Wirtschaftswissenschaft* 29(1981)4.

Schäfer, R., and J. Wahse, 1984, Die Erhöhung des Veredlungsgrades der Produktion als Grundrichtung zur Erschließung von Reserven des gesellschaftlichen Arbeitsvermögens, *Wirtschaftswissenschaft* 32(1984)7.

Schäfer, R., and J. Wahse, 1986, Zum Einfluß des wissenschaftlich-technischen Fortschritts auf die Struktur der Arbeitskräfte und ihre Qualifikation, *Wirtschaftswissenschaft* 34(1986)8.

Schäfer, R., C. Schmidt, and J. Wahse, 1982, *Disponibilität–Mobilität–Fluktuation*, Akademie-Verlag, Berlin, Germany.

Statistisches Bundesamt Wiesbaden, 1982, "Erwerbstätige im April 1982 nach Berufsordnungen, -gruppen, -ausschnitten, -bereichen und Art der überwiegend ausgeübten Tätigkeit."

Statistisches Bundesamt Wiesbaden, 1982, "Erwerbstätige nach Wirtschaftsabteilungen und Art der überwiegend ausgeübten Tätigkeit."

Tchijov, J., 1988, *CIM Application: Some Socioeconomic Aspects* WP-88-30, IIASA, Laxenburg, Austria.

Warnken, J., 1986, Zur Entwicklung der 'internen' Anpassungsfähigkeit der Berufe bis zum Jahre 2000, in MITTAB 1986, Heft 1.

Weidig, R., 1986, Soziale Triebkräfte ökonomischen Wachstums, in *Materialien des 4. Kongresses der marzistisch-leninistischen Soziologie in der DDR*, 26.–28.3.1985, Dietz Verlag, Berlin, Germany.

Zymelman, M., 1980, *Occupational Structures of Industries*, World Bank, Washington, DC.

Zymelman, M., 1986, Personal communication.

Chapter 13

Automation-Based Technologies in the USSR

Oleg Adamovic, Nadezda Mamysheva, and Sergei Perminov

Since the 1950s, the influence of technological changes on economic activity has been intensively discussed. Solow (1957) contributed one of the first works on the topic.

In the 1970s, the economies of the most advanced countries experienced fundamental changes. Among these were the development of the science-intensive areas, such as computers, microelectronics, and robots. These facilitated the dissemination of advanced manufacturing technologies in different branches of the economy. In particular, the newer technologies have been applied in the machinery industries. In the 1970s and 1980s, these sectors became the main arena of competition of international suppliers (Keyworth, 1982; Babintzev, 1989; Gorokhov, 1987) and contributed to the current interest in CIM technologies (Ayres, 1987a, 1987b; Tchijov, 1988a, 1988b; Tani, 1988; Alabyan and Ranta, 1988).

Government research and development policy is the driving force for the development in many areas of technological progress. For the most advanced countries the main purpose of R&D policy is to stimulate innovation. Innovation policy aims at creating appropriate conditions for innovation and diffusion. This consists not only in developing new technology, but also in averting the negative consequences of too rapid or too wide dissemination on society itself, with regard to human resources, the environment, natural resources, and so on.

A long-term forecast of the influence of different innovation/diffusion policy alternatives must be carried out before the decision-making process is completed.

This chapter is concerned with the impact of innovation policy on the diffusion of automation-based technology in the USSR, including the influence of different innovation strategies on economic growth. Consideration is given to the key constituents of innovation policy, to investment, and to the distribution of technical and labor resources. To accomplish this task the approach taken by Adamovic and Perminov (1988) was further developed.

13.1   Meso-Models in the Analysis of Innovation Policy in Automation

Emphasizing the role of social adaptability of CIM technologies as a matter of great concern, Tchijov (1989) remarks: "Sometimes social resistance or managerial inadequacy is much more difficult to overcome than technical problems in the diffusion process." As far as the USSR is concerned, it is necessary to note the pervasiveness of managerial/organizational resistance to the introduction of CIM technologies. Although numerically controlled machines and industrial robots are widely used in the USSR, they are mainly used as stand-alone equipment, i.e., without being linked to a system.

In general, automation influences structural change in several ways. First, it forms the historical background of structural changes. Second, it evolves rapidly in response to exogenous technological changes. Third, it makes great contributions to technological progress as such. Almost all "bottlenecks" limiting CIM introduction are due to insufficient progress in automation-based technologies. Current concerns include the development and production of current electronic components and the industrialization of software production.

Automated production is, itself, an important part of the innovative framework. Of critical importance for the acceptance of CIM technologies is the fact that comprehensive automation requires the linkage of machines within the system, as well as the flexibility of production which results.

Factors inhibiting the introduction of fully automated technologies include the following:

(1) The technical experience and knowledge of the labor force are often limited or isolated or both. These circumstances call for training and retraining of workers.
(2) The introduction of new technology demands additional expenditures. This results in the resistance of the fundamental production processes to innovation diffusion. As practical experience reveals, the most effective way to overcome stagnation and increase production levels is through competition.

However, difficulties arise not only when selecting new technologies but also when making responsible decisions. The nature of the basic technology alternatives preordains further innovative activity. It is necessary to estimate the all-around effectiveness of different production improvements.

One way to ameliorate the effect of these factors is to introduce government policy. Innovation policy aims at accelerating the reconstruction of the technostructure. One way to stimulate investment in the vanguard industries – those that are science-intensive and able to compete – is to create special "high-technology" zones in manufacturing. The Soviet government has a dominant influence on production activity through fiscal, price, tax, and other types of policies. To elaborate government strategy on innovation a multivariate long-term forecast of the new technology diffusion, in the framework of fundamental production, is needed.

To forecast the influence of the innovation process on economic growth, the meso-model was developed. The model is a neoclassic economic growth model. The differentiation of an economy by technology "clusters" is the main new element of the model. These clusters are identified by a wide range of characteristics: both qualitative and technological. These differences are mainly due to the distinction in life-cycle phases of basic technological ideas (Ayres, 1987a, 1987b). Among the most important consequences resulting from new automation-based technology dissemination is the growing share of fully automated enterprises. The movement toward a highly automated level of total production characterizes the speed of technological progress. The different technology clusters compete for all production resources. If new technologies win, a replacement process will take place. On the other hand, these clusters coexist in time as a whole economic system because of their complementarity. The new technologies can reveal their advantages only in a corresponding economic environment and in an appropriate, already developed, technostructure.

Questions arise: Which investment strategy will accelerate the progressive new technologies diffusion? Which strategy will prevent the negative consequences of technology diffusion?

In accordance with the assumptions of the model, there are four groups of main control factors of changes in technostructures:

(1) The allocation of net investment in fixed capital for expansion and renovation is based upon advanced techniques and technologies. The intensity of expansion and renovation processes in different clusters is influenced by altering the allocation pattern. Varying the pattern, one can manage the technostructure of fixed capital and, as a consequence, form and select appropriate strategies for production development at different periods of time.

(2) Changes in the rate of capital accumulation can strengthen the investment reallocation strategy. This, together with the factors of the first group, characterizes the possibility of investment policy to change the technostructure of fixed capital in the economy.

(3) The third factor consists of the following:

- Changes of fixed capital liquidation (depreciation) as a consequence of evolutionary shifts in the duration of technological life cycles, e.g., shortening life cycles of information technologies.
- Changes in the rate of renovation and reconstruction of equipment.
- Changes in the rate of capital flow to higher stages in the value-added sequence.
- Changes of fixed capital depreciation in the process of transfer to other levels.

By and large, this group of factors characterizes the influence of improvements in the technical "quality" of applied capital upon a common increase in automation levels.

(4) The process of technological transformation for all clusters is a driving force of development. The introduction of new technologies and the development of older ones are usually accompanied by changes in the growth rate of the capital–labor ratio. One question is whether higher levels of automation result in the release of labor in all cases. As a general rule it does, particularly with regard to unskilled labor. But with regard to some production processes with technological progress an opposite trend may occur. For example, at some enterprises, where sophisticated

supercomputers are substituted for personal computers, along with an enlargement in the scale of operations and employment growth, there will not necessarily be a proportional increase in the value of fixed capital. There will, however, be a substantial enhancement of labor efficiency connected with wider possibilities of including experts in the production processes, who are experienced in the subject field instead of narrow specialists.

Each group of factors is consistent with the specific group of control parameters in the model. Changing the control parameters, we model the influence of different variants of innovation policy on the transformation of the technostructure. In response we get the trajectories of development of each technological constituent (cluster of technologies). The vector of trajectories contains four basic components: investment, fixed capital, labor, and production output. Their interconnection is described in four model blocks. Each block is differentiated by the technologies involved. According to the model, the output of each cluster develops by interacting with the others: investment is the common resource for development. As a result of technical and technological renovation, "low" technologies evolve to "high" technologies. The development process is limited by the availability of labor and the problems associated with skills and training. The common value of investment "earned" in the production process is proportional to total output.

With the help of control parameters we can model situations of acceleration, recession, and even the halting of diffusion of high technologies; we can then analyze the economic consequences of these cases.

13.2 Statistical Analysis of the Diffusion of Automation

The Soviet statistics on "sectorizing" output is unique. In the USSR this type of classification yields seven sectors:

(1) Fully automated and fully mechanized production.
(2) Automated but not fully mechanized production.
(3) Mechanized but not automated production.
(4) Automated and mechanized in basic production only.
(5) Production involving some mechanized, complex automated shops.

(6) Production involving some mechanized or automated lines; automated or semiautomated equipment not built into production lines.

(7) Production not possessing mechanized and automated lines, automated or semiautomated equipment.

The statistical series are concerned with trends in the development of these sectors on the basis of the main economic indicators of output, capital, and labor. (Note that this "sectorization" is complementary to the usual classification of industries by major product.) The statistics we use are specific to these sectors. More important than this is the possibility of research and analysis of trends of innovation regarding automation-based technologies.

To use the statistic series involved in the model, the seven groups were aggregated into three sectors. Each sector includes specific types of technologies. The first one (sector A) comprises the most advanced forms of production technology; this sector includes groups (1), (2), and (3). The second cluster (sector B) comprises mature technologies with slowing or stagnating production techniques; this sector includes groups (4) and (5). The third cluster (sector C) embraces the obsolete production technologies; this includes groups (6) and (7). This division considers the material, technology, and skills used in production, as well as the quality of the goods produced.

Statistical analysis reveals that each sector is characterized by its own technical and economic performance. According to our analysis, the labor productivity for sector A very often surpasses that of the others by a considerable degree (see *Figure 13.1*). For different clusters, capital productivity is characterized by different trends of change. As a rule, capital productivity of the new technologies is growing steadily. By contrast, capital productivity declines in sectors B and C (see *Figure 13.2*), as it does in industry as a whole. As regards the difference between the rate of labor productivity and the rate of capital–labor ratio, this quantity is larger in sector A than in sector B or sector C; it tends to be positive (see *Figure 13.3*).

These trends suggest that we can expect an eventual increase in capital productivity in the economy if sector A begins to develop. We can also expect a reduction of the amount of labor required as comprehensive automation substitutes for obsolete production technologies.

The process of technical renovation in the last two decades was mainly directed toward converting low-tech plants to an intermediate level of technology. The increments of capital and labor shares of sector B were larger

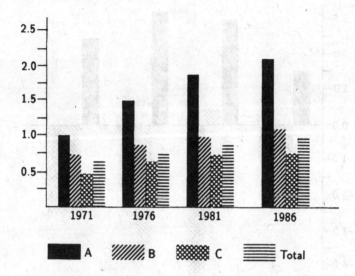

Figure 13.1. Labor productivity trends in Soviet industry by sector (1971–1986).

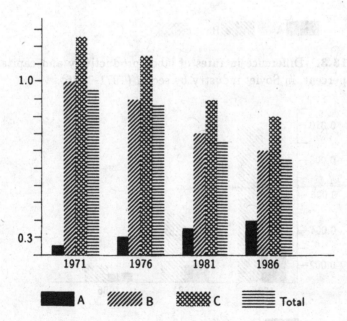

Figure 13.2. Capital productivity trends in Soviet industry by sector (1971–1986).

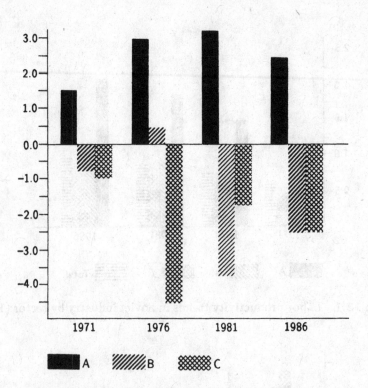

Figure 13.3. Difference in rates of labor productivity and capital–labor ratio, in percent, in Soviet industry by sector (1971–1986).

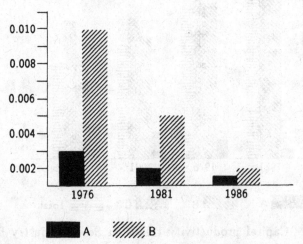

Figure 13.4. Annual increment of capital share of sectors A and B in Soviet industry (1976–1986).

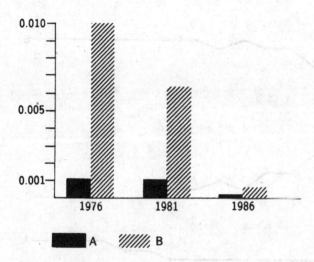

Figure 13.5. Annual increment of labor share of sectors A and B in Soviet industry (1976–1986).

than those of sector A (see *Figures 13.4* and *13.5*). As a consequence the output share of sector A remains small. It is only about 40% of the output of sector C and about 12% of the output of sector B (see *Figure 13.6*).

The low rate of diffusion of advanced (CIM) technology is particularly evident in the machine-building industry where the share of capital in sector A, according to our estimates, is very small (less than 1%), and the share of output is even smaller (see *Figure 13.7*).

In summary, at present the bulk of industrial output is produced by enterprises where automation and mechanization of production exist as "islands of automation" rather than as fully computer integrated factories. Such a strategy aggravates the trend toward decreasing capital productivity in the economy and does not create the adequate environment to introduce the basically new labor-saving CIM technologies. All of this emphasizes the necessity for radical changes in innovation policy. Undoubtedly new policies will be needed.

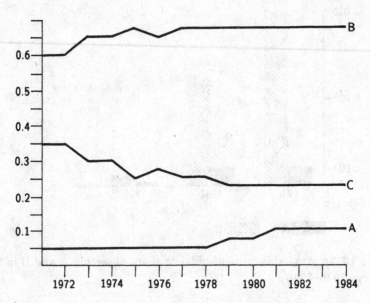

Figure 13.6. Product indicator share of sectors in Soviet industry (1971–1984).

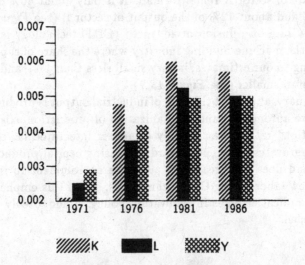

Figure 13.7. Trends in the Soviet machine-building industry, sector A (1971–1986): capital (K), labor (L), and product (Y) indicators.

13.3 Forecasts

On the basis of USSR statistics, forecasts have been made (using the model) for industry as a whole and for the seven main branches. The results suggest that it is possible to distinguish three types of innovation/diffusion strategy:

(1) The case of beneficial and rapid diffusion.
(2) The case of gradual effects of diffusion.
(3) The case of negative economic consequences of new technology diffusion.

13.3.1 Rapid and beneficial diffusion

As has been noted previously, the capital productivity of fully automated production, steadily increases; as a rule, though, it is still less than the average level for all of industry. As for labor productivity, it is much higher than the average level. The gap continuously increases.

The benefits of the wide introduction of fully automated technologies are best illustrated in the case of the energy industries in the USSR. Special attention was paid to these industries over a long-term period. This has resulted in satisfactory technical (and economic) levels of the equipment employed. As a consequence, the rates of development of sector A in the energy industries are rather high, and the share of sector A is the highest among all branches of industry (about 37% in production and 24% in fixed capital and labor). The strategy of accelerating the dissemination of fully automated production in this industry calls for an increase in the investment share of A from 10% to 40%, and an increase in the share of high-tech capital to 60%. During a six-year period the share of A production should rise to about 60% (see *Table 13.1*). This should result in an increase in the rates of production; the combined labor productivity will increase twofold, and the capital productivity will begin to rise instead of decline.

13.3.2 Gradual effects of diffusion

The steelworking industry of the USSR is still needed in the process of national reconstruction. The technology mostly used at present is of the open-hearth type. This technology is not only outdated, but ecologically harmful and unhealthy. The implications of its complete replacement by basic oxygen furnaces and continuous casters were explored using the model, and several conclusion were made. The productivity (and numbers) of new

Table 13.1. The forecast of sector A shares in total product (SYA) and total capital (SKA) as the result of the changes in sector A share in total investments (SIA) in the Soviet fuel and energy industry.

Year	SIA	SYA	SKA
1989	0.10	0.368	0.267
1990	0.15	0.420	0.335
1991	0.20	0.471	0.400
1992	0.25	0.515	0.465
1993	0.30	0.550	0.500
1994	0.35	0.575	0.556
1995	0.40	0.600	0.588

furnaces and converters in the USSR steel industry is not sufficient to permit a rapid replacement of open-hearth facilities while supporting desired output levels. If the policy of complete and fast replacement is followed, the combined output would fall over the next 10 years (see *Figure 13.8*). The faster the process of replacement, the lower will be the minimum level of production. However, after the diffusion of advanced technologies is completed, rapid rates of growth would be possible.

The different external factors involved in this decision should be taken into account. The possibilities of producing advanced construction materials to replace steel should be considered.

13.3.3   Negative economic consequences of rapid technological diffusion

In the USSR a group of technologies or products considered as "progressive" often turns out to consist mainly of obsolete technologies or products. The number of really advanced equipment units within the cluster is rather small. This is the case of sector A in the USSR machine-building industry. As was noted earlier, the actual share of sector A is very small (see *Figure 13.7*).

As regards efficiency indicators, we can see that labor productivity for sector A did not exceed the level of sector B (see *Figure 13.9*). As a rule, neither did capital productivity. This indicator was decreasing in sector A as well in sector B (see *Figure 13.10*). At the same time the consumption of production resources by A is more intensive than in sectors B and C (see *Figures 13.11* and *13.12*). Given this experience, therefore, the forecast for the machine-building industry indicates that the most favorable scenario

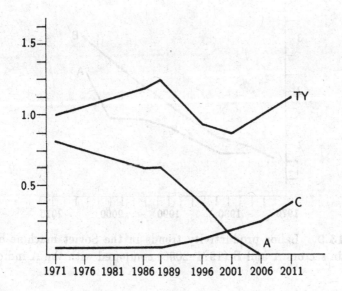

Figure 13.8. Forecast of total output (TY) trend compared with the level in 1971 as a result of changes in share trends of sectors A and C in the Soviet steelworking industry (1971–2011).

Table 13.2. The forecast of trends of shares of sectors A and B in total production in the Soviet machine-building industry.

Year	Sector A	Sector B
1989	0.00549	0.772
1996	0.00615	0.811
2001	0.00640	0.796
2006	0.00630	0.797

is one that concentrates investment at intermediate technological levels – sector B – (see *Table 13.2*).

As a consequence, a strategy of slow diffusion of the newest technologies is recommended. The first step would be to equip sector A with advanced technologies. Additional steps should be taken to accelerate its diffusion. Only in the longer run would automation in this industry achieve high levels.

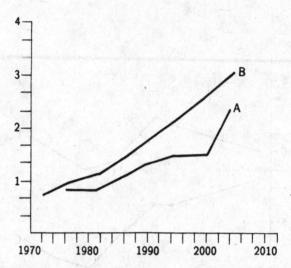

Figure 13.9. Labor productivity trends in the Soviet machine-building industry in sectors A and B (1971–2006) compared with the A indicator in 1976.

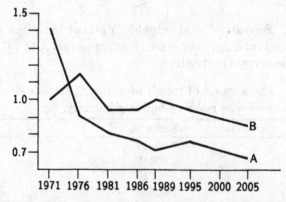

Figure 13.10. Capital productivity trends in the Soviet machine-building industry in sectors A and B (1971–2005) compared with the A indicator in 1976.

13.4   Conclusions

The basic idea is to consider the diffusion of a cluster of automation-based technologies. Three levels (A, B, C) are distinguished for this purpose. This

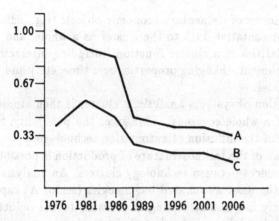

Figure 13.11. Rates of fixed capital in the Soviet machine-building industry in sectors A, B, and C (1976–2006) compared with the A indicator in 1976.

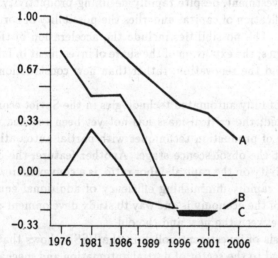

Figure 13.12. Rates of employment in the Soviet machine-building industry by sector (1976–2006) compared with the A indicator in 1976.

provides the possibility of diminishing the gap between macroeconomic and microeconomic approaches.

The proposed method of describing a cluster consists of four steps: defining the limits of a sector of the economy; developing a case study to estimate the main parameters of elementary economic objects (e.g., enterprises); extending the representative data to the cluster as a whole; and defining the statistical regularities of a cluster function (including interaction with the economic environment, changing properties over time, etc.) and basing them on the meso-model.

In the tradition of systems analysis, a cluster is then studied as a constituent part of a whole economy. This gives the possibility of estimating the final effects of the diffusion of automation technologies.

Consideration of the technostructure of production is possible because of significant differences between technology clusters. An analysis of statistics reveals that in the most automated technologies (sector A) capital productivity is relatively low but increasing in contrast to the reduction of total capital productivity; labor productivity is increasing more rapidly in sector A than in industry as a whole. In addition, the sector of partial automation and mechanization (B) is the largest in Soviet industry. It is still increasing its share of investment, despite rapidly declining productivity.

The stratification of capital underlies the possibility of correcting investment policies. The possibilities include the acceleration of the diffusion of new technologies; the extension of the share of investment in fully automated production; and the renovation, rather than new construction, of technologies.

A cluster of fully automated technologies in the Soviet economy has just been introduced; the critical mass has not yet been reached. At the same time a cluster of production techniques with partial automation and mechanization is at the obsolescence stage. Another feature, the dependence of labor productivity on the capital–labor ratio, is a convex increasing function. This means a rapidly diminishing efficiency of additional equipment. The stratification of the economy is one way to study development processes, i.e., the struggle between the new and the old.

The analysis of investment policy in the USSR shows that most investment is oriented to the sector of partial automation and mechanization (sector B). The necessary strategies to help the newest (A) technologies win acceptance have not yet been defined, except in the energy industries where sector A is developing with fairly high rates. With regard to the steel industry, the difficulty is connected with the replacement of open-hearth-based production. Forecasts show that even greater difficulties will arise if the replacement decision is delayed, but the rate of replacement must not be too rapid. An obstacle in the way of automation diffusion in this industry is that decision makers at the firm level have little interest in putting into

effect long-term policies, e.g., to invest extra funds in the short-term period
for the sake of long-term increased profits.

As far as the machine-building industry is concerned, the efficiency analysis shows that, from the point of view of the decision maker, there is little
sense in expanding sector A as it exists now. The model forecasts provide
an opportunity to analyze the ways and consequences of accelerating the
high-tech sector, and upgrading both manufacturing and the economy as a
whole.

The approach presented tries to answer the what-if question. It does not
take into account, for instance, political or social crises. Due to the current situation in the USSR, these forecasts are unfortunately being brushed
aside during a period when technical reconstruction is critically important.
Nevertheless the situation may change.

References

Adamovic, O.S., and S.B. Perminov, 1988, *Technological Structure of Economy and
    Automation: An Econometric Model to Analyze Investment Strategies in Au-
    tomation*, WP-88-99, IIASA, Laxenburg, Austria.

Alabyan, A.K., and J. Ranta, 1988, *Interactive Analysis of FMS Productivity*, WP-
    88-98, IIASA, Laxenburg, Austria.

Ayres, R.U., 1986, *Complexity, Reliability and Design: The Coming Monolithic Rev-
    olution in Manufacturing*, WP-86-48, IIASA, Laxenburg, Austria.

Ayres, R.U., 1987a, *The Industry Technology Life Cycle: An International Meta-
    Model?* RR-87-3, IIASA, Laxenburg, Austria.

Ayres, R.U., 1987b, *Complexity, Reliability and Design: Manufacturing Implications*
    (revised version), WP-87-94, IIASA, Laxenburg, Austria.

Babintzev, V.S., 1989, *USA: Priorities of the Technical Progress* (in Russian).

Dimitrov, P., and S. Wandel, 1988, *An International Analysis of Differences in
    Logistics Performance*, WP-88-31, IIASA, Laxenburg, Austria.

Fisher, J.C., and R.H. Pry, 1971, "A Simple Substitution Model of Technological
    Change," *Technological Forecasting and Social Change*, 3:75–88.

Freeman, C., 1977, *Capital Labor Substitution, Technology and Employment: Kon-
    dratiev Long Waves, Technical Change and Unemployment*, OECD, Paris,
    France.

Gorokhov, B.A., ed., 1987, *Technical Progress and Structural Changes in Current
    Economy*, AS USSR, INION, M (in Russian).

Keyworth, G.P., 1982, "The Role of Science in a New Era of Competition," *Science*,
    No. 4560.

Kleinknecht, A., 1987, *Innovation Patterns in Crisis and Prosperity: Schumpeter's
    Long Cycle Reconsidered*, Macmillan, London, UK.

Marchetti, C., and N. Nakicenovic, 1979, *The Dynamics of Energy Systems and the Logistic Substitution Model*, RR-79-13, IIASA, Laxenburg, Austria.

Mensch, G., 1979, *Stalemate in Technology: Innovation to Overcome the Depression*, Ballinger, New York, NY.

Solow, R., 1957, "Technical Change and Aggregate Production Function," *Review of Economics and Statistics*:312–320.

Tani, A., 1988, *Saturation Level of NC Machine-Tool Diffusion*, WP-88-78, IIASA, Laxenburg, Austria.

Tchijov, I., 1987, *The Cyclical Dynamics of Diffusion Rates*, WP-87-114, IIASA, Laxenburg, Austria.

Tchijov, I., 1988a, *Flexible Manufacturing Systems – Diffusion and Advantages* (Part D), WP-88-29, IIASA, Laxenburg, Austria.

Tchijov, I., 1988b, *CIM Application: Some Socioeconomic Aspects (Labor, Training, and Institutional Factors)*, WP-88-30, IIASA, Laxenburg, Austria.

Tchijov, I., 1989, "CIM Introduction: Some Socioeconomic Aspects," *Technological Forecasting and Social Change* 35:261.

Part IV

Macro and International Effects

Chapter 14

Macroanalysis of the Economic Impact of CIM: A Global Econometric Model

Rumen Dobrinsky

The diffusion of computer integrated manufacturing (CIM) appears to be one of the most outstanding technological developments of this century. Being a synthetic and interdisciplinary innovation, CIM brings about a major and complex change in human life. This explains the wide interest and the intensive research which is being carried out all over the world.

IIASA's focus on these problems within the CIM Project and the research efforts of many scientists made it possible to approach the many different aspects of CIM introduction such as the CIM diffusion process (Ayres and Ranta, 1989; Tani 1989a, 1989b; Maly and Zaruba, 1989; and *Computer Integrated Manufacturing*, Volume III: *Models, Case Studies, and Forecasts of Diffusion*); the impacts on production, organization, and management (Warnecke, 1989; Maly, 1989; Haustein, 1989); the economic aspects (Tchijov and Alabian, 1989; Ranta and Tchijov, 1989); and the impacts on labor (Mori, Chapter 4; Ayres and Mori, Chapter 6).

This chapter attempts to analyze the macroeconomic impact of the introduction of CIM technologies mainly on the general and sectoral growth process. Several studies on similar topics, which have been (or are being) carried out, are based on input–output models (Leontief and Duchin, 1986; Kinoshita and Yamada, 1989; Yamada, Chapter 10; Juszcak *et al.*, Chapter 11). In contrast, the approach taken in this chapter is inherently macroeconomic; besides, it builds heavily on the results of the Bonn–IIASA Research Project on Economic Growth and Structural Change (Krelle, 1989). This approach attempts to assess a "pure" macro effect of this technological change in a medium-term forecast, i.e., the effect on the economy due to the CIM diffusion which is measured *ceteris paribus*.

14.1 A Model for Macroeconomic Analysis of Technological Change

Consider an economy which is described by a conventional neoclassical production function with Harrod neutral technical progress:

$$Y = f[K, A(t)L] \ , \tag{14.1}$$

where Y is total output, K is capital, L is labor, and the time-dependent variable $A(t)$ is attributed to the cumulated (over time) technological change expressed in shifts in the production isoquants.[1] In the special case of the Harrod neutrality, technological progress reflects a labor-augmenting effect.

The traditional treatment of technical progress as an exogenous variable leaves little room for elaboration in the process of the actual technological development. There have been some attempts to endogenize technical progress by linking it not only to time but to variables which depend on the production process itself. We could distinguish two main lines of work in this field. The first one was pioneered by Kaldor (1957) who linked technical progress to the investment process. It was followed by Arrow's model of learning by doing (1962) and recently by Abel and Székely (1989), who broadened it by relating technical progress to logistics. The second line was started by Eltis (1971), who tried to explain technical progress by expenditures on research and development. It was developed by Krelle (1987, 1989), who introduced the latent variable "degree of activity" to explain technological change.

The approach used in this chapter diverts from these two main lines. First, we use a descriptive model of the process of technological change based on the description of the diffusion of new technologies in the economy. Second, we try to express the impact of technological change as change in labor productivity; we are concerned with total factor productivity $A(t)$

which appears in the form of changed output per unit of labor *due to* a technological change.

The aggregate labor productivity $y = Y/L$ which appears on the surface of the economy, can be presented as a weighted average of the labor productivity in the individual production units in the economy or, in a narrower sense, in the sectors which we regard:

$$y = \sum_i s_i^L y_i , \tag{14.2}$$

where $y_i = Y_i/L_i$ is the average labor productivity in sector i; $s_i^L = L_i/L$ is the share of sector i's labor input in total labor input ($\sum_i s_i^L = 1$).

Further we express each sector's labor productivity as a weighted average of the mix of individual technologies which are currently installed in that sector (again as labor productivity):

$$y_i = \sum_k s_{ik}^L y_{ik} , \tag{14.3}$$

where y_{ik} is the labor productivity of technology k in sector i and $s_{ik}^L = L_{ik}/L_i$ is the share of labor input for technology k in total labor input for sector i ($\sum_k s_{ik}^L = 1$).

We can compare the labor productivity of all technologies in the economy by rating them to a common number, e.g., a technology with a unit labor productivity $\bar{y}$:

$$y_{ik} = r_{ik}\bar{y} , \tag{14.4}$$

where r_{ik} is a productivity ratio. So, from (14.2), (14.3), and (14.4) we have

$$y = \bar{y} \sum_i s_i^L \sum_k s_{ik}^L r_{ik} . \tag{14.5}$$

Technological change in such an economy can be expressed by (a) the emerging new technologies y_{ik}^n (normally these are expected to be more productive than the average sectoral technology y_i); (b) the diffusion of the new technologies within the sectors, reflected by a structural change of employment within the sector (normally this is a drift of labor input from less productive technologies to more productive ones); and (c) sorting out of old technologies y_{ik}^o (normally these are expected to be less productive than the average sectoral technology y_i).

The rate of change of total labor productivity in this economy due to technological change is given by

$$\frac{d \log(y)}{dt} = \frac{d}{dt} \log \left(\overline{y} \sum_i s_i^L \sum_k s_{ik}^L r_{ik} \right) . \tag{14.6}$$

Now we shall make use of one result of the neoclassical growth theory, namely, that in an equilibrium growth path the rate of change of labor productivity equals the rate of technological change, provided that technical progress is of the Harrod neutral type (Krelle, 1985).

Under this assumption from (14.1) and (14.6) we shall have

$$\frac{d \log [A(t)]}{dt} = \frac{d}{dt} \log \left(\overline{y} \sum_i s_i^L \sum_k s_{ik}^L r_{ik} \right) , \tag{14.7}$$

which is an endogenized formulation for the rate of technological change in the implied diffusion process.

The assumptions of this model might at first glance seem severely restrictive. However, most of the skepticism probably can be dispersed if we consider the practical goals of its application. Taking into account the fact that we aim at analyzing medium- and long-term developments (15 years), an equilibrium growth path at the macro level is not an unreasonable approximation. As to the assumption for Harrod neutrality of the technological change, the special case of a Cobb-Douglas production function (which we use further in the model) is indifferent to the type of neutrality assumed.

However, several practical issues might make difficult the full-scale application of the approach suggested. These difficulties are mainly concerned with the necessity for a detailed description of the state of the economy, and the full set of necessary data is beyond the scope of the normal statistical practice.

We thus suggest to use this approach for the analysis of the economic impact of only one specific type of new technologies, namely, CIM. In this case we consider productivity changes induced only by the introduction of this type of new technologies and only in some economic sectors (where they are applicable).

Further if we aim at assessing *only* the future economic impact of CIM, we could assume a zero change for all other technologies and, *ceteris paribus*, analyze the pure effect of different strategies for CIM diffusion.

14.2 The Models of the Bonn–IIASA Project

The main analytical tool of the Bonn–IIASA Research Project on Economic Growth and Structural Change (Krelle, 1989) was the model of the world economy. It consists of 23 macro models of countries and country groups which are linked together through an integrated system of international trade. Omitting the details we shall just mention the main features of the models.

The central part of each country model is the production function which is specified as a Cobb-Douglas function, homogeneous to the degree one:

$$Y^* = \alpha_o \tau L^{\alpha_1} K^{\alpha_2} (IM_R)^{\alpha_3} \ , \qquad \alpha_i > 0 \ , \qquad \sum_i \alpha_i = 1 \ , \qquad (14.8)$$

where the output Y^* is defined as total production *minus* domestic intermediate inputs;[2] τ is the level of technological progress; and IM_R is the import of intermediate inputs.

Capital accumulation is described by a dynamic equation with a zero investment-lag assumption:

$$K = K_{-1}(1-d) + IF_{-1} \ , \qquad (14.9)$$

where d is the rate of depreciation and IF is gross fixed investment; IF is determined as a share of GDP by the investment ratio s:

$$IF = s \times Y \ . \qquad (14.10)$$

The import of intermediate goods IM_R is taken as a share μ_R of total imports IM

$$IM_R = \mu_R \times IM \ , \qquad (14.11)$$

and IM as well as EX (export) are explained by export and import functions (Krelle, 1989).

Total gross investment (accumulation) I is set in proportion h to gross fixed investment

$$I = h \times IF \ , \qquad (14.12)$$

and total consumption C is determined residually from the national accounts identity:

$$C = Y - I + IM - EX \ . \qquad (14.13)$$

The variables τ, l, d, s, μ_R, and h are treated as exogenous driving forces of economic growth and structural change and are used to design the forecast scenarios. By far the most important among them is the level of technological progress τ (or its rate of change W_τ) followed by the investment ratio s. Further details concerning the model of the world economy and the results of the scenario runs can be found in *The Future of the World Economy: Economic Growth and Structural Change* (Krelle, 1989).

Now we shall turn to the extension of the model which was developed for the analysis of the macroeconomic impact of CIM technologies. Because not all countries or country groups of the Bonn–IIASA Project are involved in CIM-related research, we have adopted an approach based on autonomous country models where the foreign trade variables are set to values taken from solutions of the world model. For this purpose the country models are "extracted" from the world model and are rearranged to be run in an autonomous mode. Eleven countries were selected for the CIM analysis: USA, FRG, Japan, France, UK, Italy USSR, Bulgaria, CSFR, GDR, and Hungary.[3]

The sectors within which we analyze the diffusion of CIM technologies are the five three-digit subsectors of ISIC sector 38, fabricated metal products: ISIC 381, final metal products; ISIC 382, nonelectrical machinery; ISIC 383, electrical machinery; ISIC 384, transport equipment; and ISIC 385, professional and scientific equipment. Within each sector, in accordance with the theoretical model outlined in Section 14.1, we specify four types of production technologies corresponding to four levels of automation in the industries: level 0, traditional, nonautomated, stand-alone machines; level 1, stand-alone NC and CNC machine and industrial robots; level 2, FMS and FMC; and level 3, CIM.

Within each country model we introduce a new block which simulates the diffusion of CIM technologies (levels 1 to 3) in the five sectors and, as a result, determines endogenously in accordance with (14.7) the CIM-induced change in the rate of technological progress W_τ.

For practical purposes (mainly to reduce the size of the models) we have modified the general approach outlined in Section 14.1, however, without any loss of generality. Since in all cases we start the simulation runs with a country model *given* a certain level of CIM diffusion, we consider this level as a reference point and define the CIM-induced shift of W_τ with respect to this level. If we denote y_i^0 as the labor productivity in sector i at the reference level of CIM diffusion and y_i^* as the labor productivity at some other (higher) level of diffusion, then we can write

$$y_i^* = \left\{ \left[1 + \left(r_i^{CIM} - 1 \right) s_i^{CIM} \right] \right\} y_i^o , \tag{14.14}$$

where s_i^{CIM} is the share of newly shifted (with respect to the reference state) employees to CIM technologies (the total for levels 1 to 3) and r_i^{CIM} is the productivity ratio of an average (the weighted mix of levels 1 to 3) CIM technology (as compared with the reference labor productivity y_i^o). It is easy to show that the two new variables r_i^{CIM} and s_i^{CIM} can be calculated from the basic set of variables s_{ik}^L and r_{ik}.

From (14.8) and bearing in mind that in terms of discrete growth rates we have

$$\frac{d \log y_i}{dt} = W_{y_i}^* = \frac{y_i^*}{y_i^o} - 1 , \tag{14.15}$$

it follows directly that

$$W_{y_i}^* = \left(r_i^{CIM} - 1 \right) s_i^{CIM} . \tag{14.16}$$

We can proceed in a similar manner to determine the changed level of total labor productivity y^* [from equation (14.2) and by substituting y_i^* for (14.8)], and we can then express its rate of change W_y which we substitute for the rate of technological change. Omitting the derivation which is conventional we shall only present the final result

$$W_r^* \left(= W_y^* \right) = \sum_i s_i^Y \left(W_{s_i^L} + W_{y_i} \right)$$
$$= \sum_i s_i^Y \left[W_{s_i^L} + \left(r_i^{CIM} - 1 \right) s_i^{CIM} \right] , \tag{14.17}$$

where $s_i^Y = Y_i / Y$ are the sectoral output shares and $W_{s_i^L}$ denotes the rate of change of the sectoral labor shares s_i^L. In the special case when there are no intrasectoral shifts of labor ($W_{s_i^L} = 0$), we have

$$W_r^* = \sum_i s_i^Y \left(r_i^{CIM} - 1 \right) s_i^{CIM} . \tag{14.18}$$

14.3   Data Sources and Definition of Scenarios

The extended versions of the country models were designed so that they could be made operational without imposing heavy data requirements. The data which are needed for the simulations can be divided into two categories.

The first category comprises data about the production structure of the economy (output and employment shares and, accordingly, sectoral labor productivity). In accordance with our goals we classify the economy into six sectors: five sectors in fabricated metal products and a sixth sector which is the rest of the economy. Some reference data for the countries under consideration are presented in Appendix I (*Tables 14A.1* to *14A.6*).

The output structure of the five sectors is fairly stable (*Tables 14A.1* and *14A.2*), especially if we consider the aggregate of the five sectors (the ISIC two-digit 38). However, the same is not true for the employment structure (*Tables 14A.3* and *14A.4*), which is rather country specific. We can grade the countries putting the UK on one extreme with a drastic reduction of the employment share of the five sectors during the period 1970–1985, followed by countries with a moderate decrease of the employment share such as the USA, France, and Hungary. Most of the countries lie in the middle with fluctuating shares; the USSR and the GDR indicating some modest increase in their shares.

It is of interest to consider also the relative labor productivity in the sectors (as compared with the average total labor productivity in the countries) which is illustrated by the figures in *Tables 14A.5* and *14A.6*. Despite the great variation among sectors and countries there is one stable trend indicated by these figures: the positive change in the relative position of the five sectors taken together (sector 38), the FRG in 1980 being the only outlier. This development could well be attributed to the rapid diffusion of automation during this period.

In *ex ante* simulations the output shares are determined endogenously whereas the employment shares should be supplied exogenously. We have kept the simulation runs reported here constant at their last observed level.

Actually, this is a result of one assumption and one simplification that we make. The assumption is that the process of CIM diffusion will not cause unemployment (through displacement) in the corresponding sectors. The findings of Yamada (Chapter 10) and Ayres and Mori (Chapter 6) also suggest such a development. As to the simplification, it is a fact that we disregard the intrasectoral movement of labor caused by CIM diffusion and displacement of workers. In terms of GDP growth, which we now analyze, such movements could cause an effect due to the differences in productivity among the sectors. Obviously such an effect would be negligible as compared with the CIM-induced effect.

The second category of data is the data which characterizes the productivity of the different technologies and the diffusion process. Within this category we have to build two sets of data: (a) the productivity ratios for the different levels of automation and (b) the CIM diffusion scenarios for the future (*Table 14.1* and *Table 14.2*).

To fix appropriate values for the productivity ratios we have used empirical analyses and surveys as well as experts' judgments. Thus the Sixth Survey of Machine Tools and Production Equipment in Britain (1987) suggests the following ratios for adjusting the productivity of NC machine tools of non-NC equivalents: 1971, NC M/C = 2 non-NC; 1976, NC M/C = 2-1/2 non-NC; 1981, NC M/C = 3 non-NC; and 1986, NC M/C = 3-1/2 non-NC.

The survey of about 200 FMS analyzed by Tchijov (1991) showed that, on the average, the companies reported reduction in personnel of the order of two to four times which could be interpreted as an equivalent productivity increase. The worker displacement ratio for robots reported by Ayres and Mori (Chapter 6) is about two workers per robot.

Using these and other findings as well as the opinions of experts from the CIM Project and its collaborating network, we have adopted the following scale of productivity ratios for the three levels of automation: level 1, $y_1 = 2y_0$ (1984) and $y_1 = 3y_0$ (2000); level 2, $y_2 = 5y_0$ (1984) and $y_2 = 6y_0$ (2000); and level 3, $y_3 = 30y_0$ (1984) and $y_3 = 30y_0$ (2000). For practical reasons we show only one definition – that in output shares, though these figures were derived in a complex manner.

For automation level 1 we started from national figures for the shares of numerical control machine tools (NCMT) in the total population of machine tools. We then extrapolated these figures by a logistic curve to arrive at the densities in the year 2000. As to the industrial robots we used the robot population forecast of Tchijov and Norov (1989) as well as the findings of Tani (1989a). These findings indicate that, on the average, Japan is about five years ahead of the other major industrialized countries in robot penetration. We assumed another five-year lag for the East European countries and the USSR.

For automation level 2, we used the FMS diffusion forecasts prepared by Tchijov (1991).

The figures for level 3 are mainly based on expert opinions because CIM per se is still in its embryonic stage.

Table 14.1. CIM diffusion scenarios, developed market economies (DMEs): shares in sectoral value-added by levels of automation, in percent.

Country	Level	1984					2000				
		381	382	383	384	385	381	382	383	384	385
USA	1	9	13	11	15	15	27	37	40	40	48
	2	0	0	0	0	0	9	18	24	25	23
	3	0	0	0	0	0	2	3	2	2	3
FRG	1	5	15	8	8	9	25	40	33	35	33
	2	0	0	0	0	0	8	23	15	20	17
	3	0	0	0	0	0	1	3	2	3	2
Japan	1	20	24	24	24	20	47	39	43	40	44
	2	0	1	0	2	2	23	32	28	31	25
	3	0	0	0	0	0	2	4	3	6	6
France	1	7	10	10	12	8	25	37	35	35	34
	2	0	0	0	0	0	8	20	12	20	18
	3	0	0	0	0	0	1	2	3	2	3
UK	1	6	9	9	10	9	23	33	31	37	35
	2	0	0	0	0	0	7	15	12	18	16
	3	0	0	0	0	0	1	1	2	1	2
Italy	1	7	9	10	8	8	26	32	38	38	36
	2	0	0	0	0	0	7	16	14	22	17
	3	0	0	0	0	0	1	2	2	2	2

The tables presented in Appendix II illustrate the procedures we used to design the CIM diffusion scenarios.

14.4   The Macroeconomic Effect of CIM: Results from Simulation Runs

First of all we would like to discuss briefly the possible ways to use the models for simulation purposes.

In our judgment the best use of the models is to assess the *relative* effect of the introduction and diffusion of CIM technologies in some sectors, i.e., with respect to some other simulation run in which this effect is not present. We have selected for this purpose the model solutions of simulations with zero rate of growth of technical progress. As it was pointed out this allows one to assess the "pure" effect of this technological innovation.

Table 14.2. CIM diffusion scenarios, East European countries and the USSR: shares in sectoral net output by levels of automation, in percent.

Country	Level	1984					2000				
		381	382	383	384	385	381	382	383	384	385
USSR	1	3	6	5	6	4	16	28	25	30	25
	2	0	0	0	0	0	6	13	8	15	12
	3	0	0	0	0	0	0	1	1	1	1
Bulgaria	1	5	8	7	8	5	19	30	28	30	20
	2	0	0	0	0	0	7	16	10	16	13
	3	0	0	0	0	0	0	1	2	1	1
CSFR	1	6	9	7	8	6	20	31	31	33	29
	2	0	1	0	0	0	7	16	13	18	15
	3	0	0	0	0	0	0	1	2	1	1
GDR	1	5	10	7	8	7	22	33	29	31	27
	2	0	0	0	0	0	8	20	12	17	14
	3	0	0	0	0	0	0	1	2	2	1
Hungary	1	4	6	5	6	4	19	29	30	30	24
	2	0	0	0	0	0	7	15	12	16	15
	3	0	0	0	0	0	0	1	2	1	2

Table 14.3. Estimates of the "pure" macroeconomic effect of the CIM diffusion: results from *ex ante* simulation runs.

Country	1990			1995			1999		
	Δ^a	Δ^b_{84}	$\Sigma\Delta^c_{84}$	Δ^a	Δ^b_{84}	$\Sigma\Delta^c_{84}$	Δ^a	Δ^b_{84}	$\Sigma\Delta^c_{84}$
USA	0.42	0.47	1.86	0.58	0.71	4.95	0.60	0.88	8.22
FRG	0.46	0.48	1.81	0.74	0.80	5.17	0.94	1.07	9.05
Japan	0.65	0.72	2.73	0.98	0.19	7.74	1.78	1.54	13.38
France	0.28	0.29	1.12	0.44	0.45	3.05	0.54	0.56	5.14
UK	0.21	0.22	1.04	0.33	0.34	2.30	0.42	0.43	3.88
Italy	0.30	0.32	1.25	0.45	0.49	3.37	0.53	0.61	5.65
USSR	0.37	0.44	1.62	0.56	0.76	4.77	0.69	1.01	8.42
Bulgaria	0.38	0.44	1.67	0.55	0.71	4.70	0.62	0.86	7.92
CSFR	0.69	0.78	2.94	1.00	1.24	8.25	1.47	1.55	14.00
GDR	0.64	0.73	2.69	0.98	1.26	7.95	1.17	1.65	13.97
Hungary	0.46	0.51	1.87	0.73	0.87	5.49	0.90	1.13	9.65

[a] $\Delta = \frac{Y^{CIM} - Y^o}{Y^o} \cdot 100\%$, where Y^{CIM} is GDP (NMP) in the model with CIM diffusion; Y^o is GDP in the model with zero rate of growth of technical progress.

[b] $\Delta_{84} = \frac{Y^{CIM} - Y^{84}}{Y^{84}} \cdot 100\%$, where Y^{84} is GDP (NMP) in 1984.

[c] $\Sigma\Delta^T_{84} = \sum_{t=1984}^{T} \Delta_{84}$.

Table 14.4. Relative increase of sectoral output due to CIM diffusion[a] in DMEs, in percent: results from *ex ante* simulation runs.

Country	Year	Sector 381	382	383	384	385
USA	1990	7.5	14.8	20.9	19.6	25.1
	1995	16.9	34.3	48.4	45.7	58.0
	1999	26.1	53.4	75.3	71.3	90.2
FRG	1990	6.1	15.2	10.5	14.2	11.1
	1995	14.9	38.6	26.3	35.7	27.9
	1999	24.5	64.5	43.5	59.5	46.3
Japan	1990	18.6	18.9	18.2	20.5	21.2
	1995	44.8	46.4	44.7	50.3	50.8
	1999	71.9	75.8	72.5	81.5	81.4
France	1990	5.7	14.2	9.5	12.5	12.1
	1995	13.7	35.1	23.2	41.1	30.0
	1999	22.3	57.7	37.9	51.1	49.3
UK	1990	5.0	9.7	8.2	12.3	11.1
	1995	10.5	24.1	20.3	30.4	27.6
	1999	19.9	39.7	33.5	50.2	45.7
Italy	1990	6.0	10.8	12.3	17.3	13.7
	1995	14.2	26.3	29.5	42.4	33.1
	1999	22.7	42.6	47.5	68.9	53.6

[a] Calculated as percentage differences between sectoral value-added in the model with CIM diffusion and the corresponding value-added in the model with zero rate of growth of technical progress.

On the other hand, if a reference scenario with a nonzero rate of growth of technical progress is used, it inevitably will incorporate also (at least partially) effects of CIM diffusion. If we then design a new, CIM-based scenario of technological development, it would be difficult to distinguish what is the actual effect of CIM.

In our simulation-based analysis we have tried to evaluate two main items: the CIM-induced changes in the levels of GDP (NMP for Eastern Europe and the USSR) shown in *Table 14.3* (the "pure" macroeconomic effect of CIM diffusion) and the relative changes of output within each sector (*Tables 14.4* and *14.5*).

In general the estimated "pure" economic effect of CIM diffusion (*Table 14.3*) can be considered as quite substantial. The cumulative effect of total

Table 14.5. Relative increase of sectoral output due to CIM diffusion[a] in East European countries and the USSR, in percent: results from *ex ante* simulation runs.

Country	Year	Sector				
		381	382	383	384	385
USSR	1990	3.7	8.5	6.3	10.0	7.9
	1995	8.9	20.8	15.3	24.6	19.4
	1999	14.5	34.1	24.8	40.3	31.9
Bulgaria	1990	4.0	9.6	7.5	9.6	6.1
	1995	9.6	23.6	18.6	23.6	15.2
	1999	15.6	38.8	29.9	38.8	25.0
CSFR	1990	9.2	10.0	10.1	12.1	10.1
	1995	9.9	24.2	24.1	29.4	24.4
	1999	15.9	39.3	39.1	47.8	39.5
GDR	1990	5.7	14.2	9.5	12.5	12.1
	1995	13.7	35.1	23.2	41.1	30.0
	1999	22.3	57.7	37.9	51.1	49.3
Hungary	1990	5.0	9.7	8.2	12.3	11.1
	1995	10.5	24.1	20.3	30.4	27.6
	1999	19.9	39.7	33.5	50.2	45.7

[a] Calculated as percentage differences between sectoral net output in the model solution with CIM diffusion and the corresponding net output in the model with zero rate of growth of technical progress.

GDP (NMP) increase for the 15 years of simulation ranges from 4% and 5% to 13% and 14% of the countries' GDP (NMP) in the reference year 1984.

Of course in interpreting these figures one has to bear in mind the simple assumptions that we have made. If one takes into consideration the structural adjustments and the repercussions which they cause throughout the economy (e.g., see Yamada, Chapter 10), the picture might look more complicated.

Anyway, the evaluation of the "pure" economic effect has its own specific purpose, namely, the estimated figures indicated what would happen to the economy due to CIM diffusion if no other significant changes take place. We could point out that our figures compare quite favorably with similar evaluations performed by Yamada.

The "pure" effect of CIM diffusion is more clearly revealed if we analyze the results concerning the metalworking industries (*Tables 14.4* and *14.5*). On average, at the end of the simulated period, the growth of sectoral output owing to CIM diffusion is of the order of 40% to 60% in the developed market economies and of the order of 30% to 40% in the East European countries and the USSR. Japan is the obvious leader gaining from CIM diffusion, 70% to 80% of sectoral output growth.

We would like to emphasize the fact that the cross-country differences in the results are due not only to the diffusion scenarios in the countries but also to the different weights of the sectors in the countries and to the differences in the productivity levels (see Appendix I). Thus the relative increase of sectoral output in East European countries and the USSR, in general, is smaller than that in the developed market economies (*Tables 14.4* and *14.5*) but the total effect in Eastern Europe and the USSR is largly due to the fact that the metalworking industries have a greater weight in the economies of these countries.

Following this line, however, we would like to suggest that the results concerning the general effect of CIM diffusion for the East European countries (*Table 14.3*) might be overestimated in light of the major changes occurring in the economies of these countries. The present high share of the metalworking industries in the economies of the East European countries is to a large extent due to the long-pursued policy of favoring heavy industries. Another factor was the Soviet market which was capable of absorbing a large share of the other East European countries' production. Now with the turning of these countries to market mechanisms the results of this overindustrialization become obvious. The markets for the metalworking products, which are typical for the East European countries (bulky, unsophisticated products of generally low quality), will shrink drastically, and so will (obviously) the metalworking sectors in those countries. Such a structural change would lead to a reduction in the total effect estimated in *Table 14.3*. Or, on the other hand, such a negative development might be prevented or reduced by a faster technological reconstruction and, in particular, by wider introduction of automation and CIM technologies which could support the production of higher-quality products.

With regard to the estimated effect of CIM diffusion on labor employment (*Table 14.6*), we would like to underline that these figures, and especially the estimated future share of labor employment in automated technologies, indicate a "pure" effect – i.e., assuming no change in the structure of labor employment. So these shares indicate what would be the weight of industries directly or indirectly applying automated technologies if the employment share of the corresponding sector in total employment stays constant. Or,

Table 14.6. Estimation of the effect of CIM diffusion on labor employment: results from simulation.

Country	Cumulative shift of labor to automated technologies (levels 1 to 3), 1985–1999, by sectors (1,000s of employed persons)						Share of labor employed in automated technologies (sector 38 total %)	
	381	382	383	384	385	38 Total	1984	1999
USA	178.6	542.6	721.3	652.7	271.7	2367	6.3	29.8
FRG	56.2	256.8	151.9	186.7	24.5	676	5.1	24.9
Japan	265.2	405.5	568.0	318.8	81.7	1639	13.4	42.7
France	18.7	106.9	67.5	98.6	12.4	304	5.4	23.6
UK	31.6	106.1	83.9	133.7	16.4	372	4.6	20.2
Italy	18.6	66.0	60.0	88.0	9.8	243	4.5	23.3
USSR	56.2	1406.3	122.9	239.0	156.7	1981	2.9	15.8
Bulgaria	3.3	21.6	12.9	11.8	1.6	51	3.8	16.8
CSFR	11.7	89.2	26.7	50.1	2.4	180	4.1	17.1
GDR	14.3	103.8	53.8	49.9	12.1	234	4.3	19.6
Hungary	3.6	18.3	22.7	15.3	6.8	67	2.7	16.9

to put it in other words, this is an intersectoral change in the case of no intrasectoral changes.

Labor must shift to automated technologies to deal with the demand from newly created occupations. As in the first case the figures include occupations both for the operators and for personnel indirectly involved in the new technologies (service workers, engineers, programmers, managers).

14.5 Concluding Remarks

In this chapter we discuss the possibilities of using a neoclassical growth model to analyze the economic impact of the introduction of CIM technologies. We do this by describing the diffusion process and endogenizing (partly) technological progress. The practical implementation of the approach was performed on the basis of the Bonn–IIASA model of the world economy.

The model-based analysis of some future implications of CIM diffusion was performed for 11 countries from the West and East. Our analysis shows that the diffusion of CIM and related technologies are likely to have a significant economic impact on the world economy at the end of this century:

higher productivity, new organization of production, highly skilled operators, new occupational structure. These factors taken together constitute a
major technological change which leads to accelerated growth in the sectors
where CIM is implemented and in the economy as a whole.

Certainly, our models are limited in scope and can cover only the most
important features of the macroeconomic development and the impact of
CIM diffusion. They can only be considered as a starting point for further
analysis of the socioeconomic consequences of this technological change.

Notes

[1] In this specification $A(t)$ is also referred to as "total factor productivity" since
not all production shifts can be attributed to technological changes.

[2] This is the same as the sum of GDP (denoted as Y) and IM_R, in real terms:

$$Y^* = Y + IM_R .$$

[3] At the time of the study German unification had not occurred; therefore the
FRG and the GDR are treated as separate countries.

Appendix I: Production Structure of the Fabricated Metal Products Sectors

Table 14A.1. Output shares by sectors in DMEs in percent.

Country	Year	Share of value-added in total GDP by sectors					
		381	382	383	384	385	38
USA	1973–1975	1.56	2.65	1.90	2.54	0.69	9.34
	1979–1981	1.57	3.04	2.16	2.47	0.82	10.06
	1983–1985	1.36	2.57	2.44	2.73	0.88	9.98
FRG	1973–1975	1.73	4.31	3.66	3.29	0.64	13.63
	1979–1981	1.76	4.21	3.68	3.94	0.75	14.34
	1983–1985	1.68	4.09	3.77	4.15	0.64	14.33
Japan	1973–1975	2.05	3.09	2.67	2.59	0.41	10.81
	1979–1981	1.91	3.32	3.35	2.80	0.49	11.87
	1983–1985	2.04	4.07	4.91	3.44	0.57	15.03
France	1973–1975	1.33	3.47	2.07	2.54	0.32	9.73
	1979–1981	1.32	3.75	2.03	3.31	0.29	10.70
	1983–1985	1.16	3.44	1.92	3.09	0.28	9.89
UK	1973–1975	2.04	3.27	2.30	3.10	0.51	11.22
	1979–1981	1.62	2.98	2.15	2.64	0.36	9.75
	1983–1985	1.71	2.66	2.27	2.36	0.32	9.32
Italy	1973–1975	1.56	2.58	2.47	2.87	0.60	10.08
	1979–1981	1.78	2.96	2.63	3.23	0.63	11.23
	1983–1985	1.63	3.70	2.68	2.85	0.35	11.21

Source: UN, 1988.

Table 14A.2. Output shares by sectors in Eastern Europe and the USSR, in percent.

Country	Year	Share of net production in total NMP by sectors					
		381	382	383	384	385	38
USSR	1983–1985	0.65	9.81	1.08	1.48	1.20	14.22
Bulgaria	1983–1985	1.56	5.96	4.35	3.38	0.74	15.99
CSFR	1979–1981	2.54	11.34	2.88	5.47	0.33	22.56
	1983–1985	2.58	11.80	3.40	5.71	0.33	23.82
GDR	1983–1985	1.66	7.27	5.01	4.04	1.24	19.22
Hungary	1973–1975	1.35	3.05	3.20	2.32	1.16	11.08
	1979–1981	1.29	3.10	4.00	3.03	1.60	13.02

Source: UN, 1988; UN, 1985.

Table 14A.3. Employment shares by sectors in DMEs, in percent.

Country	Year	Shares in total employment by sectors					
		381	382	383	384	385	38
USA	1970	1.85	2.69	2.30	2.17	0.59	9.60
	1975	1.50	2.38	1.80	2.03	0.58	8.29
	1980	1.57	2.62	2.07	1.99	0.64	8.89
	1985	1.26	1.94	1.88	1.76	0.56	7.40
FRG	1970	2.28	4.73	4.30	2.95	0.66	14.92
	1975	2.03	4.42	3.97	2.83	0.60	13.85
	1980	2.07	4.41	3.73	3.46	0.67	13.67
	1985	2.31	4.23	3.72	3.52	0.58	14.36
Japan	1970	1.66	2.41	2.62	1.67	0.38	8.74
	1975	1.56	2.21	2.27	1.74	0.38	8.16
	1980	1.40	2.02	2.34	1.56	0.42	7.74
	1985	1.33	2.12	2.95	1.56	0.39	8.35
France	1970	1.21	2.47	1.79	2.75	0.33	8.55
	1975	1.24	2.64	2.09	2.76	0.34	9.07
	1980	1.13	2.44	2.04	2.74	0.33	8.68
	1985	0.96	2.25	2.00	2.35	0.30	7.86
UK	1970	3.04	4.92	3.75	4.77	0.94	17.42
	1975	2.78	4.25	3.46	4.50	0.81	15.80
	1980	2.14	3.97	3.19	3.99	0.48	13.77
	1985	1.66	2.84	2.67	2.82	0.38	10.37
Italy	1970	1.11	1.38	1.17	1.67	0.39	5.72
	1975	1.10	1.57	1.79	2.01	0.37	6.84
	1980	1.01	1.50	1.64	2.07	0.32	6.54
	1985	0.90	1.84	1.43	1.56	0.22	5.95

Source: UN, 1988.

Table 14A.4. Employment shares by sectors in Eastern Europe and the USSR, in percent.

| Country | Year | Shares in total employment by sectors | | | | | |
		381	382	383	384	385	38
USSR	1970	0.85	9.48	1.09	1.38	1.16	13.96
	1975	0.92	10.20	1.17	1.49	1.25	15.03
	1980	0.96	11.03	1.27	1.61	1.35	16.22
	1985	0.97	10.78	1.27	1.57	1.32	15.91
Bulgaria	1970	1.41	—	1.72	1.27	—	—
	1975	1.77	—	2.05	1.42	—	—
	1980	1.49	5.31^a	2.44	1.65	—	10.89
	1985	1.49	4.21	3.19	2.30	0.52	11.71
CSFR	1970	2.30	8.14	2.45	3.55	1.57	18.01
	1975	2.25	7.67	2.54	4.28	1.63	18.37
	1980	3.03	9.15	2.67	4.36	0.28	19.49
	1985	2.86	9.49	2.86	4.44	0.26	19.91
GDR	1970	2.14	6.94	4.56	3.71	1.09	18.44
	1975	2.25	7.29	5.34	3.90	1.28	20.06
	1980	2.34	7.60	5.27	4.06	1.25	20.52
	1985	2.40	7.79	5.56	4.16	1.32	21.23
Hungary	1970	1.96	3.62	3.31	3.01	1.23	13.13
	1975	1.82	3.50	3.74	2.58	1.40	13.04
	1980	1.43	3.14	3.99	2.53	1.41	12.50
	1985	1.33	3.19	3.93	2.42	1.35	12.22

^{a}Sectors 382 and 385.
Source: UN, 1985; UN, 1988.

Table 14A.5. Relative labor productivity^a in fabricated metal products sectors, DMEs (total labor productivity^b = 1.0).

Country	Year	381	382	383	384	385	38
USA	1975	1.04	1.11	1.06	1.25	1.19	1.13
	1980	1.00	1.16	1.04	1.24	1.28	1.13
	1985	1.08	1.32	1.30	1.55	1.57	1.35
FRG	1975	0.85	0.98	0.92	1.16	0.97	0.98
	1980	0.85	0.85	0.99	1.14	1.12	1.05
	1985	0.73	0.97	1.01	1.18	1.10	1.00
Japan	1975	1.31	1.44	1.18	1.49	1.08	1.37
	1980	1.36	1.64	1.43	1.79	1.17	1.53
	1985	1.53	1.92	1.66	2.21	1.46	1.80
France	1975	1.07	1.31	0.99	0.92	0.94	1.07
	1980	1.17	1.41	1.00	1.21	0.88	1.20
	1985	1.21	1.53	0.96	1.31	0.93	1.26
UK	1975	0.73	0.77	0.66	0.69	0.63	0.71
	1980	0.76	0.75	0.67	0.66	0.75	0.71
	1985	0.85	0.94	0.85	0.84	0.84	0.87
Italy	1975	1.42	1.64	1.38	1.41	1.62	1.47
	1980	1.76	1.97	1.60	1.56	1.97	1.72
	1985	1.81	2.01	1.87	1.83	1.59	1.88

^{a}Value-added per employed person.
^{b}GDP per employed person.
Source: UN, 1988.

Table 14A.6. Relative labor productivity^a in fabricated metal products sectors in Eastern Europe and the USSR (total labor productivity^b = 1.0).

Country	Year	381	382	383	384	385	38
USSR	1985	0.67	0.91	0.85	0.94	0.91	0.90
Bulgaria	1985	1.05	1.42	1.36	1.47	1.42	1.37
CSFR	1980	0.84	1.24	1.08	1.24	1.18	1.16
	1985	0.90	1.24	1.19	1.29	1.27	1.20
GDR	1985	0.69	0.93	0.90	0.97	0.94	0.91
Hungary	1975	0.74	0.87	0.86	0.90	0.83	0.85
	1980	0.90	0.99	1.00	1.20	1.13	1.04
	1985	1.07	1.22	1.30	1.42	1.44	1.29

^{a}Net output per employed person.
^{b}NMP per employed person.
Source: UN, 1985; UN, 1988.

Appendix II: Diffusion Scenarios for CIM Technologies

Table 14A.7. NC machines diffusion in Japan's metalworking industries: share of NC machines in total machine tools by sector, in percent.

Year	382	383	384	385
1981	4.57	3.64	2.56	3.12
1987	12.04	13.46	9.80	9.15
1978–1982[a]	19.09	17.20	9.26	14.53
1983–1984[a]	28.80	27.92	20.01	21.84
1985–1987[a]	36.32	34.86	31.20	24.80
2000	38.00	37.00	35.00	26.00

[a]Share of NC machines in newly installed machine tools during the corresponding periods.
Source: 1978–1987, Tani, 1989b; 2000, extrapolation based on the vintage structure of machine tools.

Table 14A.8. NC machines diffusion in US metalworking industries: share of NC machines in total machine tools by sector, in percent.

Year	381	382	383	384	385
1963	0.02	0.14	0.13	0.18	0.07
1968	0.15	0.62	0.62	0.74	0.27
1973	0.33	1.37	0.92	1.26	0.80
1978	0.89	3.09	1.39	2.52	1.89
1983	2.79	6.52	4.83	5.04	4.45
1989	5.62	10.54	7.10	12.81	12.02
2000	7.40	12.60	19.00	35.50	25.80

Source: 1963–1989, *American Machinist*, issues for the corresponding years; 2000, logistic curve extrapolation.

Table 14A.9. NC machines diffusion in UK metalworking industries: share of NC machines in total machine tools by sector, in percent.

Year	381	382	383	384	385
1971	0.5	0.6	0.3	0.6	0.4
1977	0.7	1.6	1.7	2.0	1.3
1982	1.1	3.3	3.0	3.0	3.5
1987	4.3	7.4	6.9	8.1	7.9
1982–1987[a]	11.4	16.4	15.5	20.2	18.4
2000	12.6	19.5	17.3	22.0	20.5

[a]Share of NC machines in newly installed machine tools during 1982–1987.
Source: 1971–1987, III to VI Survey of Machine Tools and Production Equipment in Britain; 2000, extrapolation based on the vintage structure of machine tools.

Table 14A.10. Industrial robot penetration in Japan's metalworking industries: units of IR per thousand employees.

Sector	Share of total IR shipments in 1985[a] (%)	1985[b,d] (Total shipments of IR: 178,800 units)	1995[b,c] Low scenario (Total shipments of IR: 280,000 units)	High scenario (Total shipments of IR: 450,000 units)
381	4.1	9.1	14	22
382	9.1	12.6	19	31
383	33.9	33.8	52	83
384	25.3	47.7	73	117
385	5.1	38.5	59	95
38	77.5	27.3	42	67

[a]Tani, 1989a.
[b]Tani, 1987.
[c]Tchijov and Norov, 1989.
[d]UN, 1988.

Table 14A.11. Number of FMS installations by country, 1987–1988.

Country	Number of FMS	Country	Number of FMS
USA	139	USSR	56
FRG	85	Bulgaria	15
Japan	213	CSFR	23
France	72	GDR	30
UK	97	Hungary	7
Italy	40		

Source: Tchijov, 1991.

References

Abel, I., and I. Székely, 1989, *Technical Progress and New Logistics Technologies*, WP-89-21, IIASA, Laxenburg, Austria.

Arrow, K., 1962, "The Economics of Learning by Doing," *Review of Economic Studies* 2:153–173.

Ayres, R.U., and J. Ranta, 1989, "Factors Governing the Evolution and Diffusion of CIM," in J. Ranta, ed., *Trends and Impacts of Computer Integrated Manufacturing*, WP-89-1, IIASA, Laxenburg, Austria.

Eltis, W.A., 1973, *Growth and Distribution*, Macmillan, London, UK.

Haustein, H.-D., 1989, "Automation and a New Organizational Mode of Production," in J. Ranta, ed., *Trends and Impacts of Computer Integrated Manufacturing*, WP-89-1, IIASA, Laxenburg, Austria.

Kaldor, N., 1957, "Model of Economic Growth," *Economic Journal* 4.

Kinoshita, S., and M. Yamada, 1989, "The Impact of Robotization on Macro and Sectoral Economies within a World Econometric Model," in J. Ranta, ed., *Trends and Impacts of Computer Integrated Manufacturing*, WP-89-1, IIASA, Laxenburg, Austria.

Krelle, W., 1984, "Waves of Entrepreneurial Activity Induced by Transfer of Information and Valuation," *Zeitschrift für Nationalökonomie*, Suppl. 4:71–92.

Krelle, W., 1985, *Theorie des Wirtschaftlichen Wachsturms*, Springer-Verlag, Berlin, Heidelberg, New York.

Krelle, W., ed., 1989, *The Future of the World Economy: Economic Growth and Structural Change*, Springer-Verlag, Berlin, Heidelberg, New York.

Leontief, W., and F. Duchin, 1986, *The Future Impact of Automation on Workers*, Oxford University Press, New York, NY.

Maly, M., 1989, "Strategic, Organizational and Social Issues of CIM: International Comparative Analysis," in J. Ranta, ed., *Trends and Impacts of Computer Integrated Manufacturing*, WP-89-1, IIASA, Laxenburg, Austria.

Maly, M., and P. Zaruba, 1989, "Prognostic Model for Industrial Robot Penetration in Centrally Planned Economies," in J. Ranta, ed., *Trends and Impacts of Computer Integrated Manufacturing*, WP-89-1, IIASA, Laxenburg, Austria.

Ranta, J., and I. Tchijov, 1989, "Economics and Success Factors of Flexible Manufacturing Systems: The Classical Theory Revisited," in J. Ranta, ed., *Trends and Impacts of Computer Integrated Manufacturing*, WP-89-1, IIASA, Laxenburg, Austria.

Tani, A., 1987, *Future Penetration of Advanced Industrial Robots in the Japanese Manufacturing Industry: An Econometric Forecasting Model*, WP-87-95, IIASA, Laxenburg, Austria.

Tani, A., 1989a, "International Comparison at Industrial Robot Penetration," *Technological Forecasting and Social Change* 34:191–210.

Tani, A., 1989b, "Saturation Level of NC Machine-Tool Diffusion," in J. Ranta, ed., *Trends and Impacts of Computer Integrated Manufacturing*, WP-89-1, IIASA, Laxenburg, Austria.

Tchijov, I., 1991, "The Diffusion of Flexible Manufacturing Systems," in R. Ayres, W. Haywood, and I. Tchijov, eds., *Computer Integrated Manufacturing*, Volume III: *Models, Case Studies, and Forecasts of Diffusion*, Chapman and Hall, London, UK.

Tchijov, I., and A. Alabian, 1989, "Flexible Manufacturing Systems (FMS): Main Economic Features," in J. Ranta, ed., *Trends and Impacts of Computer Integrated Manufacturing*, WP-89-1, IIASA, Laxenburg, Austria.

Tchijov, I., and E. Norov, 1989, "Forecasting Methods for CIM Technologies," *Engineering Costs and Production Economics* 17:323–329.

UN, 1985, *Industrial Statistics Yearbook, 1985*, United Nations, New York, NY.

UN, 1988, *Handbook of Industrial Statistics, 1988*, UNIDO, Vienna, Austria.

Warnecke, H.-J., 1989, "Integration of Information and Material Flow in a Pilot Plant – Development and Experiences," in J. Ranta, ed., *Trends and Impacts of Computer Integrated Manufacturing*, WP-89-1, IIASA, Laxenburg, Austria.

Chapter 15

Macroeconomic Impact of Structural Changes in the 1990s: Consequences for CIM

Robert Boyer

15.1 The Need for a New Analytical Approach

During the past 30 years it has been common practice to emphasize that technical change has rarely played a major role in the history of economic theory. While specialists in technical change have attributed the bulk of total factor productivity to advances in the technical and scientific knowledge base (Solow, 1957; Denison, 1987), macroeconomists have explained short- and medium-term adjustments as productivity and labor-force trends generated by external forces. With possibly the sole exception of R&D, there has been an almost complete dichotomy between the technological and economic spheres.

The economic growth in OECD member countries was strong and relatively stable as they progressed along the technological path laid down after World War II: the approximation provided by these theories appeared to be adequate. However, the situation changed in the 1980s. The decline of certain branches that were formerly powerhouses for development, the flowering of innovations with the potential to initiate an overall restructuring of the productive system, and the as yet uncertain direction of the new technological pathway have brought the issue of the relationship between science, technology, economic growth, and employment to the forefront. Far from being merely an external by-product, innovation is destabilizing or reshaping a large proportion of the earlier organizational structures. The integration of innovation into macroeconomic analysis results in new structures. The integration of innovation into macroeconomic analysis is therefore a necessity, even though this task may prove to be extremely difficult.

This situation provides an opportunity to reconsider the few theoretical constructs that accord technical change a central role in the process of economic development. It should come as no surprise therefore to find that Schumpeterian intuitions have been reinstated and utilized in new lines of research. This new research addresses not only technical change, but also organizational and institutional changes. Indeed, it is clear that technologies relating to information processing play a major part in the definition of requisite skills, the role of employees within firms, and even the workings of labor markets; not to mention the fact that even government intervention seems to be increasingly subordinated to the pressing need to be competitive and to modernize industry. As a result, the interfaces between technical innovations and determinants of activity at a macroeconomic level have proliferated.

The unavoidable conclusion is that there is a damaging discrepancy between the present exigencies of positive adjustment policies and the rudimentary nature of the analytical instruments available for reviewing such policies. The diffusion of new technologies is expected to stimulate investment and consequently improve productivity and quality, both of which are considered to be prerequisites for the recovery of national competitiveness. After a period of gestation, the prospects for growth in the economy and employment should therefore improve. In contrast, it is extremely difficult to accommodate the probable consequences of such policies in macroeconomic models. Faced with an inadequate integration of technological and organizational variables, econometricians are reduced to modifying exogenously the variables determining the formal relationship between productivity, investment, and so on.

The impact of computer integrated manufacturing (CIM) might be a good example for analyzing the relevance of economic theory and econometric modeling within this domain. On one side, this emerging organization provides an opportunity to study the technological and macroeconomic foundations for its viability and diffusion. On the other side, if considered as an alternative to Fordism, CIM might have, in the long run, significant (if not exclusive) influences on the growth regime. Similarly, given the variety of methods used by the IIASA Project, a systematic comparison of their strengths and weaknesses might deliver fruitful insights for future researches about CIM as well as technical change in general.

15.2   Paradoxes of the 1980s: Researches on CIM

In recent years, numerous examples have come to light of discrepancies between the predictions made in the most widely respected theories and the actual developments observed. Three paradoxes have unerringly attracted the attention of commentators with respect to the relationship between technology and economics.

The R&D paradox may simply be expressed as follows: given the fact that R&D expenditure is accelerating, why is it that, except in one or two cases, productivity has failed to make a significant recovery? From a more general point of view, what contribution does science and technology policy make to the medium- and long-term development of productivity? It is quite conceivable that the other components of economic policy, for instance, maintenance of high interest rates and a certain budgetary austerity, may inhibit the expected benefits of technological innovation. This is one point that warrants investigation.

The employment paradox arises from the apparent conflict between the implicit model that provides a basis for numerous economic policies and the reality of sequential progressions that have been in effect for the past 15 years. Policymakers and economists basically consider that accelerating the diffusion of new technologies should boost employment by increasing competitive supply. In fact, the results paint a much more ambiguous, if not contradictory, picture: in the United States, paltry performances of productivity have not precluded the pursuit of job-creating growth, whereas in most European countries buoyant productivity does not seem to have lowered their extremely high rates of unemployment. Another question raised is that of the relationship between exposed and sheltered sectors, and between industry and services. Are macroeconomic scenarios, which are essentially based on industrial logic, still valid when the focus of employment has shifted to the tertiary sector?

The paradox of the new technical system represents a third topic for analysis. All economic agents are sensitive to the impact of innovations, in that the novelty of the innovation has a concomitant effect on product ranges, organization of work, management of firms, market operation, products, and labor practices. On the other hand, econometric estimates based on aggregate data paint a somewhat different picture in which most basic relationships remain more or less stable. As a result, many economists are tempted to conclude that technical development, in an initial approximation, has no effect on macroeconomic regulation. Is this discrepancy due to the inadequacy of review instruments (the total economy rather than the sector or firm) or, more insidiously, does its cause lie in a failure to account for technical changes?

CIM provides an opportunity to address most of these paradoxes. Of course, its implementation is by definition restricted to manufacturing activities, but the same tools could probably be applied to information and telecommunication technologies, which are transforming many service activities (insurance, banking, general administration, welfare system, engineering). But the new manufacturing principles give a very precise content to the general hypothesis about an emerging new technical system. The various researches undertaken by IIASA therefore deliver very valuable assessments about the respective relevance of alternative analytical tools.

15.3 CIM Capital Stock in Aggregate Production Functions or Technical Progress Relations

Insofar as research into new products and processes is becoming increasingly subordinated to the achievement of economic objectives, the obvious analytical approach to adopt consists in considering that technical progress is itself the outcome of the allocation of labor and equipment and that, in short, the principles of production theory are applicable to technical progress. Nevertheless, this particular approach may result in analyses that give undue emphasis to other factors. From this point of view, it might be useful to make a tentative synthesis of the main theories concerning the economic determinants of innovation, with particular reference to observations concerning the past 20 years.

First, many studies (Griliches, 1980, 1984; Mairesse and Cuneo, 1984; Soete and Patel, 1985; OECD, 1987) have assessed the capital stock associated with the past flow of R&D, and have then estimated a generalized production function that, in addition to the usual factors (labor, plant, energy, and raw materials), includes this new variable.

It would therefore be helpful to update the synthesis work already completed in the light of the econometric research and specific sectoral studies that have since been carried out. An attempt could be made to take stock of the various questions still unanswered: Are the lagged effects due to the slowdown of R&D in the mid-1970s or to a drop in the marginal efficiency of R&D expenditure? Are problems due to the volume or to the slant and methods of financing? What are the scale and speed of the diffusion of new innovations among firms and among sectors?

Second, it might be instructive to establish the relevance of theses that put forward market buoyancy as a major stimulus for innovation activity and likewise for diffusion. On the one hand, note should be taken of the correlation between patent data and the vigor of demand in the sector concerned – a factor that has already been identified with respect to an earlier industrial revolution (Schmookler, 1966). On the other hand, according to the spiritual heirs of Smith, the depth of the division of labor, and consequently the invention of new production processes, is governed by market size. This school of thought, which has been rehabilitated by Kaldor (1966), developed a meaningful interpretation of growth in the 1950s and 1960s. Even though the hypothesis fails to explain all the stylized facts characterizing the 1980s (Michl, 1985; Boyer and Petit, 1988), it does provide a meaningful interpretation whereby, *ceteris paribus*, innovation activity will be slowed down by uncertainty and less buoyant aggregate demand.

Whereas technical change is increasingly considered to be driven by R&D expenditure, it is worth investigating the possibility that it may also be partially led by the buoyancy of investment or demand or both. If so, there might be a minimum growth threshold below which the cumulative diffusion of innovations, even major innovations, would be blocked.

In the same vein, contemporary economists have perhaps acted too hastily in discarding a hypothesis that was once widely used and which postulates that the direction and even the rate of technical change respond to the relative cost of the different factors or even income distribution (Fellner, 1961).

This hypothesis was fairly widely accepted with regard to energy following the surge in its relative price in the aftermath of the two oil shocks. However, it could equally well be applied to capital–labor substitution in the new technologies, in that it would provide an explanation of why present industrial innovations are still responding to the fall in returns in the early 1970s whereas the equilibrium of income distribution has been restored. This approach would explain the increase in labor productivity as the 10- or 20-year lagged effect of pressures on income distribution (Sylos-Labini, 1984).

In adopting the taxonomy proposed by Pavitt (1984), it might perhaps be advisable to make a distinction between different types of determinants of technical change according to whether the change is the result of a basic research effort, the acquisition of equipment and materials, or innovation within firms in response to routine management problems. This raises the possibility of formulating different technical progress equations for the three categories. On the other hand, at the aggregate level, surely the above factors should be used in combination (R&D, market buoyancy, relative cost, and income distribution).

According to this typology, CIM is resulting from previous R&D expenditures – expanding technical and organizational knowledge – and delivers a new generation of manufacturing equipment: in a sense, CIM would be the modern successor to the Fordist assembly line, given the extension of product differentiation allowed by modern information and data-processing techniques. Ideally, the share of investment in modern manufacturing equipment with respect to total investment would have to be included into generalized production function (Dobrinsky, 1989). Still more, the potential performances of early CIM equipment remain quite superior to actual indices for efficiency and productivity increases: learning-by-using effects are far from absent in these modern equipments. In fact, a significant polyvalence and commitment of blue-collar workers are required to improve economic performances, as clearly shown by studies on Japanese firms (Watanabe, 1989). No doubt that this theme will be documented in Volume V on management and organization (Ranta, et al., forthcoming).

15.4   From Catching-up Models to CIM Diffusion Models

Rostow (1960) postulated that when countries industrialize they all pass through the same stages of development. Accordingly, US history should provide advance indication of the paths that are likely to be followed by Europe and Japan. This hypothesis, which is highly general in scope and also strongly contested, has several features in common with the interpretation of the slowdown in growth and productivity in OECD member countries. At the end of World War II, the United States was at the leading edge of technological and organizational innovation, and therefore enjoyed the highest productivity levels, while being the source of most of the new products that shaped contemporary modes of living. Although starting with older production techniques, the other countries have gradually caught up with the United States. Accordingly, the high productivity gains achieved in the 1950s and 1960s are considered merely to represent a transitional stage.

This line of argument clearly bears the stamp of the analyses of Gomulka (1971), whose work has been advanced by many authors including Madison (1977, 1987). Although this model sheds some light, it still falls far short of explaining all the developments observed. First, it postulates a fairly mechanical process whereby an economy modernizes its productive structures faster when it starts from a lower level of productivity. This approach effectively disregards the role of internal organizational structures which, depending upon circumstances, can either accelerate or, on the contrary, obstruct the adoption of technologies (Hodgson, 1987). Second, even if econometric adjustments were to confirm the essence of the link, the model still falls far short of accounting for the differences in national development paths (Johnson and Lundvall, 1988). The limitations of this model become clearly apparent if an analysis is made of the reasons for (or failure to achieve) industrial takeoff in Third World countries over the past 20 years. Finally, the model provides no explanation for the slowdown in US productivity, which is an issue of crucial importance.

A more microeconomic approach might consider that productivity should vary not only in accordance with inventions and innovations, but also with the speed at which they spread throughout the economic system. In the models developed by Griliches (1957) and Mansfield (1968), the diffusion of products and processes follows a logistic curve. This implies that there is a positive correlation between the speed of diffusion and the returns afforded by the innovation, and a negative correlation between speed of diffusion and the size of the investment required. In addition, an extensive literature attests to the fact that the type of competition has a definite effect, albeit complex and not easily reduced to a few simple ratios, on the speed of diffusion. At a more general level, diffusion can be brought to a standstill by management problems within firms, a lack of coordination between organizations, a lack of training, and even confrontational industrial relations.

In terms of macroeconomics, can low productivity gains be justifiably attributed to the slow speed of diffusion of new technologies (CIM for manufacturing sectors, coherent and efficient information system for the service sectors)? Were this to be the case, should the cause be attributed to low effective rates of return afforded by new processes or the size of investment, particularly that in infrastructure necessary for the investment to be fully effective? To what extent does the uneven diffusion of electronic capital goods explain the relative productivity levels and degrees of competitiveness within the different OECD economies and, by extension, their growth and external trading position (Petit *et al.*, 1987)? At a more general level, is it possible to include variables representing diffusion determinants, which are themselves conditioned by such factors as rates of return, market growth, and real interest rates, in the aggregate macroeconomic functions?

To be more specific about the impact of CIM, let us suggest two complementary approaches. On the one hand, it is important to adapt standard diffusion models to the specific features of integrated manufacturing: evolution of capital output ratio at full capacity utilization, relative unit production costs, reactions to unexpected demands in volume and composition, and consequences of an increased adjustment speed upon oligopolistic power and market share. Volume III of the CIM Series, *Models, Case Studies, and Forecasts of Diffusion*, precisely addresses this issue. On the other hand, given the potential structural change associated with CIM, the stock of related modern equipment, as well as the learning effects linked to its progressive implementation, could capture a modernization effect, to be included within either a conventional production function or a technical progress function, along the seminal proposal by Arrow (1962). Maybe IIASA's experts could benefit from the various tools elaborated by people who have been trying to capture the modernization impact of new and sophisticated equipment goods, for example, within the textile sector (Petit *et al.*, 1987).

15.5 Reconsideration of Cumulative Growth Models

These models have the advantage of offering an integrated approach to technical change and long-term economic trends in that innovation in the division of labor and the mobilization of specialization and learning effects go hand in hand with the extension of markets, with the result that dynamic returns to scale and growth are intimately connected within economies dominated by the manufacturing industry. This formulation is clearly based on concepts advanced by authors such as Myrdal (1957), Young (1928), Verdoorn (1949), and Kaldor (1966, 1972, 1975). The role of industry as an engine of growth, the existence of returns to scale in this sector, and the inhibiting effect on growth exerted by reserves of manpower or by competitiveness in the international market are the three components on which this model is based.

Although these elements are extremely simple, they provide a fairly satisfactory interpretation of the comparative economic growth of OECD member countries, or at any rate during the period from 1950 to 1973. This model may be considered to correspond to the growth characteristics described by Ford, although perhaps he underestimated the organizational factors that afforded such growth (Boyer, 1988). In fact, above and beyond the few reduced-form equations initially proposed by Kaldor, it is possible to describe the structural relations conducive to growth. This model describes, first, the medium- and long-term productivity determinants (mobilization of returns to scale, investment ratios, slant and intensity of innovations) and, second, the conditions governing the distribution of productivity gains between employees, enterprises, and the outside world from which different demand regimes will ensue, in terms of the pull exerted by consumption or investment or the relative openness and competitiveness of the economy.

However, most econometric studies addressing the past 15 years confirm the breakup of some of the basic relations underpinning this model. On the one hand, the technical system based on specialized equipment (relatively unskilled labor) and mass production of standardized products is faltering and is facing growing difficulties in terms of productivity, quality, and the need to adapt to changing markets. On the other hand, the virtuous loop linking mass production and consumption on the domestic market is being compromised by the growing externalization of production, investment flows, and financial movements. This has brought about a breakup in most of the usual macroeconomic relationships at work in the domestic economy that may be seen as the result of a mismatch between the mode of economic management and its technical bases.

On the basis of this diagnosis, what will be the macroeconomic effects of the new technical system(s) emerging and the institutional changes that are affecting, for example, industrial relations, the labor markets, and competition? One possible result might be the emergence of a variety of growth regimes, depending upon the type of synchronism established between the way in which productivity gains are achieved and the distribution of the product. As a result, the medium- and long-term trends in productivity might combine, in different proportions, returns to scale, intensity of tangible and intangible investment, volume and slant of R&D, and labor-learning curves.

This framework has been used for historical studies of long-run manufacturing growth (Caussat, 1981; Leroy, 1988) and for cross-national analyses of relative growth performance after World War II (Boyer and Petit, 1989). But now, given the large institutional and technological transformations clearly exhibited by econometric studies (Boyer and Mistral, 1988), the issue has to turn from retrospective to prospective: given the observed trends, is a coherent and stable new growth regime emerging? A previous study (Boyer and Coriat, 1986) began to address the question. For example, the electronization for equipment goods, the fulfillment of which is represented by CIM, can either deliver a model of flexible specialization (Piore and Sabel, 1984) or introduce more product differentiation within mass production, as the Japanese model would suggest (Boyer, 1989a). According to these very crude models, flexible specialization would only slow the growth pattern of modern economies, whereas flexible automation is powerful enough to trigger a renewal of growth. Fleissner and Polt (1990) extend and refine the analysis by an explicit formalization of the dynamic process of competition between Fordist and CIM manufacturing processes. We must now calibrate the related parameters in the light of the detailed data extracted from the IIASA studies.

15.6   Product Innovations in Input–Output Models and CIM

The major failing of most approaches is that they address aggregates, thereby disregarding the fact that technical change produces new products and helps to create new industries while contributing to the decline of older or more mature branches. It is therefore legitimate to question the relevance of purely aggregate models. In fact, such models assume that equipment, workers, and know-how are perfectly adaptable and can therefore be switched from one industry to another. However, the specificity of industries is an essential factor in the transition from one technological system to another. The sectoral mismatches produced by technical change may have a crucial effect upon macroeconomic performance, particularly with respect to employment.

The first steps to an approach would be to collect all the technical coefficients relating to major innovations of a period (e.g., office automation and robotics), to survey probable trends in the demand for new products, and then to collate the data using the traditional input–output matrices. In this way account would be taken of interdependences, which through intermediate consumption, investment, and final consumption may reveal the multiplier effects of new technologies. In many cases analysis focuses on the effects of both volume and skills on employment (Leontief and Duchin, 1985; Ayres *et al.*, 1987).

Bearing in mind the number, variety, and cost of such studies, it would be advisable to establish both the potential input they could provide and their limitations. In particular, how could the results obtained be directed toward a more traditional form of the macroeconomic model? Would it be possible, for example, to limit the number of sectors, to those unaffected by major innovations in information technology or to those transformed by the innovation or to new sectors built up around specific innovations? Would it then be possible to investigate how flows of workers conform to occupational trends in the labor force?

Further development along these lines might address the scale of direct and indirect multiplier mechanisms in conjunction with the predicted broad characteristics of the new technical system. It will be recalled, for example, that in the postwar model consumer durables and car manufacturing displayed tightly knit consistency which fueled a series of industrial investments and infrastructures (urban development and transport). Would it therefore be possible to devise an equivalent system by means of an adequate combination of information technologies, new materials, biotechnologies, space technology, and so on?

Starting out from the matrix of sectoral R&D and innovation flows, the analysis could be directed toward identifying sectors that have the long-term potential to become pacesetters, particularly with regard to productivity. Does CIM belong to this category? One possibility would be to use the results of dominance theory. The final objective might be to identify the major explanatory factors to be incorporated in a possible aggregate equation for productivity since inadequate account is taken of sectoral evolution in total factor productivity. The aim would be to reintroduce some of these variables into aggregate macroeconomic models.

In most cases, applied studies addressing input–output indicators show the long-term impact of a whole series of technical changes without describing the intermediate sequences. But the process by which an economy progresses from one technical system to another is itself important; innovations may initially compound employment problems without instigating a massive recovery in productivity, and this movement may only be reversed through the comprehensive diffusion of such innovations. If full employment is to be maintained, the demand for new products and productivity gains calls for proper management of aggregate demand (Pasinetti, 1981). In an open economy, this very problem has produced some interesting results (Ros, 1986): specialization that initially proved to be profitable in accordance with a Ricardo-type static criterion for comparative advantages may prove to be unprofitable in the long term once growth has fallen in the sector in question (mature or declining products) and once the economy forgoes the corresponding learning effects. Similarly, financial tensions and shortages of skilled labor may arise during the transition period and thereby jeopardize the viability of the new system (Amendola and Gaffard, 1988).

Would it be possible, on the basis of these various sectoral models, to abstract a model that would represent, preferably dynamically, the basic relationship between the old and the new technical systems to establish the time path for the adjustment of a domestic economy? By extension, would it be possible to attempt to build a Schumpeterian-type model, i.e., one that would contrast routine with innovation as the characteristics of two discrete sectors? The book by Petit *et al.* (1987) about modern textile equipments and the paper by Fleissner and Polt (1990) clearly show the interest and viability of such models. Consequently, in a Schumpeterian approach, economic dynamics is conceived as resulting from the progressive replacement of old equipments and routines by more efficient ones, capturing the more recent advances in basic and applied researches.

15.7   Long-Wave Neo-Schumpeterian Models

The spectacular reversal of productivity growth trends and the shift from quasi-full employment to long-term high unemployment have given fresh relevance to Kondratieff-type long-wave approaches. Historians and statisticians have endeavored to determine whether or not such 50-year cycles exist, but a clear consensus has not been reached. In addition, an explanation had to be found for the sequences and causality that could give rise to such long cycles, which have tended to be neglected in favor of business cycles. Nonetheless, of the hypotheses put forward, Schumpeter's construct does have its attractions. In a Schumpeterian interpretation the upward phase would be the result of the creation of a technical system and a set of appropriate institutions which would then produce a cumulative movement combining the diffusion of innovation, profit, investment, and growth (Schumpeter, 1911, 1939). However, widespread diffusion of a cluster of major innovations would itself be limited by the erosion of monopolistic advantages as a result of general competitive pressure characterizing this stage of the process. This would explain the reversal of the process and the start of a downward phase.

Specialists in technical change have therefore updated and extended Schumpeter's intuitions. According to their interpretation, the industrialized OECD economies have been in the grip of a long-term downward phase since the late 1960s. Owing to the slowdown in investment and the heightening of competition, innovative processes have apparently gained ascendancy over innovative products, with unemployment in large part caused by the swing in the technical system within the unchanged system of economic and social institutions (Freeman *et al.*, 1982). In response to these problems, they argue that an alternative technical system is emerging based on information technologies, new materials, and biotechnologies. However, this system would conflict with the earlier configuration of organizational structures, thereby resulting in mismatches in the internal organization of firms, industrial relations, cooperation between firms and government, and the status of international relations (Perez, 1981). This interpretation would appear to be borne out by a series of converging case studies.

Perhaps it might be even more important to investigate the hypothesis of a lag between technical change and socio-institutional change. To this effect, an even balance would have to be struck in the treatment of institutional innovations (new types of cooperation between enterprises of the vertical quasi-integration type, recasting of wage relations and type of wage agreements applicable in pace-setting sectors, and so on) and technological innovations. Two different strategies would, *a priori*, be feasible.

First, the institutional aspects of the model previously described could be reinforced, thereby providing a specific description for the mismatch hypothesis whereby incompatibility would generate a set of structural parameters relating, first, to the technical system and, second, to socioeconomic management. The corresponding dynamic model would therefore be structurally unstable or would at least produce a cumulative downward movement in the absence of corrective measures linked with, for example, economic policy.

Second, the recasting of cumulative growth models might open the way to a dynamic representation of considerably greater complexity than those studied to date. The mismatch in this case could be ascribed to the incompatibility of a system based on productivity with one based on demand; both are created through the combined effects of the technical and organizational components of the economy. This interpretation would by opposition allow scope for the definition of changes that would lead to the recovery of high and stable growth.

Nevertheless let us mention an important methodological caveat. Most neo-Schumpeterian models deal with macroeconomic dynamics, whereas CIM is only one part of a global technical and organizational change. Therefore it is not easy to reconcile macroeconometric *ex post* analysis with *ex ante* partial simulations dealing with only a subpart of economic dynamics. The research by IIASA treats these difficulties in an effort to capture the impact of information technologies upon employment, growth, and skills at the European level (Standaert and van Zon, 1990). A systematic comparison of both researches would enlighten some of the unsettled issues.

15.8 Evolutionary Models of Technical and Institutional Change

Most of the preceding analyses have been based upon the simplistic and comforting hypothesis that new technologies, and particularly CIM, dominate older ones, regardless of pricing systems or demand prospects. Furthermore, all economic agents have knowledge *ex ante* of returns, with the result that diffusion is reduced to an almost mechanical process of adjustment to a technical organization that is clearly more efficient. In short, technology is for all intents and purposes a public good freely available to all and at no cost in terms of assimilation, learning, or implementation.

This set of hypotheses conflicts with what is known of the stylized facts relating specifically to technical innovation. Technical innovation is risky enough when pursued as a marginal activity in an existing technical system, but becomes a highly precarious venture should there be inconsistencies in the fabric of the system. Accordingly, enterprises have the choice of either innovating or imitating the processes and products developed by others. Bearing in mind the contingencies that govern such decisions, the soundness of a strategic course can only be determined *ex post*. If this is so, then a major overhaul is needed of the way in which technology is accounted for in economic models (Nelson and Winter, 1982). Technology does not become a free good, but it is more or less appropriated in its entirety and in all cases entails learning costs. The confrontation between various strategies adopted by firms only achieves compatibility through a sequential process of selection consisting of the elimination of enterprises through bankruptcy and the strengthening of enterprises that after initial success manage to remain on a course offering long-term viability.

The configuration of the socio-technical system therefore depends upon the paths that are followed. It is possible for two enterprises or two national systems to make different initial choices and embark upon diverging paths. At this point the concept of technological trajectory must be considered (Dosi, 1982; Dosi *et al.*, 1988). The value of this approach is all the greater if account is taken of the externalities relating specifically to certain major innovations, e.g., networks. If one particular technology gains an initial advantage, it is capable of definitively eliminating a later technology that might well prove to be superior were its diffusion possible (Arthur, 1987). This type of formulation runs counter to the idea that dominant technologies will always carry the day; this is not the case when the future is uncertain and significant externalities exist. *A contrario*, technologically superior equipment might appear as inefficient from an economic point of view, both at the micro and macro levels. For example, premature integrated manufacturing (the SATURN project in the USA) might propose a too daring innovation which might turn out as increasing the capital output ratio and promoting a new built-in rigidity, facing more flexible and divisible technologies.

Despite their appeal, these models do not as yet afford a potential alternative to the traditional macroeconomic approach of a production function augmented by technical change. Since simulations are intrinsically stochastic in nature only Monte Carlo methods will provide results, contrary to the deterministic use made of most standard macroeconomic models. This raises the important issue of whether it would be possible to translate the intuitions embodied in these evolutionary models into a simple macroeconomic model.

On the basis of available models (including those already cited: Silverberg, 1985, 1988; Eliansson, 1985, 1986), would it be possible to define a standard model that would explain the historical relevance and irreversibility of technical systems characterized by significant externalities? Would it be feasible to give these models a numerical basis to obtain some indications at the macroeconomic level along the lines of the formulation of links between the behavior of firms and macroeconomic dynamics proposed by Eliansson?

Finally, the role played by learning effects in the changeover from one technical system to another is receiving wider recognition. Certain theorists have even made this factor an essential characteristic of firms (Aoki, 1984, 1987). Furthermore, case studies have shown the importance of such factors as learning by doing (introduced into economic theory by Arrow in 1962, but subsequently somewhat neglected). Specialists in business management have demonstrated just how important this factor is with respect to both shop-floor workers and management (Adler, 1985, 1986). Similarly, it would seem that in the present context the chances of success for product innovations are enhanced by ongoing dialogue between designers and final users (Freeman, 1987), which ensures that full advantage is taken of the gains to be made from learning by using (Lundvall, 1988). It is therefore essential to review the results of the models to study the pathways followed as innovation is progressively integrated into production facilities, producers' know-how, and consumers' expertise (Amendola and Gaffard, 1988). The strategy in this respect, too, should be directed toward obtaining a sufficiently compact model for comparison with the more traditional concepts embodied in standard econometric models.

15.9   Analyses of CIM: A Contribution to the Economics of Technical Change

Whatever the theoretical framework adopted, the IIASA study will probably provide a benchmark in the analyses about contemporary technological and institutional changes. Adding a capital stock variable measuring investment in CIM is a first avenue for enriching conventional production functions. At a more disaggregated level, input–output methods should consider both process innovations (CIM as a means for improving total factor productivity) and product innovations (CIM as a device for capturing oligopolistic or monopolistic rents out of choosy consumers). Still more sufficiently rich mechanisms have to be considered to give technological optimism or neutralism a chance: impacts upon real wage and profit, consumption and investment, production capacity, and external competitiveness usually have positive impacts upon growth and employment. If these mechanisms are not taken into account, technical progress will ineluctably destroy more jobs than it will create, leading to a shaky and uncomfortable technological pessimism (Boyer, 1989b).

Past researches (specially Boyer and Coriat, 1986) suggest another avenue according to the regulation approach: the issue of CIM reflects the compatibility between a new emerging productivity regime and the transformation of a demand regime. A long-run growth model can exhibit the likelihood of a renewal of rapid and stable growth, given the estimates for the capital output ratio, the learning effects, and the input coefficients associated with CIM. But the more enlightening exercise deals with the dynamic competition between marginally transformed Fordist techniques and new productive organizations linked to modern information equipment: Under what conditions will CIM replace older investments and organizations? This would conform with the neo-Schumpeterian concept of development, growth, and cycles. The first activity in this approach is stimulating indeed, even if not totally conclusive (Fleissner and Polt, 1990). Of course, manufacturing is only a declining part of aggregate productive activities, but similar models could be done for the diffusion of information technologies within financial and administrative services (Standaert and van Zon, 1990).

It is to be hoped that the impressive amount of research now available from IIASA's CIM Project will stimulate many follow-up and innovative researches on contemporary organizational and technical change.

References

Adler, P., 1985, *Technology and the Future of the Firm: A Schumpeterian Research Agenda*, Mimeo, Department of Industrial Engineering and Engineering Management, Stanford University, Stanford, CA.

Adler, P., 1986, *When Knowledge is the Critical Resource, Knowledge Management is the Critical Task*, Mimeo, Department of Industrial Engineering and Engineering Management, Stanford University, Stanford, CA.

Amendola, M., and J.L. Gaffard, 1988, *The Innovation Choice: An Economic Analysis of the Dynamics of Technology*, Blackwell, Oxford, UK.

Aoki, M., 1984, *The Co-operative Game Theory of the Firm*, Clarendon Press, Oxford, UK.

Aoki, M., 1987, *A Microtheory of the Japanese Economy: Information, Incentives and Bargaining*, Mimeo, Kyoto Institute of Economic Research, Kyoto, Japan.

Arrow, K., 1962, "The Economic Implications of Learning by Doing," *Review of Economic Studies* 29.

Arthur, B., 1987, "Competing Technologies: An Overview," in G. Dosi *et al.*, eds., 1988, *Technical Change and Economic Theory: The Global Process of Development*, Frances Pinter, London, UK.

Ayres, R.U., 1985, "A Schumpeterian Model of Technological Substitution," *Technological Forecast and Social Change* **27**.

Ayres, R.U., H.-U. Brautzsch, and S. Mori, 1987, "Computer Integrated Manufacturing and Employment: Methodological Problems of Estimating the Employment Effects of CIM Application on the Macroeconomic Level," Paper presented at Colloquium GERTTD AMES, Paris, 2–4 April, IIASA, Laxenburg, Austria.

Boyer, R., 1988, "Assessing the Impact of R&D on Employment: Puzzle or Consensus?" Paper prepared for the International Conference on New Technology: Its Impact on Labor Markets and the Employment System, 5–7 December, Berlin, Germany.

Boyer, R., 1989a, *The Eighties: The Search for Alternatives to Fordism, A Very Tentative Assessment*, Working Paper CEPREMAP, Paris, France, No. 8909.

Boyer, R., 1989b, "New Directions in Management Practices and Work Organisation: General Principles and National Trajectories," Paper prepared for the OECD Conference on Technical Change as a Social Process: Society, Enterprises and Individual, December 11–13, Helsinki, Finland.

Boyer, R., and B. Coriat, 1986, "Technological Flexibility and Macro Stabilization," *Ricarche Economiche* **XL**(4), October–December: 771–835.

Boyer, R., and J. Mistral, 1988, "Le bout du tunnel? Stratégies conservatrices et nouveau régime d'accumulation," Paper presented at the Colloque International sur la Théorie de la Régulation, 16–18 June, Barcelona, Spain.

Boyer, R., and P. Petit, 1988, "The Cumulative Growth Model Revisited," *Political Economy* **1**(4).

Boyer, R., and P. Petit, 1989, "Technical Change, Cumulative Causation and Growth: Accounting for the Contemporary Productivity Puzzle with Some Post-Keynesian Theories," Paper presented at the OECD International Seminar on Science, Technology and Growth, Workshop IV, 5–8 June, Paris, France.

Caussat, L., 1981, *Croissance, emploi, productivité dans l'industrie américaine (1899–1976)*, Mimeo, CEPREMAP, September, Paris, France.

Denison, E.F., 1987, "Growth Accounting," in J. Eatwell *et al.*, eds., *The New Palgrave: A Dictionary of Economics*, Macmillan, Basingstoke, UK.

Dobrinsky, R., 1989, *Macro Analysis of the Economic Impact of CIM Technologies: An International Comparison Based on a Global Econometric Model*, Unpublished paper, IIASA, Laxenburg, Austria.

Dosi, G., 1982, "Technological Paradigms and Technological Trajectories: A Suggested Interpretation of the Determinants and Directions of Technical Change," *Research Policy*.

Dosi, G., C. Freeman, R. Nelson, G. Silverberg, and L. Soete, eds., 1988, *Technical Change and Economic Theory: The Global Process of Development*, Frances Pinter, London, UK.

Eliansson, G., 1985, *Dynamic Micro-Macro Market Coordination and Technical Change*, Working Paper No. 139, Industrial Institute for Economic and Social Research, Stockholm, Sweden.

Eliansson, G., 1986, "Innovation Change, Dynamic Market Allocation and Long-Term Stability of Economic Growth," Paper presented at the Conference on Innovation Diffusion, 17–22 March, Venice, Italy.

Fellner, W., 1961, "Two Propositions in the Theory of Induced Innovations," *The Economic Journal*, June.

Fleissner, P., and W. Polt, 1990, "Switching from a Fordism to a CIM Accumulation Regime: A System Dynamics Model," Paper presented at the Conference CIM: Revolution in Progress, July 1–4, 1990, IIASA, Laxenburg, Austria.

Freeman, C., 1987, "Structural Unemployment," in J. Eatwell *et al.*, eds., *The New Palgrave: A Dictionary of Economics*, Macmillan, Basingstoke, UK.

Freeman, C., J. Clark, and L. Soete, 1982, *Unemployment and Technical Innovation: A Study of Long Waves and Economic Development*, Frances Pinter, London, UK.

Gomulka, S., 1971, *Inventive Activity, Diffusion and the Stages of Economic Growth*, Institute of Economics, Aarhus, Denmark.

Griliches, Z., 1957, "Hybrid Corn: An Exploration in the Economics of Technological Change," *Econometrica*, October: 501–522.

Griliches, Z., 1980, "R&D and the Productivity Slowdown," *American Economic Review* 70, May: 434–448.

Griliches, Z., 1984, *R&D, Patents and Productivity*, NBER, University of Chicago Press, Chicago, IL.

Hodgson, G., 1987, *Institutional Rigidities and Economic Growth*, Mimeo, Newcastle upon Tyne, UK.

Johnson, B., and B.A. Lundvall, 1988, "Institutional Learning and National Systems of Innovation," Paper presented at the Conference on Strategies of Flexibilization in Western Europe: Techno-Economic and Socio-Political Restructuring in the 1980s, April 6–10, Roskilde Universitetcenter, Denmark.

Kaldor, N., 1966, *Causes of the Slow Rate of Growth of the United Kingdom*, Cambridge University Press, Cambridge, UK.

Kaldor, N., 1972, "The Irrelevance of Equilibrium Economics," *Economic Journal* 82, December; reprinted in *Further Essays on Economic Theory*, 1978, Duckworth, London, UK.

Kaldor, N., 1975, "Economic Growth and the Verdoorn Law: A Comment on Mr. Rawthorn's Article," *The Economic Journal*, December.

Leontief, W., and F. Duchin, 1985, *The Future Impact of Automation on Workers*, Oxford University Press, New York, NY.

Leroy, C., 1988, *Un modèle de croissance de longue période de l'industrie manufacturière américaine*, Mémoire de D.E.A., Mimeo, E.H.E.S.S., October.

Lundvall, B., 1988, "Innovation as an Interactive Process: From User-Producer Interaction to the National System of Innovation," in G. Dosi *et al.*, eds., *Technical Change and Economic Theory: The Global Process of Development*, Frances Pinter, London, UK.

Madison, A., 1977, "Phases of Capitalist Development, Banca Nazionale del Lavoro," *Quarterly Review*, December.

Madison, A., 1987, "Growth and Slowdown in Advanced Capitalist Economies: Techniques of Quantitative Assessment," *Journal of Economic Literature*, June: 649–698.

Mairesse, J., and Ph. Cuneo, 1984, "Productivity and R&D at the Firm Level in French Manufacturing," in Z. Griliches, ed., *R&D, Patents, and Productivity*, NBER, University of Chicago Press, Chicago, IL.

Mansfield, E., 1968, *Industrial and Technological Innovation: An Econometric Analysis*, W.W. Norton, New York, NY.

Michl, Th., 1985, "International Comparisons of Productivity Growth: Verdoorn's Law Revisited," *Journal of Post-Keynesian Economics* III(4), Summer.

Myrdal, G., 1957, *Economic Theory and Underdeveloped Regions*, Duckworth, London, UK.

Nelson, R., and S.G. Winter, 1982, *An Evolutionary Theory of Economic Change*, Harvard University Press, Cambridge, MA.

OECD, 1987, *Diverses Notes du Groupe de Travail No. 1 du Comité de Politique Economique*, Paris, France.

Pasinetti, L., 1981, *Structural Change and Economic Growth*, Cambridge University Press, London, UK.

Pavitt, K., 1984, "Sectoral Patterns of Technical Change: Towards a Taxonomy and a Theory," *Research Policy* 6: 343–374.

Perez, C., 1981, "Structural Change and Assimilation of New Technologies in the Economic and Social Systems," *Futures* 15(5), October.

Petit, P., Ch. Antonelli, and G., Tahar, 1987, "Technological Diffusion and Firm's Investment Behavior: The Case of Textile Industry," *International Journal of Industrial Organisation*.

Piore, M., and Ch. Sabel, 1984, *The Second Industrial Divide: Possibilities of Prosperity*, Basic Books, New York, NY.

Ranta, J., J.E. Ettlie, and R. Jaikumar, forthcoming, *Computer Integrated Manufacturing*, Volume V: *Fewer and Faster: A Story of Technological, Organizational, and Management Innovation in the Manufacturing Enterprise*, Harvard Business Monograph Series, Harvard Business Press, Cambridge, MA.

Ros, J., 1986, "Trade, Growth and the Pattern of Specialization," *Political Economy* 2(1): 55–72.

Rostow, W., 1960, *The Stages of Economic Growth*, Cambridge University Press, London, UK.

Schmookler, J., 1966, *Invention and Economic Growth*, Harvard University Press, Cambridge, MA.

Schumpeter, J., 1911, *Théorie de l'évolution économique*, Edition Française Dalloz, (1935) Paris, France.

Schumpeter, J., 1939, *Business Cycles*, Procupine Press, Philadelphia, PA.

Silverberg, G., 1985, "Technical Progress, Capital Accumulation and Effective Demand: A Self-Organisation Model, Institute for Social Research," in D. Batten, ed., *Economic Evolution and Structural Change*, Springer-Verlag, Berlin, Heidelberg, New York.

Silverberg, G., 1988, "Modelling Economic Dynamics and Technical Change: Mathematical Approaches to Self-Organization and Evolution," in G. Dosi *et al.*, eds., *Technical Change and Economic Theory*, Frances Pinter, London, UK.

Soete, L., and P. Patel, 1985, "Recherche-développement, importations de technologie et croissance économique," *Revue Economique* 5, September: 975–1000.

Solow, R.M., 1957, "Technical Change and the Aggregate Production Function," *Review of Economics and Statistics*, August.

Standaert, S., and A. van Zon, 1990, *A First Attempt to Model the Employment Effects of the Use of "New" Information Technologies within the Context of the European HERMES Model*, Mimeo, Maastricht, Netherlands.

Sylos-Labini, P., 1984, *The Forces of Economic Growth and Decline*, MIT Press, Cambridge, MA.

Verdoorn, P.J., 1949, "Fattori che regolano lo sviluppo della produttivitá del lavoro," *L'Industria*: 3–10.

Watanabe, S., 1989, "The Diffusion of New Technologies, Management Styles and Work Organisation in Japan: A Survey of Empirical Studies," Paper presented at the International OECD Conference Technological Change as a Social Progress: Society, Enterprises and the Individual, 11–13 December, Helsinki, Finland.

Young, A.A., 1928, "Increasing Returns and Economic Progress," *The Economic Journal*, December.

Chapter 16

Switching From Fordism to CIM: A System Dynamics Model

Peter Fleissner and Wolfgang Polt

To bridge the gap between microanalysis and macroanalysis is a difficult task, especially when dealing with new technologies that are still in their infant stage such as CIM technologies. For the assessment of macroeconomic impacts of technical change different methods (as econometric techniques, input–output modeling, and system dynamics) exist, each with its own advantages and shortcomings.

Here we present a system dynamics approach that makes it possible to overcome the comparative-static view often inherent to econometric and input–output analysis and to describe the breakdown of an accumulation regime and the adjustment process to another. An accumulation regime is a long-run macroeconomic basic configuration of technological, social, and institutional variables that maintain among themselves specific rules of economic adjustment. System dynamics is a helpful tool in constructing dynamic models with feedback loops. It helps to organize one's hypothesis in a consistent and operational way.[1] Recently this approach has been used to construct a national economic model of Austria (see Bruckmann *et al.*, 1989).[2]

In our model we distinguished two sectors: a Fordist sector and a CIM sector as representatives of different technological paradigms determining the economy's accumulation regime. Then we tried to simulate the breakdown of the Fordist accumulation regime and the emergence of a CIM production regime by configuring the respective sectors with some "stylized facts." This is difficult to do because one of the main features of CIM, the growing importance of diversity and economies of scope, could not be reflected *directly* in the macroeconomic variables as they are measured in aggregated terms. So we had to capture this effect indirectly in several macroeconomic variables:

(1) A fading out of the economies-of-scale effect (and a related rise of the importance of economies of scope) could be seen in changes in the Verdoorn relation (that is, the change of productivity following a change in output). In the CIM sector we expected the Verdoorn effect to be lower due to the fact that productivity growth could be held stable despite fluctuations in demand as a result of economies of scope ("delinking hypothesis").

(2) Economies of scope could also be embodied by decreasing inventory rates in the long run. This is due to better coordination and handling of uncertainty.

(3) They can be reflected by a lower price elasticity of demand because enterprises could change their production program rather than lower their prices.

(4) Further we used a lower capital output ratio to capture the CIM sector's ability to respond to changes in demand without heavy new capital investment.

16.1   The Model of a Fordist Economy

16.1.1   General structure

Our model was constructed to characterize the most important technical features of a technological regime that could be embodied in macroeconomic variables, but was kept as simple as possible. First we built a general structure of the Fordist economic regime. In general this structure was reproduced for the CIM sector, of course with some modifications, and is described in detail in this section. Only the main economic indicators should be included.

Many relevant sectors like the public or financial sector were neglected. The economy was designed as a closed one to avoid complications with foreign trade. We started with a one-commodity economy without prices. Fixed capital was used as the key variable of production. By means of two marginal coefficients (marginal capital intensity and marginal capital output ratio), the technology of production can be defined. The quotient of marginal capital intensity and marginal capital output ratio is equal to marginal labor productivity.

Instead of the usual neoclassical production function we used a different approach: the main indicators of the production sector – fixed capital, labor, and full capacity – are increased by capital investment and labor and capacity increases and decreased by capital allowances and labor and capacity reductions. The increments are governed by gross capital investment as the key variable. Capital allowances are determined by fixed capital stock over a time constant, which means a fixed life span of fixed capital. Whereas gross labor increment is defined by gross capital investment over marginal capital intensity and gross capacity increase by gross capital investment over marginal capital output ratio, respectively, the corresponding decrements are defined by average coefficients, and not by marginal coefficients. One could take this model as the most simple vintage model available because it differentiates between two types of capital only, between actual and recent parts of fixed capital and capital stock in place.

To get the model running we included two feedback loops, considering the formation of demand: the consumption loop and the capital investment loop. Consumption is financed out of the wage sum. It is equal to the product of the wage rate times the number of workers. In our first experiments we determined gross capital investment by means of expectations about future capacity, which in turn were based on the past behavior of total demand. The difference between expected capacity and actual capacity in place divided by the marginal capital output ratio was used as a measure for net capital investment. Gross capital investment resulted by adding capital allowances. Unfortunately this experiment was not a success; it showed high instability. Changing the marginal propensity of private consumption by a very small amount led to a change in long-term behavior; all the important variables became negative instead of moving toward plus infinity as before. The reason for this behavior may be seen in the lack of negative feedback. Positive loops as the capital investment loop and the consumption loop reinforce any disturbance (be it negative or positive with respect to a theoretical equilibrium) and finally lead to explosions. Thus we had to look for a negative feedback loop, which should stabilize the economy.

Finally the inclusion of a consumer price index brought the desired result. Consumer prices are determined by an adaptation mechanism; they are incrementally increased if the actual inventory rate (inventory stocks over total demand) is lower than a desired fixed value. An increase in prices reduces demand for consumer goods. Decreased consumption reduces total demand, which in turn increases inventory rate. As the adaptation of the prices needs some time, oscillations may result by overcompensation.

If management is confronted with excess demand, price adaptations occur. In addition, production capacity increases. If – as before – the desired inventory rate surmounts actual values, gross capital investment is increased. A time delay of 1.5 years between the decision to invest and the effects of investments on capacity was taken into account.

16.1.2 Matching initial values and parameters

To start with a more or less equilibrium situation it is necessary to match the initial conditions and parameter values. To make this task easier we postulated the equality of marginal and average parameters first. We skipped this assumption in later models.

Fixed capital stock was given a numeric value of 2000, in billions of currency units (BCU) at constant prices. With a marginal capital intensity (i) of value 1, which is assumed to be equal to the average capital intensity, and is of dimension BCU/millions of persons, we arrive at the same numerical initial value for labor, 2000, as for capital stock. A marginal (average) capital output ratio of four leads to an output (production capacity) of 500 BCU. This output has to feed consumers and firms. As gross capital investment was set to 100 BCU, resulting in a net accumulation rate of 2.5% because of capital allowances of 50 BCU, initial consumption has to be 400 BCU. The real-wage rate was set to 70% of the productivity of labor, thus

$$0.7 \times 500/2000 = 0.175 \ ,$$

in thousands of currency units. Thus the wage sum is equal to 350 BCU. A Keynesian consumption function with a marginal propensity to consume of less than one could be represented by

$$\text{Consumption} = 60 + 0.97 \times \frac{\text{Wagesum}}{\text{Price}^{0.9}} \ .$$

Because of a price index of one at starting time, the value of consumption is precisely 400 BCU.

By this function all the necessary initial values are defined with one exception, the stocks of inventory. By assuming an inventory rate (inventory stocks/output) of 0.4 the initial inventory value has to be 200 BCU.

The values of these marginal coefficients will be changed in later models. Some will become functions of other variables to meet the special needs of a definite economic regime.

16.1.3 The Fordist model

Boyer (1988a) characterized the Fordist regime mainly by the following properties:

- Economies of scales, reflected by a high Verdoorn effect.
- A rather large "multiplier" (change of investment induced by changes in consumption) as the main driving force for investment.
- A Fordist distribution, linking real wages closely to productivity.

These have been the main building blocks for the stable but gradually eroding growth mechanism in the postwar period as described in Ayres and Zuscovitch (1989). In the model we could take into account only five factors:

(1) Real wages are highly correlated with the growth of the productivity of labor (thus increasing consumption and creating additional demand). In our model this relationship was reflected by using the following equation for the real-wage rate:

$$\text{Wagerate}_{\text{FORD}} =$$
$$0.7 \times \text{Labor productivity}_{\text{FORD}} \{\text{wage rate, Fordist sector}\} \ .$$

(2) As a substitute for Boyer's multiplier we took the marginal capital output ratio. In the long run this coefficient is constant over time, although it varies with the growth rate of the economy by means of economies of scale (see *Figure 16.1*).

(3) There is an exponentially increasing capital intensity. In the model we assumed the marginal capital intensity to grow by 3% per year:

$$\text{marginal capital intensity} =$$
$$e^{0.3t} \{\text{marginal capital intensity, Fordist sector}\} \ .$$

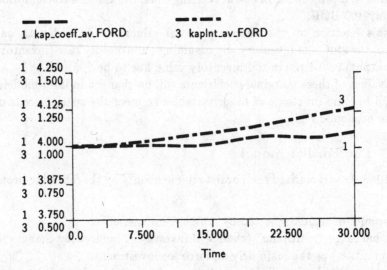

Figure 16.1. Capital output ratio and capital intensity, average values: Fordist sector.

In the result, the average capital intensity is growing, too, but with a lower rate than the marginal capital intensity (see *Figure 16.1*).

(4) As we had found in the literature and also empirically for the FRG and Austria, there is a Verdoorn effect (*Figure 16.2*), in which economic growth, which is due to economies of scale, leads to an increase in the productivity of labor. In our model this effect can be included by letting the marginal capital output ratio be dependent on the growth rate of final demand for the Fordist sector; increased growth of final demand will reduce the marginal capital coefficient. We used the following formula:

kap_coeff_marg_FORD =

$\quad$ 4 [1 − (gY_d_FORD/100 − 0.02) × 1.5]

marginal capital output ratio =

$\quad$ {marginal capital output ratio, Fordist sector} ,

$\quad$ 4 [1 − 1.5 (growth of demand$_{\text{FORD}}$ − 0.02)] .

gLab_prod_d_FORD vs gY_d_FORD

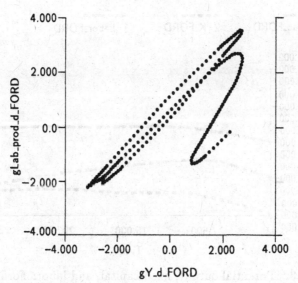

Figure 16.2. The Verdoorn effect: growth rates of labor productivity over final demand in the Fordist sector.

If growth of demand in the Fordist sector is greater than 2% the marginal capital output ratio is reduced. The strength of the influence can be controlled by a numeric constant. In accordance with some of our empirical investigations in the economies of the FRG and Austria we defined it as 1.5.

(5) After some periods of expansion the Fordist regime runs into stagnation. Our model (*Figure 16.3*) shows a stagnating production capacity (1), a slowly growing capital stock (2), and – after some periods of expansion – a decreasing labor force (3). There are two reasons for this stagnative behavior. First, the marginal propensity to consume is smaller than 1, thus leading to a relative slowdown in consumer demand, compared to the growth of the wage sum. Decreasing consumer prices are partly compensating this effect, but do not outweigh it. Second, because of a growing discrepancy between the desired and the actual inventory rate, capacity expansion is reduced having a slowdown effect on final demand via lower gross capital investment.

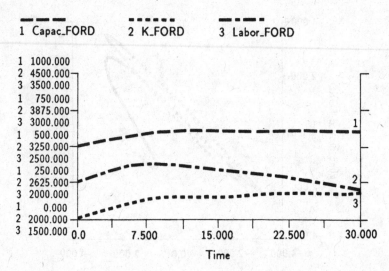

Figure 16.3. Potential output, fixed capital, and labor: Fordist sector.

16.2   The CIM Sector

16.2.1   Different initial and parameter values

The CIM sector differs from the Fordist sector in some parameters and initial values. The main differences are:

- We expected the share of gross investment directed to the CIM sector to follow a logistic diffusion curve. This share of 60% maximum should not be exceeded.
- In the consumption function we tried to reflect the fact that the comparative advantage of CIM is not mainly price competitiveness. Instead we assumed that there is a "flexibility premium" by the flexibility to produce differentiated high-quality products, thus resulting in a lower-price elasticity of consumer demand. This can be seen as a rent from "monopolistic competition."
- Another difference related to the ability for flexible production is the (necessary) speed of investment change in response to perceived gaps between actual and desired inventories. In the CIM sector we assumed this need to respond to be much smaller than in a Fordist regime. This

is due to the reprogrammability of CIM equipment, by which it can be adapted to changes in demand.

- Then we introduced a higher (but constant) marginal capital intensity to the CIM sector than that used in the Fordist sector.
- The Verdoorn effect is one-third of the Fordist sector (following our "delinking hypothesis"). The function kap_coeff_marg_CIM was lowered in comparison to that in the Fordist sector because of the capital savings that could be achieved by CIM (see Ranta and Tchijov 1989).
- While in the Fordist sector life of fixed capital was held constant at 40 years, in the CIM sector it was annually reduced by 1 year down to 30 years. This reduction should indicate the high speed of technical progress especially in the early stages of diffusion of a new technology, making fixed capital obsolete faster. Later we allowed for a subsequent rise in life expectancy of CIM, thus assuming that once the technology has been standardized, its flexibility advantage (the ability to cope with changes in the structure of demand without the need for heavy new investment) will outweigh the tendency for increased obsolescence. In the end, the lifetime of CIM equipment was higher than in the rest of the economy (45 and 40 years, respectively).
- Real-wage formation is less influenced by labor productivity than under the Fordist regime (see Boyer, 1988a, for a similar assumption).
- The desired inventory rate (reflected by the variable Inv_rate_CIM_des) is lower according to the empirical findings in the literature (Tchijov, 1991a, 1991b). The function generates an annual fall of 1% of the desired inventory rate.
- A 10% higher capital intensity was built in, but growth rates of marginal capital intensity are equal.

16.2.2 Connecting the CIM sector to the Fordist economy

The main question arising here is the way in which the sum of total demand has to be allocated to the sectors. Fordist and CIM sector are competing for their shares of final demand which are determined by the following mechanisms:

- It is assumed that the share of investment demand that the CIM sector is able to attract is equal to its production capacity, and thus depends only on the speed of diffusion of CIM.
- The respective shares of consumption are *roughly* distributed according the relative capacities, but codetermined by prices and price elasticities

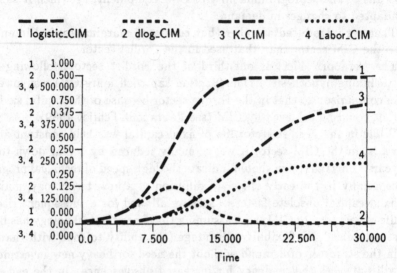

Figure 16.4. Diffusion of CIM, capital, and labor: CIM sector.

of demand. By this way it was ensured that the CIM sector could show over or underproduction, although it seemed reasonable to rule this out for the share of investment demand that the CIM sector is able to attract because of the pervasiveness of the new technology. The residual demand is left to the Fordist sector.

16.3 Simulation Results: The CIM Sector and Its Effect

By assuming a diffusion speed of 0.5, the CIM sector reaches its saturation level after roughly 20 years, the point of inflection is reached after 9.5 years.[3] At the end of the simulation period the share of capital stock is approximately 20%, roughly corresponding to the figures of the diffusion of CNC and FMS given by Ranta and Tchijov (1989) and Tchijov (1991b), see *Figure 16.4*.

After the initial phase of diffusion the capital output ratio is steadily increasing. This could be interpreted as the normal effect once a new technology is established because of market saturation (see *Figure 16.5*).

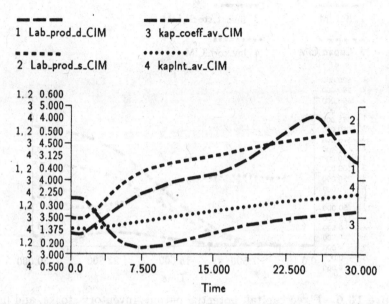

Figure 16.5. Labor productivity, capital output ratio, and capital intensity, average values: CIM sector.

Inventory stock of CIM is fairly low during the simulation period; inventory rate decreases to very low levels because of increasing output (see *Figure 16.6*).

Price levels decrease in both sectors. In the first phase the Fordist price is lower than the CIM price; during the saturation period its the other way round (see *Figure 16.7*).

There is an overall increasing trend of labor productivity in the CIM sector, but – caused by a lack of demand – a cyclical reduction occurs in the final phase. Wage sum and labor show the expected patterns (see *Figure 16.8*).

Over the phase of diffusion and establishment, the CIM sector exerts very different influences on the Fordist sector and on the behavior of the economy as a whole (see *Figures 16.9* and *16.10*, which show the differences between the simulations with and without CIM).

During the initial growth phase (up to year 17) investment demand increases in the Fordist sector, while private consumption in the Fordist sector decreases. The reason is additional capital investment demand for the CIM sector and increased competition between the sectors on consumer demand.

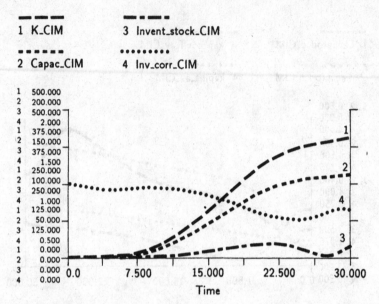

Figure 16.6. Fixed capital, potential output, inventory stocks, and inventory corrections: CIM sector.

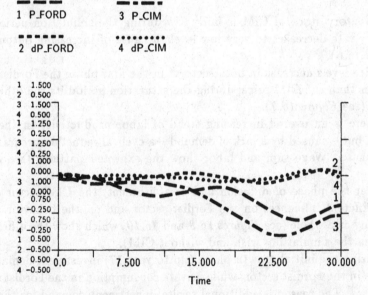

Figure 16.7. Price changes.

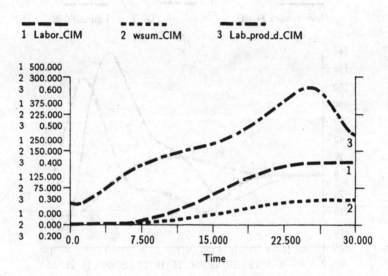

Figure 16.8. Labor, wage sum, and labor productivity: CIM sector.

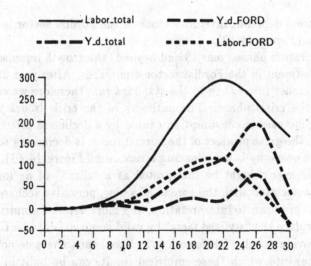

Figure 16.9. Deviations from standard run: labor and demand in total economy, Fordist sector.

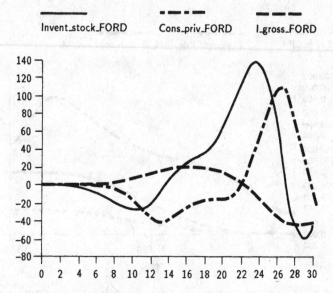

Figure 16.10. Deviations from standard run: inventory stocks, consumption, and gross investment in the Fordist sector.

Because of excess demand, inventory stock of the Fordist sector is lower in the first phase (*Figure 16.10*).

In the saturation phase (year 18 and beyond) the growth impulse of CIM to capital investment in the Fordist sector diminishes. After year 22 capital investment becomes lower than in the standard run; therefore we call it the beginning of the crisis phase. The outbreak of the crisis is delayed by a compensating increase in consumption caused by a decline in price level. In this last phase the gross product of the Fordist sector is decreasing so rapidly that the whole economy is experiencing a recession (*Figure 16.10*). To sum up, this development might be interpreted as a "climb" of a "long roller coaster ride" associated with the spread of a new, pervasive technology.

The results are open to interpretation, and more detailed empirical work is needed to replace the "stylized facts" by valid empirical data on CIM. Our task was to provide a tool, a framework (which we hope to have demonstrated as a useful one) into which these empirical results can be built in to get a picture of the long-term dynamic macroeffects of CIM.

Notes

[1] For a comparison of different modeling concepts see Meadows and Robinson (1985); for special system dynamics software see Richardson and Pugh (1981, DYNAMO) and Richmond et al. (1987, STELLA).

[2] Rounded figures of the Bruckmann and Fleissner model were taken as parameter values of the model discussed in this chapter.

[3] A caveat should be kept in mind while interpreting the results of the simulation: at the beginning the initial values of the CIM sector are very small, thus resulting in strange effects on the quotients like capital output ratio and so on.

References

Ayres, R.U., and E. Zuscovitch, 1989, "Information, Technology and Economic Growth: Is there a Viable Accumulation Mechanism in the New Paradigm?" *Technovation*.

Boyer, R., 1988a, "Formalizing Growth Regimes," in G. Dosi, et al., eds., *Technical Change and Economic Theory*, Frances Pinter, London, UK.

Boyer, R., 1988b, "Technical Change and the Theory of 'Regulation'," in G. Dosi et al., eds., *Technical Change and Economic Theory*, Frances Pinter, London, UK.

Bruckmann, G., et al., 1989, *Am Steuerrad der Wirtschaft. Ein ökonometrisch-kybernetisches Model für Österreich*, Springer-Verlag, Berlin, Heidelberg, New York.

Meadows, D.H., and J.M. Robinson, 1985, *The Electronic Oracle – Computer Models and Social Decisions*, John Wiley & Sons, New York, NY.

Polt, W., 1989, "Some Considerations on the Possible Macroeconomic Effects of Computer Integrated Manufacturing Automation," in J. Ranta, ed., *Trends and Impacts of Computer Integrated Manufacturing*, WP-89-1, IIASA, Laxenburg, Austria.

Ranta, J., ed., 1989, *Trends and Impacts of Computer Intergrated Manufacturing*, WP-89-1, IIASA, Laxenburg, Austria.

Ranta, J., and I. Tchijov, 1989, "Economics and Success Factors of Flexible Manufacturing Systems: The Classical Theory Revisited," in J. Ranta, ed., *Trends and Impacts of Computer Integrated Manufacturing*, WP-89-1, IIASA, Laxenburg, Austria.

Richardson, G.P., and A. Pugh, 1981, *Introduction to System Dynamics Modeling with Dynamo*, MIT Press, Cambridge, MA.

Richmond, B., S. Peterson, and P. Vescuso, 1987, *An Academic User's Guide to STELLA*, High Performance System, Lyme, NH.

Tchijov, I., 1991a, "The Diffusion of Flexible Manufacturing Systems," in R.U. Ayres, W. Haywood, and I. Tchijov, eds., *Computer Integrated Manufacturing*, Volume III: *Models, Case Studies, and Forecasts of Diffusion*, Chapman and Hall, London, UK.

Tchijov, I., 1991b, "International Diffusion Forecasts," in R.U. Ayres, W. Haywood, and I. Tchijov, eds., *Computer Integrated Manufacturing*, Volume III: *Models, Case Studies, and Forecasts of Diffusion*, Chapman and Hall, London, UK.

Chapter 17

CIM Impacts on Logistics: Macroeconomic Aspects

Pavel Dimitrov

The introduction of computer integrated manufacturing (CIM) technologies on a broad scale entails fundamental changes both in the organizations implementing these technologies and in the economy as a whole. These changes and their consequences should be anticipated as early as possible, and the implications for management and policymaking should be studied.

Manufacturing and logistics activities are highly interrelated and interdependent. In this relationship manufacturing plays the dominant role, since only what has been produced can be stored, handled, and transported. On the other hand, logistics developments are a prerequisite and a catalyst for manufacturing improvements. Due to the natural interrelations, the implementation of CIM technologies will have a major impact on logistics, which in turn will exercise a reverse impact on CIM.

Logistics comprises all activities that are necessary for overcoming disparities of a temporal, spatial, or quantitative nature, ranging from the procurement of materials to the distribution of finished products to the customer. These activities incorporate the planning, implementation, and monitoring of transport, handling, and storage, as well as the information flow necessary for these operations. Logistics integrates activities that traditionally have been located in different functions of the business (procurement, materials management, and distribution) and different sectors of the economy (production sectors, transport, trade, and communication). The individual activities that are included in logistics have always been an essential feature of the manufacturing process. Logistics links the distribution network, the manufacturing process, and the supply activity in such a way that customer service is at a high level and costs are low. In this respect, logistics can be viewed as a prerequisite and continuation of the manufacturing process in the supply and the distribution spheres, since from the wider economic perspective the production process is actually completed when the product reaches the customer.

The developments in manufacturing have always been accompanied by corresponding changes in logistics activities. Several common factors have contributed to the restructuring of both manufacturing and logistics and have triggered the development of CIM and radical improvements in logistics. The following are among the most important of them:

- Saturation of the markets, leading to growth in the demand for high-quality customized products and raised standards for customer service.
- Growing product complexity and diversification resulting in increased complexity of the intracompany economic links and increased vertical integration.
- Shortening of the products life cycles.
- Globalization of production, markets, and competition.

To compete successfully in today's marketplace requires total cooperation and coordination of all activities. It is no longer possible to manufacture in isolation. The current situation in most markets – characterized by short product life cycles, increased complexity of economic links (due to product complexity and product diversification), severe competition, and fast-changing demand – requires short lead times (production and delivery including the time for designing new products).

The introduction of CIM technologies, and especially flexible manufacturing systems, contributes to the growth of production flexibility and the ability of manufacturing systems to respond quickly to fluctuations in demand. CIM contributes to higher market performance of the system by assuring high standards and increasing production flexibility (by shortening lead times, decreasing production lots, increasing product assortment). IIASA's FMS data bank, as well as other studies, provides considerable evidence in this respect. The analysis of IIASA's FMS data bank shows that the implementation of FMS reduces the lead time (the time from order to delivery) on average by a factor of four in comparison with conventional technology. Simultaneously, product variety is increased and production lots are decreased.

On the other hand, many experts and adopting firms seem aware that to use the increased production flexibility the whole material-flow system (including the supply and the distribution stages) should be flexible (de Vaan, 1989). It is of little advantage to employ flexible production (through CIM) to manufacture products that will sit for weeks in the distribution network, or that will require large inventories to compensate for an inflexible supply system. In other words, the successful implementation of CIM requires flexible supply and distribution systems. There are signs that management is aware of the importance of logistics for the successful implementation of CIM. Case studies and surveys (Shapiro and Heskett, 1985; Mortimer, 1986, 1988; Voss, 1987; Jansen and Warnecke, 1988) show that the successful adopters of CIM (FMS, in particular) either have made the required changes in their supply and distribution systems (usually through implementing the just-in-time concept) to match the requirements of CIM or have used existing flexible supply and distribution systems. They indicate that this has become a conventional approach of the leading Japanese and Western companies. The competitive advantage of the Japanese firms to a great extent can be explained by the far-better synchronization of their manufacturing and logistics systems (Schonberger, 1982). It is also often the case that the introduction of CIM in one company not only triggers developments in its logistics system, but forces increased flexibility in its suppliers (eventually implementation of CIM) and is a prerequisite for implementation of CIM by the buyer companies.

Common managerial concepts and strategies are being applied in the manufacturing and logistics fields. The most popular and quickly spreading is the just-in-time philosophy. According to this philosophy only what is needed (in amount and quality) is produced, transported, distributed, etc. The JIT philosophy is supported by a wide range of physical improvements (changing plant layout, cutting setup times, etc.) and socio-managerial rationalizations (introducing total quality control, creating a flexible work force, close cooperation with vendors and customers) aimed at increasing the flexibility and the responsiveness of the whole manufacturing logistics chain (Hutchins, 1988; Lubben, 1988). The manufacturing resource planning (MRP) system has been developed and modified into the distribution resource planning (DRP) system to serve the needs of the management in the distribution processes.

These new technologies and managerial strategies along with market pressures and greater awareness of the potentials for growth are providing an impetus to manufacturing and logistics interaction and development. It is quite obvious that in the efficient harmonization of manufacturing and logistics activities there is a great potential for accelerated and competitive development. There are enough reasons and sufficient evidence to believe that under the influence of CIM, the role, the structure, and the organization of logistics activity will radically change in the near future. It may also be expected that these changes will exercise a reverse impact on manufacturing.

The importance of the analysis of CIM impacts on logistics is emphasized by the fact that logistics activities constitute a good deal of the economic activity of a nation. Logistics activities contribute some 20% to 30% of the gross domestic product of industrialized countries. Different studies reveal a similar magnitude of logistics activity in the various countries. Thus, costs of logistics activities amount to 20% to 30% of the total national expenditure. In relation to the GDP it was estimated that logistics costs amount to 21.2% in the USA (Kearney, 1984), 20.3% in Sweden (Agren, 1983), 19.5% in France, 22.0% in the United Kingdom (McKinnon, 1989), and 26.1% in Bulgaria. A study carried out by A.T. Kearney, Inc. (1984) covering 500 companies in Europe showed that logistics costs constitute from 8% to 22% of sales of the companies investigated, depending on the sector of the economy in which the companies operate. The logistics costs of the companies operating in the machine tool industry and the electronics-equipment industry are reported in this study to be 7.82% and 8.74% of the sales value, respectively. Similar orders of magnitude of the logistics costs are reported in several case studies (Mortimer, 1986).

Based on data from the input–output tables, estimates of the logistics costs for the machinery industries (the industries belonging to ISIC 38) of the USA and Japan for 1970 and 1980 showed that the logistics costs in the industries investigated range between 9% and 12% of the output. These figures are quite consistent with the results of other studies. The greater portion of the logistics costs is represented by transportation and inventory costs. Energy crises have contributed to the increase in transportation costs, and soaring interest rates have greatly increased the costs of holding stocks.

Logistics activities employ 20% to 30% of the working population of the industrialized nations: 23.13% in Austria; 20.25% in Finland; 20.6% in the FRG (prior to October 1990); 22.2 in the Netherlands; 20.5% in Sweden; and 30.7% in the USA. The trend in most of the countries is toward an increase in the share of those employed in logistics in the total number of the employed in the economy. These figures are quite similar to the estimates of the logistics costs in the respective sectors. Manufacturing-logistics interaction can be studied at different levels of aggregation – micro (shop floor to factory), meso (company or organization to industry), and macro (national or world economy). Since most studies have concentrated on the microeconomic aspects of the CIM-logistics relationship, the focus in this chapter is on macro and meso levels. The micro level is addressed to reveal the micro foundations of the macroeconomic impacts of CIM on logistics. In this sense, the results in this study can be treated as a part of a more global issue concerning the macroeconomic impacts of CIM.

The problem of analyzing manufacturing and logistics interaction at the macroeconomic level has not been sufficiently addressed. The lack of theories and models as well as examples of such analysis made it necessary to develop hypotheses of the major impacts of CIM on logistics and their socioeconomic consequences, supported as much as possible by facts and representative cross-country data.

The scope of the analysis of CIM impacts on logistics at the macro-economic level depends upon the definition of logistics. For the purposes of this chapter, logistics is treated in a broad sense to encompass all activities related to the flow of goods in the national economy (e.g., transportation, handling, and storage) as well as the related managerial and information processes. From the perspective of the national economy as a whole, these activities are carried out both by specialized sectors (logistics or logistics-related sectors such as freight transport, wholesale and retail sales, and communication) and by the production and nonproduction sectors. The basic function of logistics consists in assuring the continuity of the production process by smoothing the time, space, and volume of production and consumption.

There are several ways to analyze CIM impacts on logistics. The focus of this chapter is on the expected changes in distribution patterns, transportation, inventories, and logistics labor.

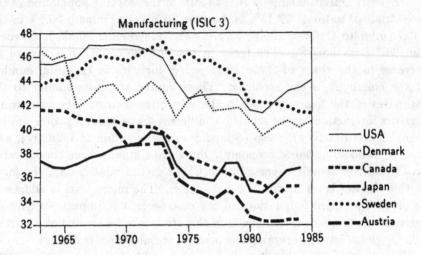

Figure 17.1. Value-added–output ratio in manufacturing (ISIC 3), in percent. Source: *United Nations Industrial Statistics Yearbook*, various years.

17.1   CIM Impacts on Distribution Patterns

The literature and statistical sources provide examples and supporting data on the growing complexity and diversity of products (see, for example, Yamashina and Masumoto, 1989; Mortimer, 1988). Ayres (1991) has presented convincing evidence of the growing complexity of products, measured in terms of the required parts and components. The growing product complexity and diversification along with Fordist-type (highly specialized and large-scale) production have contributed to the increased complexity of intracompany economic links and the growing trend of vertical integration. It is typical for a big manufacturing company nowadays to maintain a large network of suppliers, vendors, and subcontractors. One illustration regarding the growing rate of vertical integration, measured by the value-added–output ratio in the manufacturing industries of six countries (the USA, Japan, Denmark, Canada, Sweden, and Austria) for the 1963–1985 period is presented in *Figure 17.1*. The figure shows a clear trend of growing vertical integration in the manufacturing industries of these countries. However, this trend is not uniform among all countries and industries.

The prevailing organization of manufacturing activity of big multinational companies nowadays is characterized by strong specialization of the different plants belonging to the company. In such a way they are exploiting economies of scale.

The introduction of CIM provides the possibility to manufacture a greater variety of products in small volumes, closer to the raw materials or consumer markets or both. The potential of large-scale production for productivity increases has been largely exploited by more and more companies to increase productivity and competitiveness with CIM. These companies will turn to small-scale customer-oriented production. This, in turn, will lead to the spread of small-scale manufacturing outlets capable of producing products according to customer orders. This may result in a completely different territorial distribution of production and will contribute to a trend away from vertical integration.

The implication of this phenomenon for distribution patterns is quite remarkable. In the first place, the logistics (the supply and the distribution) chains will be shortened and several intermediaries in the flow of goods will be eliminated. Statistical studies already indicate a decline in the role of wholesaling. Their business is contracting as a consequence of the absorption of their logistical functions by producers and multiple retailers. A comparison of the structures of the UK consumer-goods marketing channels in 1938 and 1983 revealed that the flow of consumer goods (by value), which was passing through the wholesalers, had decreased from 47% in 1938 to 36% in 1983 (McKinnon, 1989). However, there exist large differences among countries and trades regarding the role of the wholesalers. *Figure 17.2* depicts the participation of the wholesale sphere in the flow of goods in the national economy (measured by the wholesale sales–manufacturing output ratio) in nine industrialized countries during the mid-1980s. The data show similar magnitude of the role of wholesaling in the USA, the FRG, the UK, Sweden, Finland, and Austria; the wholesale sales–manufacturing output ratio is between 0.6 and 0.8. The lowest wholesale sales–manufacturing output ratio is manifested by Canada (0.38), while the participation of the wholesale sphere in goods flows in Denmark (1.15) and especially in Japan (1.71) is much higher.

The differences between the countries in respect to the role of the wholesale sphere determine the scope of the possible influence of the restructuring of the distribution patterns as the result of a wider application of CIM. In this respect, Japan deserves special attention. Japan, the leading country in the implementation of new manufacturing and logistics technologies, has a very complicated multilayered distribution system. The analysis of the structure of goods flows (in value) of a typical Japanese regional wholesaler (in

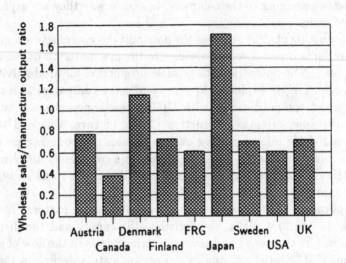

Figure 17.2. Wholesale sales–manufacturing output ratio. Source: IIASA logistics data base.

Okinawa prefecture) showed that in 1985 about 64% of the incoming goods of the wholesaler originated from other wholesalers and 31% of the outgoing goods were delivered to other wholesalers (MITI, 1985). These figures indicate a complicated distribution pattern in Japan. However, recent studies indicate that this system is in a process of reconfiguration; many producers have started to develop their own distribution systems. A good example in this respect is the Kao Company – the biggest sanitary goods manufacturer in Japan, which produces 5,000 different items (as of 1988). Kao has a head office in Tokyo, eight operation centers, and 111 distribution centers which supply 300,000 retailers. Kao has established a computer-based distribution system which claims to be able to meet any order within 12 hours. This example indicates that the role of wholesalers will decline in the future. It might well be the case that some may find a niche as organizers of new partnerships and networks. In any event the wider application of CIM will contribute to the declining role of wholesaling in the economy, which in turn will have social and economic impacts.

In addition, the wide application of CIM is a factor facilitating new organization of intrafirm economic links and relationships. CIM will increase the shift from multiple to single sourcing, from competitive to long-term cooperative intrafirm relations. New concepts of intracompany relations (the so-called value-added partnership), based on the intensive use of advanced communication and information technologies (electronic data interchange), are emerging and spreading rapidly (Johnston and Lawrence, 1988).

17.2   CIM Impacts on Transportation

The transportation system is one logistics area which will experience major impacts from the wider application of CIM. Developments in transportation have contributed to the fast growth of large-scale, highly specialized production. Again, the development of the transportation sphere made it physically and economically possible to penetrate and exploit territories and regions that are situated far from the raw materials and consumer markets but which can provide low-cost labor. At the same time the transportation system had always had an extensive impact of manufacturing, since a major demand for transportation services comes from manufacturing.

The analysis of the transport systems for several countries shows that during the last several decades dramatic changes have taken place in transportation systems; one example has been the shift to fast and reliable modes. *Figure 17.3* shows the development of transport modal split in some OECD countries over a period of about 25 years. The demand for rapid and reliable transport has triggered the enormous development of road transport. Correspondingly, dramatic changes in the transport infrastructure have taken place. This trend is accompanied by several technological innovations in all modes of transport resulting in a decisive increase in speed and loading capacity. Among the most important common trends are:

- Faster growth of GDP than the transport work (in tons per kilometer).
- Similar modal split changes, especially for high value goods (the shift from rail to road and air transport).
- Changes in the criteria for the evaluation of the transport performance from price and cost toward quality, punctuality, flexibility, reliability, and minimization of goods that are damaged or lost.
- Fast development of combined transport.
- Integration with production and trade (the emergence of the so-called production–transport chains).
- Unification of the transport, packaging, and loading–unloading equipment.

The implementation of CIM on a broad scale will give further impetus to these general trends. CIM will decrease the transportation work (in tons

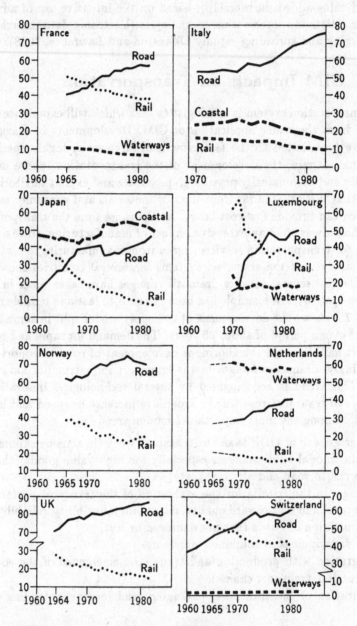

Figure 17.3. Development in transport modal split. Source: OECD, 1986.

Table 17.1. Modal split (of tons per kilometer) of inland transport in some countries in 1986, in percent.

Country	Rail	Road	Inland waterways
Western Europe			
UK	13.1	86.6	0.3
Belgium	23.3	62.4	14.3
FRG[a]	24.6	54.8	20.6
France	33.5	61.4	5.1
Eastern Europe			
Hungary	52.3	28.2	19.5
CSFR	71.8	23.2	5.0
Poland	76.8	22.0	1.2
GDR[a]	76.8	20.0	3.2

[a]Based on data collected prior to German unification.
Source: IIASA, NLT data base.

per kilometer); this will be due to the closeness of production to the raw materials and supply market. On the other hand, the process of production to customer order will require reliable supply and distribution systems which will contribute to a further shift to more reliable and flexible modes of transport (road and air); this in turn may result in increased costs for transportation. Japan is an example of this phenomenon. Estimates based on input–output tables for a 10-year period (1970–1980) in Japan show that the transportation services purchased by each machinery sector (ISIC 38) have visibly grown in relation to the sector's output.

Within the general trends, common for all countries investigated, there exist large differences, regarding the technological level of the different transport modes and their performance, the prevailing modal split, and the state of development of the transport infrastructure. The most striking difference in this respect is the difference in the transport modal split between Western and Eastern European countries (where, in the latter, rail transport still dominates over freight transportation; *Table 17.1*).

The strong correlation between transport modal split and logistics performance (measured as value-added–inventory ratio) indicates that the state of development of the transport system is a decisive factor for the strategic logistics advantage in some nations but not in others; this may become important for the difference in the penetration rates of the CIM technologies in the various countries.

17.3   CIM Impacts on Inventories

CIM's impact on inventories is one of the strongest and most visible impacts of CIM on logistics. In the first place, CIM implementation directly affects inventories in adopting firms and sectors. The implementation of CIM results in several significant improvements contributing to inventory reduction. The dramatic setup-time reduction and the entire reconfiguration of the in-plant logistics system as a consequence of installing automated storage and transportation systems, able to store and handle a variety of parts in the sequence required, lead to a dramatic reduction of the in-process waiting time and work-in-progress, respectively. The buffer stock component of work-in-progress is drastically reduced due to decreased uncertainty in the production process. CIM assures standard quality of production which reduces the scrap and rework, resulting in less work-in-progress. The flexible production system, as a rule, is complemented by flexible supply and distribution systems which in turn reduce the inventories of raw materials and finished goods. The magnitude of inventory reduction as a result of CIM (FMS, in particular) depends upon the implementation strategy (e.g., the capacity-flexibility dilemma). According to IIASA's FMS data bank, the typical inventory reduction (both raw materials and finished goods) as a result of FMS implementation ranges between 25% and 75% (or by a factor of 1.5 to 4). In about half of the observed flexible manufacturing systems, inventories (both indicators) were reduced by a factor of two. The analysis of IIASA's FMS data base shows a trend of increasing reduction of inventories over time, e.g., the more recent systems cause higher inventory reduction than the earlier FMS. This would imply that the direct CIM impact on inventories is expected to grow in the future.

In addition, CIM's flexibility and short production lead time allow the system to work directly on customers' orders. This could have a chain effect on the input inventories of the customers of the CIM manufacturer. Because the sizes of inventories along the material flow are interdependent, reduced inventories (through small batches and lead times) at the producer level reduces the cycle and the buffer inventories (the safety stock) at the customer level. Similarly, CIM systems would trigger increased flexibility in the subcontracting and supplying networks and thus reduce their inventories. Unfortunately, there is not sufficient data to prove this statement. However, the existing data regarding some Finnish cases indicate that FMS has triggered the introduction of JIT supply and distribution systems. The result was an almost 100% reduction of raw materials and finished goods inventories and a corresponding reduction of suppliers' and customers' inventories. Most of the just-in-time adopters claim that this is what happens with their suppliers' and customers' inventories (Mortimer, 1986).

As suggested earlier, CIM will induce changes in the distribution patterns, the result of which will be further decline of the role of the wholesalers and reduction of their inventories. The inventories held by the wholesaling sector constitute a great deal of the total inventory stock – 32% in Japan, 22% in the USA, 20% in Sweden, and 23% in the UK (in 1983–1984). Similarly, production to customers' orders along with fast and reliable distribution systems will decrease inventories in the retail sector. This would imply that CIM's impacts on inventories go far beyond the firms and sectors implementing CIM and can be felt by suppliers and customers (upstream and downstream in the logistics chain).

Looking for the possible influences of CIM implementation on inventories an analysis has been carried out using IIASA's data base. This data base covers 15 countries with different economic systems over a period of 10 years. The sample is limited to those countries for which reasonably harmonized data could be obtained. The data are structured according to the three-digit level of the International Standard Industrial Classification of All Economic Activities (ISIC, 1968). For each country data were collected for about 25 ISIC sectors of the manufacturing industry. Additionally, data for inventories of the wholesale and the retail sales sectors of some countries were collected.

This data base provides broad analytical possibilities. For the purposes of studying CIM's impact on inventories the analysis is restricted to the manufacturing sector as a whole, as well as to the sectors of the machining industries (ISIC 38): metal products (ISIC 381), machinery n.e.c. (ISIC 382), electrical machinery (ISIC 383), transport equipment (ISIC 384), and motor vehicles (ISIC 3843). These are the sectors that are reported to be the most intensive users of the new manufacturing and logistics technologies. The value-added–inventory ratio and inventories in terms of days of stock-holding have been used as measures for the relative inventory levels.

The analysis of inventories over a 10-year period reveals that, despite cyclical variations, in most countries (with the exception of the FRG, Portugal, and the East European countries) there is a trend of steady improvement of the value-added–inventory ratio (*Figures 17.4, 17.5, 17.6,* and *17.7*). At the same time the gap between Japan and the other countries is widening. The upward trend in the value-added–inventory ratio, especially during the last years of the period, may be partially attributed to the developments in manufacturing and logistics. Similarly, the gaps between Japan and the

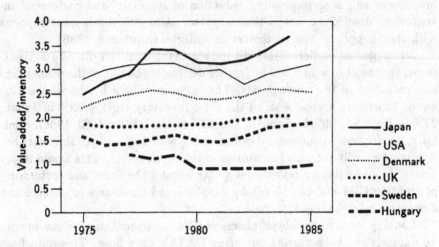

Figure 17.4. Value-added–inventory ratio in manufacturing (ISIC 38). Source: IIASA logistics data base.

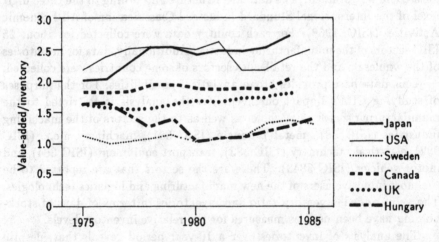

Figure 17.5. Value-added–inventory ratio in machinery n.e.c. (ISIC 382). Source: IIASA logistics data base.

USA, between these two countries and the rest of the countries, and between Western Europe and Eastern Europe can also be partially attributed to these developments.

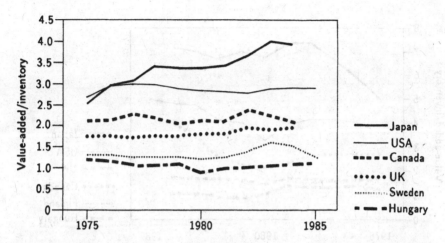

Figure 17.6. Value-added–inventory ratio in electrical machinery (ISIC 383). Source: IIASA logistics data base.

Ceteris paribus, the implementation of CIM should lead to observable changes upstream in the value-added–inventory ratio. However, the data do not show such a clear trend. One reason is that the current rate of CIM application cannot lead to measurable changes in inventories at the sectoral level. Another reason is that inventory levels depend upon several interrelated and interdependent factors. An extended list of these factors and their analysis by country is presented in Dimitrov (1984) and Chikan *et al.* (1986). Correspondingly, when interpreting the inventory trends in the context of macro inventory impacts of CIM technologies one should bear in mind the numerous contradictory factors that influence inventory levels. Just to name a few: the equilibrium of the market, the participation of and the dependence on foreign trade, the state of the money market, the changes in the interest rates, and the development of the transportation system. In many cases the influence of these factors can be so large that their effect on stocks can offset the impacts of CIM. It may also be the case that the combined action of the inventory factors may result in a trendless value-added–inventory ratio for many of the sectors and countries studied. To obtain a quantitative measurement of the CIM impacts on inventories at the macroeconomic or sectoral level it is necessary to isolate the influence of the most important inventory factors. Without direct measure of the diffusion of CIM technology in each sector the influence of CIM on inventories could

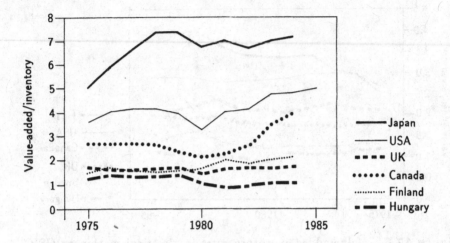

Figure 17.7. Value-added–inventory ratio in motor vehicles (ISIC 3843). Source: IIASA logistics data base.

be distinguished from the trend after statistically controlling for the other factors. This could be the topic of future research.

An analysis of the value-added–inventory ratio, as well as inventories in terms of days of stock-holding, reveals the existence of stable differences in the value-added–inventory ratio between the countries (*Figures 17.8* and *17.9*). This is true for the whole manufacturing industry as well as for the different manufacturing sectors. Japan is the leader as a rule followed by the USA, several West European countries (the FRG, Denmark, the UK, Ireland), Canada, the Scandinavian countries (Sweden, Finland, and Norway), Austria, the East European countries (Poland, Bulgaria, and Hungary), and Portugal.

The difference between Japan and Hungary in the whole manufacturing sector in terms of value-added–inventory ratio is a factor of 4.0; for the nonelectrical machinery industry and the motor vehicle industry the factors are 2.5 and 8, respectively. The analysis also reveals completely different patterns of inventory formation between West European countries (with the exception of Finland) and the East European countries regarding the inventory structure, e.g., the breakdown of inventories into raw materials, work-in-process, and finished goods (see *Figures 17.8* and *17.9*). The greatest part of the inventory stock in the East European countries is kept as raw

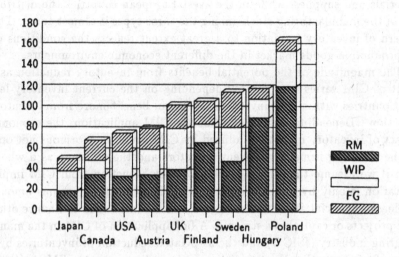

Figure 17.8. Number of days stock is held in inventory in the manufacturing industry (ISIC 3) for 1980–1985: raw materials (RM), work-in-progress (WIP), and finished goods (FG). Source: IIASA logistics data base.

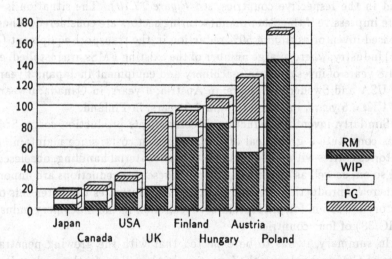

Figure 17.9. Number of days stock is held in inventory in the motor vehicles industry (ISIC 3843) for 1980–1985: raw materials (RM), work-in-progres (WIP), and finished goods (FG). Source: IIASA logistics data base.

materials and supplies, while in the West European countries the distribution of the manufacturing stock among the three types is almost equal. This pattern of inventory formation to a great extent reflects the conditions on the production goods market in the different economic environments.

The magnitude of the potential benefits from inventory reduction as a result of CIM varies considerably depending on the current inventory levels. Countries with higher inventory levels can benefit more from inventory reduction. Depending on the state of the CIM application, the economic impact of inventory reduction induced by CIM could be enormous not only for the adopting firms, but for entire sectors and the economy as a whole. Even if we assume that the average inventory reduction from CIM implementation is only by a factor of two, this will mean an enormous amount of released financial and other resources, which can be used to finance other CIM projects or capital investments. A full application of CIM in the manufacturing industry (ISIC 3) and the associated reduction of inventories by a factor of two (note that the average inventory reduction due to FMS, without the extreme cases, is by a factor of 2.8) will release financial resources equal to from nearly 2 (in the case of Japan) to 5 (in the case of Hungary) years of investment in machinery and equipment in the different countries (under the current inventory levels and rates of investment in machinery and equipment in the respective countries, see *Figure 17.10*). The situation is even more impressive if the other manufacturing sectors are considered. Thus the released inventories from a 50% reduction in the transport equipment (ISIC 384) industry, where a large number of the existing FMSs are installed, equal to 1.8 years of investment in machinery and equipment in Japan; 3 years, in the USA and Sweden; 3.5 years, in Austria; 4 years, in Canada; 6 years, in the UK; 6.5 years, in Hungary; and 7.5 years, in Finland.

Similarly, inventory reduction directly affects production costs. Storage costs constitute a great deal of the production costs, since maintaining inventories carries with it the costs of storage, material handling, obsolescence, and so on, as well as interest payments. These cost reductions are important to increase production efficiency. *Table 17.2* illustrates the potentials of inventory reduction for production cost reduction in the machinery industries (ISIC 38) of four countries.

In summary, it has to be expected that with the growing penetration rates of CIM technologies the inventory levels will be further reduced.

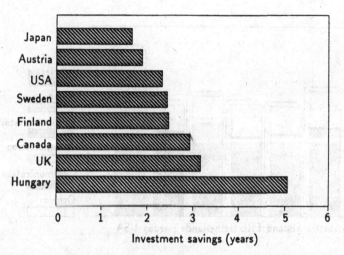

Figure 17.10. Potential investment savings in the manufacturing industry (ISIC 3) from a 50% inventory reduction. Source: IIASA logistics data base.

Table 17.2. Inventory storage costs in percent of value-added.[a]

ISIC	Japan (1984)	Finland (1984)	UK (1984)	USA (1985)
381	6.3	13.7	14.7	9.9
382	11.1	15.5	16.6	12.0
383	7.6	15.8	15.1	10.4
3843	4.2	13.7	16.6	6.0

[a]Storage costs are assumed to be 30% of the value of stocks.
Source: IIASA NLT data base.

17.4 CIM Impacts on Logistics Labor

The wider application of CIM will exercise major impacts on workers in logistics activities. Logistics activities provide occupations for some 20% to 30% of the working population of the industrialized nations. This is illustrated in *Figure 17.11*, derived from the internationally comparable occupation-by-sector labor matrices prepared by Brautzsch (1988) for the CIM Project.

Those employed in logistics activities in the machinery industries (ISIC 38) occupy 15.5% of the total number of the employees in Austria; 10.7%, in Finland; 8.9%, in the FRG; 14.8%, in the Netherlands; 11.9%, in Sweden; and 11.7%, in the USA. The implementation of CIM technologies (FMS,

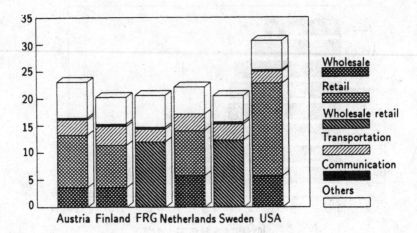

Figure 17.11. Logistics labor of total labor in six national economies, in percent.

in particular) exercises a great impact on a major logistics-related occupational group – material-handling and related equipment operators (ISCO 9–7). IIASA's FMS data base shows that personnel reduction due to FMS is approximately a factor of four. A good deal of this personnel is related to in-plant logistics – warehousing, transportation, and handling. This occupational group constitutes 41.4% of logistics labor in the production sectors in Austria (1981); 39.8%, in the Netherlands (1985); 48%, in Sweden (1980); 47.4%, in Finland (1980); 28.8%, in the USA (1984); and 22.4%, in the FRG (1982). Nicol and Hollier (1985) have found that the work force employed in materials-handling occupations (including persons employed full time on handling, storage, and transport duties, as well as those who include handling as part of their duties) constitutes 12% of the companies' total labor costs (33 UK companies investigated).

In the context of the discussion on CIM impacts on the distribution patterns (e.g., the declining role of the wholesaling sector, changes in transportation modal split, and the decline of railways in transport), it is to be expected that the wider application of CIM technologies will cause visible labor impacts on the transportation and the trade sectors as well as on occupations related to warehousing, handling, and in-plant transportation. The magnitude of these impacts will depend on the penetration rates of the CIM technologies.

17.5   Conclusions

Undoubtedly, logistics is one of the economic spheres that will be strongly affected by CIM. The major impacts of CIM on logistics consist in the following: entire reconfiguration of the supply and the distribution network, based on long-term collaborative intrafirm relations; fast growth of high-quality transportation services at the expense of the decreased amount of transportation work; a drastic reduction of inventories along the logistics chain; and labor-force shifts.

From the other side, logistics activities will exercise a reverse impact on CIM. Logistics can facilitate the fast development of CIM by providing the framework of future computer integration beyond the boundaries of the factory. The first steps in this respect have already been undertaken by the so-called value-added networks (VANs). The developments of CIM and logistics technologies require constant monitoring to be able to identify the emerging new trends and possible impacts.

References

Agren, B., 1983, *Cost for Transport, Handling and Storage in the Various Branches: Measurement of Efficiency and Rationalization Possibilities*, Research Report 130, Linköping University, Sweden (in Swedish).

Ayres, R., 1991, *Computer Integrated Manufacturing*, Volume I: *Revolution in Progress*, Chapman and Hall, London, UK.

Brautzsch, H.-U., 1988, "The Occupational Structure by Sectors in Selected Countries," Report to IIASA, Laxenburg, Austria.

Chikan, A. *et al.*, 1986, "Macroeconomic Factors Influencing Inventory Investments: An International Analysis," in A. Chikan, ed., *Inventory in Theory and Practice*, Akademiai Kiado, Budapest, Hungary.

de Vaan, M., 1989, *FMS Magazine* 7(4):183–188.

Dimitrov, P., 1984, "Classification and Analysis of Factors Influencing Aggregate Inventories," in A. Chikan, ed., *New Results in Inventory Research*, Akademiai Kiado, Budapest, Hungary.

Hutchins, D., 1988, *Just in Time*, Gower Press, Hampshire, UK.

IIASA Logistics Data Base, Compiled on the basis of *United Nations Industrial Statistics Yearbook*, *United Nations Statistical Yearbook*, *Annual Bulletin of Transport Statistics for Europe*, United Nations, New York, NY; and contributions from collaborators in 12 countries.

ISIC (International Standard Industrial Classification of All Economic Activities), 1968, *Statistical Papers*, Series M4, Rev. 2, New York, NY.

Jansen, R., and H. Warnecke, 1988, *Just in Time Manufacturing*, IFS Publications/ Springer-Verlag, Berlin, Heidelberg, New York.

Johnston, R., and P.R. Lawrence, 1988, "Beyond Vertical Integration: The Rise of the Value-Added Partnership," *Harvard Business Review*, July–August: 94–101.

Kearney, A.T., Inc., 1984, *Measuring and Improving Productivity in Physical Distribution*, National Council of Physical Distribution Management (NCPDM), Chicago, IL.

Lubben, R., 1988, *Just-in-Time Manufacturing: An Aggressive Manufacturing Strategy*, McGraw-Hill, New York, NY.

MITI (Ministry of International Trade and Industry), 1985, *Census of Commerce*, Tokyo, Japan.

McKinnon, A.C., 1989, *Physical Distribution Systems*, Routledge, London and New York.

Mortimer, J., ed., 1986, *Just-in-Time: An Executive Briefing*, IFS Publications/Springer-Verlag, Berlin, Heidelberg, New York.

Mortimer, J., ed., 1988, *Logistics in Manufacturing*, IFS Executive Briefing, IFS Publications/Springer-Verlag, Berlin, Heidelberg, New York.

Nicol, L., and R. Hollier, 1985, "Plant Layout in Practice," in V. Bignel *et. al.* eds., *Manufacturing Systems: Context, Application and Techniques*, Open University, Milton Keynes, UK.

OECD, 1986, *The Future of European Transport*, Paris, France.

Schonberger, R., 1982, *Japanese Manufacturing Techniques, Nine Hidden Lessons in Simplicity*, Free Press, New York, NY.

Shapiro, R.D., and J.L. Heskett, 1985, *Logistics Strategy, Cases and Concepts*, West Publishing Company, St. Paul, New York, Los Angeles, San Francisco.

United Nations Industrial Statistics Yearbook, various years, United Nations, New York, NY.

Voss, C., 1987, *Just-in-Time Manufacture: International Trends in Manufacturing Technology*, Springer-Verlag, Berlin, Heidelberg, New York.

Chapter 18

Flexible Automation in Less Developed Countries

Pentti Vuorinen

18.1 Global Industry at the Crossroads

The global industrial system is undergoing major changes. They appear in many economic and social issues; severe recessions, exceptionally strong cyclical movements, and labor market mismatches are common in most industrialized countries. National economies are facing deep financial problems, and the crisis in the world monetary system has grown worse. New protectionist barriers to trade are being introduced.

In this chapter, the effects of introducing flexible production technologies in less developed countries (LDCs) are considered. The problems of industrial development in LDCs, especially in newly industrializing countries (NICs), are approached by studying the results from previous research and by evaluating competing theories of technological development in LDCs.

In Section 18.2, a brief history of the development of some industrial branches in various LDCs is presented. The effects of multinational corporations, the development of R&D activities, and the supply of skilled labor are discussed. The features of LDCs are compared with each other and with some industrial countries.

Two contradicting hypotheses are presented in Section 18.3. Both are criticized for being too simplistic and too general. The most common theoretical explanation for international industrial development, the theory of comparative advantages, is examined. As an overall explanation for LDC development or policy guideline it is rejected, though accepted as a normative tool for planning sequential paths for industries in LDCs.

The problems of producing and applying flexible automation technologies in LDCs are summarized and formulated in Section 18.4.

18.1.1   Two contradicting hypotheses

If we examine industrial transformation from the less developed countries' viewpoint, the East Asian NICs seem to have benefited from the rapid growth of electronics in the world economy. For many other LDCs the opportunities and challenges of modern technology are only beginning to manifest themselves. While manufacturing automation is diffusing quickly in the developed countries, the future success of the NICs is not self-evident. Two contradictory hypotheses are often proposed (Hoffman, 1986):

(1) The microelectronics industry offers an opportunity for LDCs to leapfrog over the long-standing problems which have constrained their development.
(2) The technological gap between the developed and developing countries is irrevocably widening. Automation in the developed world will undercut the LDCs ability to compete in international markets. On the other hand, an open embrace of new technology without sensible policies to mitigate negative social impacts would be likely to have little beneficial effect on the development process of less developed countries.

Production technology is the focal point of the latter hypothesis. It is based on the assumption that the prevailing trend toward worldwide sourcing and global production chains is about to change because patterns of relative advantages, costs, and benefits are changing. The new relative advantages for developed countries rise from the diffusion of flexible manufacturing automation. This poses major problems for less developed countries.

The underlying line of argument, according to Roobeek and Abbing (1986), is that the relative advantage for LDCs – and the basis for their industrialization – has been low-cost labor. Now, when benefits from automation outweigh low-cost labor benefits, they are losing this advantage. Labor costs are no longer critical, and the urge to automate outweighs the urge to localize production near cheap labor resources. Modern industry will, from now on, be located near markets which are mainly within the so-called Triad region – Europe, North America, and Japan. Ohmae (1985) finds three main causes for this change in locational patterns:

(1) New production structures of FMS, CIM, and other microelectronics-based technologies and new manufacturing ideologies.
(2) Large markets and common consumption patterns in the Triad region.
(3) The protectionist barriers that are increasingly used in developed countries to protect domestic production against import competition.

There are also arguments opposing this reasoning. The most common is that LDCs now have the opportunity to narrow the technology gap by jumping over intermediate levels of technological development and directly implementing more advanced techniques. Their relatively modest level of industrialization would even assist this process, since the existing capital stock to be replaced by modern microelectronics is not so valuable.

18.1.2 Dangers of simplifying

The whole picture may not be so simple, however. Neither of the hypotheses can be taken as a complete description of the possibilities. The history of industrialization in some LDCs so far shows that no theoretical scheme can give an *ex ante* prediction of the future (c.f. Harris, 1987). The success stories of East Asian NICs have not followed the predictable paths of development based on obvious relative advantages. The simplifying assumptions can be questioned. Are all LDCs really so dependent on multinational companies (MNCs)? Is cheap labor their only resource? Are the locational decisions of foreign companies only based on labor costs? Some preliminary considerations need to be taken up before beginning a more elaborate discussion:

(1) The remarkable industrialization in some LDCs during the 1960s and 1970s has not only been due to cheap labor. Other countries, with very modest industrial development, have more obvious relative advantages of cheap labor, though labor can still be a critical supply factor (Harris, 1987; Hoffman, 1986).
(2) The NICs or other industrializing LDCs are not totally dependent on multinational corporations or on imported technology. The growth of domestic industry has been rapid in many countries, and the importance

of foreign manufacturing plants varies between LDCs. The relationships between domestic and foreign companies are also many-sided and differ in various industrial branches and countries (c.f. Grunwald, 1985; Woronoff, 1986).

(3) Locational choices are not based on one or two simple factors. Industrial development is a more complicated process. This is especially true for diffusion of new technology. It is the outcome of evolutionary processes whereby the interaction between agents induces changing incentives, selection mechanisms, and learning processes (Dosi *et al.*, 1986).

We are dealing with a change in the whole global manufacturing system with diverse and contradictory trends and varying developments in national economies. A unified approach to theory and empirical study is needed.

This is not an easy topic to investigate; the research approach should include development theory, theory of technological change, the study of diffusion of innovations, and various issues of international relations and trade. On the one hand, there is no way to reveal a common path of evolution or the trends of change based on plain theoretical reasoning. On the other hand, empirical studies, as such, do not contribute much more than new cases and examples.

18.2   History of LDC Industrialization

Among the most important industrial sectors in the history of LDC industrialization have been:

- Electronics (semiconductors, computers, and consumer electronics): This sector is the core of the new technology. Semiconductor assembly has been one of the main fields for multinationals' offshore assembly. Even many domestic firms in LDCs have achieved success.
- Mechanical and engineering industry (machines, machine tools, and other metal products): The capital goods for other sectors are mostly produced within this industry. Together with electronics this sector forms the core of technologically dynamic industries.
- Textiles and clothing industry: This sector is the most traditional branch of manufacturing, based on cheap labor resources. Production technology is being modernized much slower than in the mechanical industries.
- Motor vehicle manufacturing: This is the sector where advanced flexible automation technologies are used widely. Vehicle production is quite

remarkable in several LDCs. Cars are mainly produced for the domestic market in LDCs; however, some NICs have also had noticeable success in exporting vehicles.

- Manufacture of plastic products may not yet be among the most important industrial branches in LDCs, but it is likely to grow in importance. Products from this industry are substituting in many fields for metal goods.

18.2.1 Electronics

The electronics complex is at the heart of the new technological systems. The electronics industry and its products will be the most important carrier of technological change in LDCs on a short- to medium-term basis. Its importance is, according to Hoffman (1986), based on many factors:

(1) Electronic consumer products are affecting consumer patterns even in the poorest countries. Microcomputers offer enormous scope for immediate applications which could yield substantial social benefit.
(2) This sector – because of the pervasive character of microelectronics – will increasingly play a role in economic development akin to that attributed to the capital goods sector.
(3) Electronics-related skills will have wide applicability throughout the economy: first to adapt imported technologies to local conditions and eventually to develop indigenous technologies. They are also necessary to implement industrial automation techniques.

The annual growth rate of LDC exports in seven categories of electronics products exceeds that of world exports by a factor of two to three. Hong Kong, Singapore, and South Korea are exceptionally strong in virtually every product category. Taiwan, Malaysia, the Philippines, Mexico, and Brazil form another important group.

Consumer Electronics

During the initial expansion of the consumer electronics industry in Asia, the multinational corporations involvement in offshore assembly for export was a crucial element. LDCs have mainly been successful exporters in low-technology products, which entail only limited local linkages. Several countries have succeeded in developing a strong local industry in certain mature products where design has standardized and process changes have been incremental. Even in Thailand, where most components are imported, there are seven large TV and radio assemblers (Hoffman, 1986).

The emergence of Asian countries as the dominant world source of consumer electronics products was a principal feature of electronics industry during the 1960s and 1970s. This persuaded many others to follow the same path. However, where the export market in more sophisticated products is the objective, a whole new set of difficulties has arisen. This suggests that some form of mutually beneficial cooperation with the MNCs will be necessary. For example, in Southeast Asia dependence on Japanese MNCs is strong for product design, know-how, and components. Japanese firms are extremely reluctant to provide product and process technology, preferring to reserve production of these products for their domestic facilities where they can quickly exploit scale economies to achieve market dominance. This can cause problems. Hoffman (1986) stresses that:

> In the past Japanese producers directed their attention to Asian countries as a place for overseas production but there is no move to divert investment to developed countries in North America and Europe. No substantial expansion beyond the current fairly active situation is expected.

Computers

The computer industry is still dominated by US firms. Small firms have emerged with the introduction of microcomputers, and high growth rates have led many established firms from other parts of the electronics complex to enter the market. This has led to intense competition in an already fierce and crowded market.

PCs are, however, extremely important for the development of LDCs. They can be used as basic tools for modern technological systems. For example, even advanced manufacturing technologies rely on PCs. They will increase domestic and commercial use of stand-alone units, which can integrate to systems by local area networks. These can further merge by connecting domestic terminals to subscription-based interactive information systems.

Production of computers and peripherals for export has grown fast in some LDCs. The sourcing strategies of foreign firms and export activities of local firms mean that a small group of NICs have become important forces in the world computer market.

Some LDCs have even imposed policies to protect their domestic computer manufacture. For example, Brazil and Mexico have used a market reserve strategy with some success. But problems have occurred: domestic PCs are often produced with very high costs, and they are often inferior. It also is an extremely hard task to shift domestic start-up firms toward a self-sustaining growth path defined by local innovation. The market reserve strategy is strongly opposed by foreign firms barred from what they see as extremely lucrative markets.

Computer Software

Software may be even more important than the manufacture of computer hardware, at least in regard to export prospects. Software costs are rising, and demand is increasing. Throughout the developed countries there is a shortage of trained software personnel. For LDCs themselves software is necessary: without a software capability there can be no real indigenous electronics or mechatronics production capacity in the country, nor can the country go far in adapting systems to its needs.

Development of applications may be one of the best and most cost-effective ways through which even small and poor developing countries can begin to build up a capacity in electronics. This capability is crucial for introducing all kinds of advanced industrial automation techniques into the manufacturing processes. Software capability will, in fact, determine a country's ability to develop an independent capacity in electronics and other modern technologies.

According to Hoffman (1986), developing countries have good prospects for exporting software and computer services to the developed countries, because of the following:

- Large demand of products and services outstrips the supply capacity of the industry in the advanced countries.
- The highly fragmented market for products means that there are many market niches where small firms can gain entry provided they have a reliable product.
- The skill barriers to entry are really quite low.
- Capital costs are low as well.

In LDCs unit costs of software development can be three to ten times less than in developed countries. But, there are countervailing factors as well:

- A wide variety of programming tools is being developed, and this will lead to substantial cost reductions in developed countries.
- LDCs software exports are almost all tied to the operations of MNCs who subcontract only relatively simple processing tasks to their offshore locations. There are also possible benefits from this entry route; subcontracting may provide a springboard – like in hardware – which NIC firms can use to launch their independent capacity.
- The distance problem is real, but it can be partially overcome by telecommunications or subsidiaries in the main markets.

Semiconductors

In the semiconductor industry the barriers to entry are currently too high for new entrants. Investment costs are high. Chip manufacturers must stay near the forefront of technology or else rapidly lose their market share. Therefore the level of R&D needed is beyond the capabilities of most LDCs.

Hoffman (1986) argues that the trends toward rapidly moving technological frontiers, regional concentration of MNC investments, and expanding national capabilities within the NICs, which are evident in the semiconductor industry, parallel the developments in the machine tools and clothing industries. Smaller LDCs are presently excluded from gaining access to the most rapidly growing parts of the electronics markets. This situation would only change when (or if) the rate of technological change in the sector slows down.

The present conditions governing entry into the integrated circuit (IC) market differ significantly from those of consumer electronics, software, and computers. For these three sectors there are some common factors that are not – at least not yet – present for the semiconductor sector:

- Because of the rapid diffusion of microelectronics within the electronics complex, various product niches are emerging with characteristics that could allow much greater participation of Third World firms.
- The successful exploitation of these product niches depends much more on product design capabilities than on process technology.
- In spite of MNCs' major role, small firms enjoy distinct advantages in responding to or anticipating specific or changing market demands in many product categories.

- For several products, efficient scales of production are quite low; domestic market opportunities can be more easily exploited to nurture the development of small firms without forcing them to move to export markets too directly.

Two main questions remain unanswered: Will the trends of technical change that are dominant in the semiconductor industry expand to other segments of the electronics complex? Is a sphere of "appropriate technology" suitable for LDCs to produce circuits to be used in their own products emerging?

18.2.2 Machines, machine tools, and metal products

In many NICs the capital goods sector has been a target area for governmental policy and support measures. Production of capital goods is often regarded as the backbone of a nation's industrial structure. Some LDCs have had remarkable success in gaining industrial strength (Erber, 1986; Chudnovsky, 1986).

The sector is also of great importance for the development and diffusion of modern automation technologies in LDCs. In particular, producers of advanced electronics or metal products supply not only the needed machinery but also skilled labor to install and maintain complex imported machinery and systems.

However, it is not always advisable to try to produce the necessary capital equipment domestically. The make-or-import decision is important when determining the national strategy of industrial development. It may often be a more successful strategy to import most modern technology, and then adapt, maintain, and develop the imported equipment.

On the other hand, the sector itself is an important user of flexible automation technologies. In many LDCs the same multisector corporations – with the core in engineering industry – often produce and use the flexible automation in the respective country.

The most important feature in the international machine tool market has been the swift rise to dominance of the Japanese. Hoffman (1986) attributes this to the following:

- Major domestic users of machine tools such as the automobile industry undertook an intensive innovative effort to develop these tools for their own use.

Table 18.1. Production of and demand for CNC lathes in some countries, in units.

Country	Production	Demand
Argentina	10 (1983)	45 (1981)
Brazil	120 (1982)	150 (1982)
India	15 (1983)	101 (1983)
South Korea	268 (1984)	248 (1984)
Taiwan	347 (1984)	250 (1984)
Sweden	200 (1984)	NA
UK	816 (1984)	1,449 (1984)
Italy	830 (1984)	494 (1983)
France	616 (1984)	1,001 (1984)
FRG	2,356 (1984)	1,661 (1984)
USA	1,524 (1984)	4,575 (1984)
Japan	16,555 (1984)	10,551 (1984)

Source: Jacobsson, 1986.

- Producers set out to capture scale economies in machine tool production based on extensive use of automation technologies and via product standardization so that unit costs were considerably reduced.
- The Japanese identified particular market niches at the lower end of the cost-complexity scale and designed superior products to fill these niches.
- The producers established an extensive worldwide network for marketing and after-sale service which served to cultivate demand among users ignored by other firms. Now the Japanese network covers more than 130 overseas locations.
- Japanese machine tool producers established close design links with suppliers of CNC units and were able to reap substantial savings in purchasing the control systems by buying in bulk – achieving unit reductions of up to 35%.

The last point is the most important: the CNC unit accounts for about 25% of total costs, so this gave an important boost to its price competitiveness compared with conventional producers who manufacture machine tools in small batches.

The use of NC machine tools has grown quite fast in some NICs (see *Table 18.1*). In South Korea the share of CNC lathes in total lathe investment grew from 2.4% in 1977–1978 to 34% in 1981–1982. In Taiwan the shares were 7% in 1977–1978 and 20% in 1981–1982. On the other hand, the overall diffusion of NC machines into LDCs is still very modest. In Argentina, for example, NC tools accounted for only 6% to 9% of capital goods imported between 1978–1982 and NC lathes accounted for 38% of all imported lathes. In Brazil, there were 834 NC machine tools in 1983 of which 422 where domestically produced.

18.2.3 Textiles and clothing

Textiles were another important field for multinationals. Textile plants in LDCs are usually either domestically owned or joint ventures with foreign firms and not subsidiaries of multinationals. However, the industry is not knowledge intensive, and products are usually designed in the developed countries.

Since the manufacturing of garments is still labor intensive, production seems to be moving from the original NICs to countries with greater relative advantages of cheap labor resources. The most important among these are China, the Philippines, and Malaysia. These second-tier NICs are now increasing their textiles and garments production. In particular, the Chinese garments industry has shown a remarkable growth in exports (Ballance and Sinclair, 1983).

In the long run, it is evident that flexible automation techniques will be developed for garment assembly and that they will be quickly adapted, particularly in firms operating in countries with high labor costs. Computerized technologies are already widely used in other functions – cutting, drafting, and designing.

This means that manufacturers in LDCs should be prepared to modernize their production technologies. But, to succeed in this, they need to develop the domestic machine tools and electronics industries. More research and development in the manufacturing technology of garments is also needed, as well as advanced software and systems for developing the production systems.

18.2.4 Motor vehicles

Motor vehicle production is one of the leading sectors using flexible manufacturing technologies. Many of the world FMS are installed in car or car-parts factories. This sector has the best data for comparing the national and cultural approaches to flexible manufacturing. For example, American, European, and Japanese car manufacturers have rather different work organizations; they have adopted flexible manufacturing technologies with

Table 18.2. Selected indicators for expenditure for R&D.

Country	GDP (%)	Per capita (USA = 100)	Per R&D person (USA = 100)
USA (1983)	2.7	100	100
Brazil (1982)	0.6	1	14
India (1982)	0.7	1	NA
Japan (1983)	2.6	68	50
South Korea (1983)	1.1	5	21
Philippines (1982)	0.2	1	10
Singapore (1981)	0.3	5	44
Finland (1984)	1.5	43	59
France ((1979)	1.8	29	66
FRG (1981)	2.5	65	97
Sweden (1983)	2.6	76	103
UK (1981)	2.3	44	NA
Yugoslavia (1981)	0.8	2	7

differing expectations and based on different types of investment calculations.

Even the targets of flexibility seem to differ. Also the relationships of production organization and production technology in implementing flexibility are not the same. The Japanese most often begin with redesigning the organization; on the other hand, manufacturers in the USA and Europe tend to install new machines, then educate the work force, and finally introduce the changes in the production organization when the system meets with difficulties.

Both trucks and cars are produced in many LDCs; however, most of them are produced for the domestic market only. Recently various countries (notably Taiwan, South Korea, and Yugoslavia) have also attempted to enter the world market.

18.2.5   R&D

Domestic R&D is an important issue from the point of view of developing manufacturing technologies and the industrial structure. Some indicators showing the level of R&D investments in a few developed countries and the most important LDCs are collected in *Table 18.2*.

If measured as a percentage of GDP, the share of R&D investments is usually much higher in the developed countries than in LDCs. However, in

Table 18.3. Total expenditure for the performance of R&D by source of funds, in percent.

Country	Govern. funds	Priv. ent. & spec. funds	Foreign funds	Other funds
USA (1983)	46.9	49.6	—	3.5
Brazil (1982)	66.9	19.8	5.3	8.0
India (1982)	86.1	13.9	—	—
Japan (1983)	24.0	75.9	0.1	—
South Korea (1983)	27.3	72.5	0.2	—
Philippines(1982)	76.8	14.9	7.6	0.7
Singapore (1978)	37.6	55.9	5.2	1.3
Finland (1983)	42.3	55.6	0.9	1.2
FRG (1981)	41.3	57.0	0.9	0.8
Sweden (1983)	39.3	59.0	1.5	0.2
France (1979)	51.2	42.9	5.2	0.7
UK (1981)	47.7	42.7	6.9	2.7
Yugoslavia (1981)	31.8	57.2	1.8	9.2

France and Finland the share is smaller than in other industrialized countries and only a little larger than in South Korea. On the other hand, in most LDCs the share is still very modest or insignificant.

When compared to the USA, the rankings between countries vary depending on the method. According to R&D expenditure per capita, the USA was clearly first in 1983; Sweden, was second; followed by Japan and the FRG. The LDCs with huge populations are far behind the developed countries in R&D.

The list looks different when the ranking is made according to the expenditure per person employed in R&D (scientist or engineer). Sweden, the USA, and the FRG are at the same level approximately, followed by France and Finland, and Japan and Singapore third. In this list the position of most LDCs is not so far behind.

Finance for R&D can come from government funds, from private enterprises and specific funds (foreign or domestic), or from both sources (see *Table 18.3*). Examples with the bias of the first type are India and Philippines; of the second, Japan and Korea. USA serves as an example of mixed financing. As a rule, both sources are relevant in developed countries.

Foreign funds are significant not only in LDCs but also in some developed countries like the UK and France. It is quite interesting to notice that the structure of sources for R&D is almost identical in Japan and South Korea: a

Table 18.4. R&D expenditures by sector of performance in some countries.

Country	Productive[a] sector	Higher[b] education	General[c] service
USA (1983)	73.0	12.2	14.8
Brazil (1982)	30.1	16.5	53.4
India (1982)	24.7[d]	3.3[d]	72.0[d]
Japan (1983)	64.8	23.0	12.2
South Korea (1983)	60.4	10.3	29.3
Philippines(1982)	9.8[d]	10.6[d]	79.6[d]
Singapore (1981)	54.2	30.0	15.8
Finland (1983)	60.3	19.4	20.3
France (1979)	61.3	15.5	23.2
FRG (1981)	68.3	16.8	14.9
Sweden (1983)	64.5	30.2	5.3
UK (1978)	64.1	10.6	25.3
Yugoslavia (1981)	56.4	18.9	24.7

[a]Productive sector: The industrial and trading establisments which produce and distribute goods and services for sale.
[b]Higher education sector: Establishments of education at third level including research institutes serving them.
[c]General service sector: Various public or government establisments serving the community as a whole.
[d]Estimate.

quarter from government and three-quarters from private sources, no foreign or other funds. Other funds are important in both Yugoslavia and Brazil.

The governments of Brazil and Mexico may soon have considerable difficulties in financing R&D, mainly because of the high accumulation of debts. To a lesser extent this may also cause problems for the Philippines and Yugoslavia. High foreign debts may introduce two important implications. First, other sources are likely to increase their share in financing R&D; second, the total expenditure in R&D will increase slower than previously and slower than in other countries.

In developed countries, the productive sector is responsible for about 60% to 70% of all R&D activities measured in costs. The figure is largest for the USA, and exceptionally low for LDCs like the Philippines, India, and Brazil. The share of the higher education sector ranges from 3% to 30%. The figure is highest for Sweden and Singapore, and lowest for India, the Philippines, and South Korea (see *Table 18.4*).

The average expenditure for R&D by sector of performance varies from country to country. In India, Japan, the USA, and the UK the average

Table 18.5. Relative expenditure per R&D scientist and engineer by sector in some countries.

Country	Productive sector	Higher education	General service
USA (1983)	100	90	117
India (1982)	100	13	132
Japan (1983)	100	65	122
South Korea (1983)	100	21	93
Philippines (1982)	100	NA	67
Singapore (1981)	100	152	42
Finland (1983)	100	57	64
France (1979)	100	71	86
FRG (1981)	100	81	94
Sweden (1983)	100	98	72
UK (1978)	100	NA	150
Yugoslavia (1981)	100	47	73

expenditure is highest in the general service sector; in Singapore it is highest in the higher education sector; and in other countries, in the productive sector. In LDCs the differences between sectors are much clearer than in developed countries. Extreme examples of uneven average expenditures between sectors are India, South Korea, and Singapore. In the first two the figure for higher education is exceptionally low; in Singapore, it is extremely high (*Table 18.5*).

The actual number of scientists and engineers in R&D as a share of the population is clearly largest in Japan. If we consider R&D potential, Japan's performance does not seem likely to weaken. The strong position is an outcome of a long-term educational policy.

On the other hand, the potential R&D manpower in Japan is not as concentrated in R&D as it is in South Korea or the USA, which are extreme examples. The figure for Japan is at about the same level as for other developed countries. The concentration level is surprisingly low in Singapore; only 2% of potential scientists and engineers are actually working in R&D (*Table 18.6*).

However, it is not at all evident that a high concentration of scientifically educated persons in R&D occupations is necessarily the most productive from the point of view of industrial development. Their skills are

Table 18.6. Scientists and engineers in R&D as percent of potential scientists and engineers.

Country	Actual (per million pop.)	Potential (per million pop.)	Actual (as % of pop.)
USA (1983)	3,111	14,777	21
Argentina (1982)	360	18,970	2
Brazil (1982)	256	11,231	2
China (1984)	NA	7,129	NA
Hong Kong (1981)	NA	19,137	NA
India (1982)	131	2,829	5
Japan (1984)	4,436	59,636	7
South Korea (1983)	801	2,486	33
Philippines (1982)	101	NA	NA
Singapore (1981)	296	15,846	2
Austria (1981)	894	20,506	4
Finland (1983)	2,265	35,789	6
France (1979)	1,364	23,747	6
FRG (1981)	2,084	37,001	6
Sweden (1983)	2,292	40,597	6
UK (1978)	1,545	NA	NA
Yugoslavia (1981)	1,109	17,918	6

also needed in production, commercial, managerial, and educational occupations. Focusing on these functions may explain, e.g., the exceptional success in Singapore.

18.2.6　Conclusions

Success in economic development during the last 20 years is obviously dividing LDCs into two groups. The first group (NICs) now approaches the most advanced countries in many industrial fields. For the second group, the gap seems to be widening quite dramatically. Difficulties met by countries in the second group are often multiplied because of a growing burden of debt.

However, the increasing weight of Japan in the banking world may open some possibilities for LDCs with close economic relations with Japan. Within the group of LDCs with tighter connections to the USA, the threat of a debtor's cartel may give debtors some leverage. This threat has not been effective up to now.

Because of a long-term educational policy some LDCs will be reaching the educational level of DCs by the end of this century. In most LDCs, labor-skill levels for flexible manufacturing will be weak for a long time. In some big LDCs the general skill level is low and quite static.

The sources for financing R&D in LDCs are twofold. Small Asian NICs rely on the private sector; most others count more on governmental funds. Accordingly, the expenditure for R&D will mainly be spent either within the productive sector or within the general service (i.e., government) sector. The share of the higher education sector is usually small in LDCs. Also, the average costs of R&D are usually smallest in the higher education sector.

18.3   Comparative Advantage in LDCs

The concept of comparative advantage can be approached from two angles, the positive and the normative. As a positive concept, it can be used to explain the specialization of production and trade: to explain how the things are. As a normative concept, it can be used as a guideline for government policies on resource allocation and trade: to outline what should be done.

When investigating the possibilities of LDCs using flexible automation, we are primarily interested in comparative advantage as a normative concept. There are two main reasons. First, comparative advantage as a positive concept is getting weaker year by year. The information included in market prices is not, in the long run, comprehensive enough. When focusing on flexible automation it is necessary to endogenize technology and productivity. Second, it is difficult to find LDCs where steady state could be regarded as a positive development. For most LDCs a dynamic process of economic growth and development is a necessity.

From this point of view the important questions are: How could the structure of comparative advantages be changed in the long run? How should the physical and human resources be used to take advantage of flexible automation and CIM? Conversely, which of the old comparative advantages are strong enough to be used as the basis of new development?

18.3.1   Three approaches to comparative advantage

In economic theory, three approaches to comparative advantage can be found: the static approach common in economics textbooks; the dynamic-growth approach and the process of industrialization (Bruno, 1970); and the industry-specific sequential approach (Cline, 1982).

Besides these, the concept could be approached from other angles: for example, from those of the dynamic advantages, adaptation advantages, co-operation advantages, and cumulative effects (Kozma, 1982). However, from the normative point of view the division into three approaches seems to be the most useful one.

In the static approach, the factor proportions and comparative advantages over time are given. Countries have their static, predetermined places in the global division of labor.

The dynamic – or quasi-dynamic – approach is based on a theory of different stages of development. The first stages are often seen as periods of interdependent processes of economic growth within the economy, largely independent of the international environment and often aided by different measures of protection. These measures cause distortions in factor prices and changes in factor supplies, which in time support the development of economies of scale in the production of commodities. In a way, the comparative advantages develop with the aid of general support measures. In the later stages of development, the economy can turn to more open international relations now based on the fully developed relative advantages.

According to the third approach, domestic industrial development is regarded within the context of the international economic environment, but differs from the second approach in a selective way. The comparative advantages are actually created and shaped by protective measures aimed at some industries chosen beforehand. In this approach, the role of governmental policy is important.

When promoting growth and strengthening international competitiveness are major goals for industrial policy in a country with modest industrial development, the case for intervention is fairly strong. In LDCs, industrialization strategies are strongly influenced by the position of the country within the international economy. An implication of this is that prevailing comparative advantages are starting points for formulating policy measures, and that a renewal – or even a re-creation – of such advantages is a central goal for policy.

18.3.2   Changes in comparative advantage

Although the concept of comparative advantage is somewhat ambiguous, it is a relevant basis for policy discussions to find out the possible paths for industrial change and success in the international market. It is necessary, however, to know the factor endowments in the countries under study.

To ascertain preferences in several countries, UNIDO (1986a) compared capital and skill technology with total capital. The results on some LDCs as well as some developed countries regarding the industries most relevant from the FMS/CIM point of view are presented in *Table 18.7*. The figures show that some relevant changes in comparative advantage have taken place in the 1980s in a number of LDCs. The corresponding changes in DCs are quite marginal.

In South Korea calculated comparative advantage has become positive – industry has grown "competitive" – in all industries considered except non-electric machinery (which appears to be on the way to becoming competitive). In other LDCs the results are ambiguous; while some industries seem to be going up, some are declining. For instance, in Brazil manufactures of metal and nonelectric machinery seem to be going down but electrical machinery and transport equipment are up. The transport-equipment industry of Brazil seems to be already competitive.

According to these results, the comparative advantage for India and Mexico seems to be best in manufactures of metal (for India nonelectric machinery is also rising); for Singapore, electrical machinery (all the others rising); and for the Philippines, transport machinery.

A static analysis of comparative advantage can only serve as a starting point for a dynamic and sequential analysis. In the next phase of investigation, the development in average labor costs in CIM-sensitive industries has to be considered. This is to be done to find out the comparative advantage in labor costs, which is of central importance when a firm makes locational decisions.

In the 1980s the purchasing power of one US dollar in the Southeast Asian countries with low-cost labor has decreased. In other words, the manufacture of "CIM-sensitive" goods in the fast-growing Asian NICs is getting more expensive compared with manufacture in other countries, for instance, the developed European countries (*Table 18.8*).

Although the figures are not representative, it is reasonable to argue that the average labor costs are rising very rapidly in Singapore and South Korea. This seems to be true especially in ISIC 385 in South Korea and in ISIC 383 in Singapore.

From the perspective of MNCs with offshore assembly in these NICs, the trend implies an increase in production costs in the short run. The urge to substitute capital for labor is obvious, and the incentive for further operations in this area depends on other advantages in the area and the characteristics of alternative locations.

Table 18.7. Index of revealed comparative advantage (RCA) in some CIM-sensitive industries.

	1981–1983	Change		1981–1983	Change
USA			*Brazil*		
SITC 69	−0.2	0.0	SITC 69	−1.2	−1.1
SITC 71	0.8	−0.1	SITC 71	−2.4	−2.0
SITC 72	0.0	−0.2	SITC 72	−0.7	0.9
SITC 73	−0.3	−0.3	SITC 73	0.6	1.2
Japan			*Mexico*		
SITC 69	1.4	−0.4	SITC 69	−0.7	0.3
SITC 71	1.4	0.9	SITC 71	−1.3	0.0
SITC 72	2.2	0.0	SITC 72	−1.2	0.6
SITC 73	2.8	0.6	SITC 73	−2.0	−0.5
Finland			*Hong Kong*		
SITC 69	0.2	0.8	SITC 69	−0.4	−0.7
SITC 71	−0.3	0.5	SITC 71	−0.4	0.1
SITC 72	−0.2	0.5	SITC 72	−0.5	−0.3
SITC 73	0.0	0.5	SITC 73	−0.3	0.1
France			*India*		
SITC 69	0.3	0.2	SITC 69	0.7	0.2
SITC 71	0.0	0.1	SITC 71	−0.7	0.6
SITC 72	0.1	0.0	SITC 72	−0.3	0.5
SITC 73	0.5	−0.2	SITC 73	0.2	0.0
FRG			*South Korea*		
SITC 69	0.8	−0.2	SITC 69	1.7	2.6
SITC 71	1.0	−0.3	SITC 71	−0.8	0.8
SITC 72	0.5	−0.3	SITC 72	0.2	1.2
SITC 73	1.0	0.0	SITC 73	0.7	1.7
Sweden			*Philippines*		
SITC 69	0.3	−0.2	SITC 69	−1.3	0.2
SITC 71	0.4	0.0	SITC 71	−1.6	0.4
SITC 72	0.2	0.2	SITC 72	−0.9	0.3
SITC 73	0.8	0.0	SITC 73	−0.5	0.6
Yugoslavia			*Singapore*		
SITC 69	0.8	1.2	SITC 69	−0.7	0.2
SITC 71	−0.8	0.6	SITC 71	−0.5	0.3
SITC 72	0.4	0.6	SITC 72	0.0	0.8
SITC 73	0.2	0.4	SITC 73	−0.3	0.2

SITC 69 = manufactures of metal, N.E.S.
SITC 71 = nonelectric machinery.
SITC 72 = electrical machinery, apparatus, and appliances.
SITC 73 = transport equipment.
Source: UNIDO, 1986a.

Table 18.8. The change in average labor costs (in US dollars) per annum in the 1980s in manufacture of fabricated metal products, machinery, and equipment (ISIC 38), in percent.

Country	381	382	383	384	385
Mexico (1981–1984)	−13	−16	−15	−17	NA
Hong Kong (1981–1984)	1	5	3	4	3
South Korea (1981–1984)[a]	8	7	8	6	11
Japan (1981–1984)	2	2	3	2	3
Singapore (1981–1984)	14	12	17	9	16
Finland (1981–1984)	− 2	− 1	− 2	− 2	− 1
France (1981–1984)	− 4	− 3	− 4	− 2	− 4
FRG (1981–1984)	− 4	− 3	− 3	− 2	− 4
UK (1981–1984)	− 5	− 4	− 3	− 5	− 3

381 = manufacture of fabricated metal products, except machinery and equipment.
382 = manufacture of machinery except electrical.
383 = manufacture of electrical machinery apparatus, appliances, and supplies.
384 = manufacture of transport equipment.
385 = manufacture of professional and scientific and measuring and controlling equipment not elsewhere classified, and of photographic and optical goods.
[a] According to average wages.

According to the latest developments in labor costs, the best alternatives are to be found in Latin American countries, such as Mexico, and in Europe! Of course, characteristics of labor force, work climate, and marketing must be considered besides wages. In manufacture based on advanced flexible automation technologies, characteristics other than wages are decisive.

Even based on these figures, some changes are to be expected in world trade and industrial locations. They will largely depend on the MNCs location decisions concerning the CIM-sensitive industries. We can speculate on some possible trends:

- If the offshore plants are to be moved nearer to the markets, LDCs with low levels of demand are likely to lose their advantage in adoption and implementation of new technology. These LDCs are usually in an early stage of industrialization. In these circumstances there is not much room for export-oriented policies. The only possibility left for financing industrial advancements is to take loans from richer countries or multilateral agencies.
- If offshore industries still prefer low wages, the expansion of assembly plants in Latin America may be expected.

- The most developed NICs have already passed the initial phases of industrialization and are quite independent of the actual low-wage assembly plants. The multinationals' decisions to locate in these countries have been based for some time on other reasons.

18.3.3   Conclusions

As noted above, comparative advantage cannot be taken as a central guideline for policy measures or more than a starting point in an investigation. But as a normative concept, it can be useful for planning the sequential development path for industry in a country. From this preliminary study, some conclusions on comparative advantage can be drawn.

Flexible automation is changing the structure of comparative advantage of countries within the international division of labor. To be successful in this process, LDCs need to shape their industrial structure of comparative advantage, industry by industry. Export promotion depends on the country's stage of industrialization, factor endowments, and policy measures.

When a country is orienting its industrial production mode toward flexible automation, the role of human capital is important. Because it is difficult for LDCs to reach the technological frontier developed in DCs, learning by doing should probably be preferred to science-based learning in industries not on the technological frontier, like the general machinery industry.

Planning industrial change is a process in which information is gathered stage by stage according to the strategy adopted. In this process, the last phase (implementation of the new technology) is the most difficult one, and with regard to industrialization often the most important.

According to recent statistical surveys, the structure of comparative advantage in industry has changed more in LDCs than in DCs, at least in the industries relevant from the point of view of flexible automation. The pattern of change, however, is quite different depending on the individual LDC.

18.4   LDCs and Industrial Automation

Flexible automation is presenting severe challenges to LDCs at various levels. Already, implementing automatic stand-alone machines is changing the picture of their industrialization process; in particular, the more advanced and integrated automation systems can have a much deeper effect on several aspects of the LDCs.

At the most general level, as a question of national development, the challenge of flexible automation and CIM consists of many diverse problems. These are mainly structural problems that are common to all technical change, both process and product innovation:

(1) The national system for research and development: capabilities for research and development, resources granted to R&D, the relations of academic research to industry, etc. It is obvious that the higher the average technological capabilities of the country – and of the potential adopters of new technology – the faster the rate of diffusion of the new techniques. High-technological capabilities include the ability to evaluate the properties of the technology, to use them effectively, and to improve them (Dosi *et al.*, 1986).
(2) The educational system and its capability to produce qualified researchers, engineers, managers, and workers needed to plan, implement, develop, repair, maintain, and operate modern technology.
(3) National attitudes to new technologies.

The second group of problems is specific to industrial automation, although related to the more general conditions for technological change stated above. They also depend on the industrial structure of the country and the composition of firms, companies, and plants in various industrial branches. The problems of introducing new technology to manufacturing processes could be divided into three types:

(1) Problems connected to production, implementation, and adoption of stand-alone machines, diverse automation apparatus, and single techniques.
(2) Problems connected to producing and implementing whole flexible manufacturing systems, changes in whole factories, and implementing new automated production.
(3) Changes in the mode and the institutional or organizational setting of manufacturing, within large companies, intercompany production chains, and the whole industrial culture of the country or region.

18.4.1 Production of manufacturing automation technology in LDCs

The manufacture of equipment for automation is quite modest in most LDCs. However, the capital goods sector has grown quite substantially in some

NICs. But modernizing the sector is another problem. Only a few LDCs have had any notable success in manufacturing numerically controlled machine tools (see Chudnovsky, 1986).

The machine tool industry began using CNC technology much earlier than other sectors. Earlier machine tools were relatively mature products with a low pace of technical change. They were precisely the products that have their best chances of gaining export markets by building up the necessary capabilities through the learning process. Now the incorporation of microelectronics into these products adds a whole new dimension to the learning process.

The principles of microelectronics to operate machinery are simply not clear. The design of the machine is based on different principles, and the operational relationship between different components can no longer be perceived from mere observation and the application of seat-of-the-pants innovativeness. The LDCs are absolutely dependent upon the availability of specialists and sufficient R&D resources. This has implications for the innovation and training policies for LDC firms and government policies in the area of higher education and R&D subsidies.

Domestic production is by no means absolutely necessary for adopting and using flexible manufacturing technologies. However, it helps build up the capabilities. And when a country is aiming toward a wider diffusion of more advanced flexible manufacturing, it cannot avoid the necessity of some domestic production. The sector not only supplies the economy with the machinery but also provides it with the necessary skills in installation and maintenance of imported equipment in any manufacturing sector. An up-to-date machine tool industry is needed for the adoption and development of more complex imported flexible automation systems.

18.4.2   Manufacturing automation technologies in LDCs

The diffusion of microelectronics technology has been quite rapid within the electronics complex, but much slower than was originally anticipated in other sectors. The textile industry is one example with design and other pre-assembly activities being radically transformed, but the assembly process has rarely been affected as yet.

This uneven process of diffusion suggests that in some sectors LDCs will enjoy a "breathing space." But this space is likely to diminish as microelectronics penetrates even the most technologically stagnant industries. In sectors where the future really lies with microelectronics, firms clearly need to enter the learning curve as quickly as possible, conceivably in an incremental fashion to stay in a race that becomes increasingly intense.

The digital nature of microelectronics facilitates the attainment of a degree of systems integration not possible with analogue control devices. The economic and technical advantages of integration in turn compel producers and users to pursue the process. Due to the inherent logic of the technology, the locus of technical change and innovation has increasingly begun to take on a "systemic" character. Isolated islands of automation are being linked together at progressively higher levels of integration. Whereas stand-alone applications can be relatively easily absorbed by firms, the movement toward higher degrees of systems integration will necessitate basic changes in the organization of production, in the social relations of production, in the structure of the firm, and in its relations with suppliers and end-users.

Difficulties in adopting manufacturing automation technologies are accelerating in a hierarchical order: implementing a single stand-alone NC machine usually represents the simplest case; the most difficult one is planning and building a factory based on modern manufacturing concepts. The problems are hierarchical also from the viewpoint of learning: the experience acquired by both the organization and the work force from stand-alone machines and flexible systems is usually a necessary condition for implementing larger and more advanced systems. Among the hierarchy of problems are:

- Implementation and use of single techniques: stand-alone NC-machines, robots, etc.
- Implementation of complete systems from small FMCs to complex FMSs.
- Adaptive work to install and modify imported equipment.
- Renewal of old plants: the decision between a total renewal and a slower process based on modular approaches (from NCs and robots, to FMCs and FMSs).
- Organizational problems at the process of implementation.
- The division of labor adopted with the new production technologies.
- Problems connected with the maintenance of modern technologies.
- The development of the production processes.

Computer aided design (CAD) is not directly involved in the production process, but is important for flexible manufacturing automation and the new industrial production mode. How could LDCs use CAD in an innovative way to renew the whole production system?

CAD has profound impacts on the design function in almost every industrial sector. When CAD is used in manufacturing, a major obstacle to automation in subsequent stages of production is removed. All the subsequent manufacturing activities can be based on the information generated and stored at the design stage. CAD is a basic precondition for flexible systems consisting of various producers and quickly changing models.

Since the late 1970s the diffusion of CAD has been quite rapid. In 1982 there were approximately 10,000 CAD installations in the world (expected to grow to 27,000 by 1986). Not surprisingly, very few were in LDCs – according to one study only 32 out of a total of 8,000. Further, most CADs studied in LDCs were in subsidiaries of multinational corporations. Doubtless, there are many more in domestically owned firms, especially in the NICs not reached by the study. Market reserve policies certainly inhibited the adoption of CAD.

Interest in selling CAD to LDCs has been growing. For example, several US manufacturers have distributors in Singapore and Hong Kong. CAD in LDCs is, as yet, limited to the NICs and to a few central sectors such as shipbuilding, automobile and component manufacturing, metal works, and consultant engineering. For example, out of six systems installed in Brazil in the last three years, four are used by engineering firms and two by electronics firms.

The rate of diffusion will probably be well below its potential until capital costs decline or until the suppliers penetrate markets in the LDCs. Of course, the problems associated with implementing CAD systems in LDCs are enormous. The lack of software capabilities is among the most acute. This is especially so when the suppliers do not necessarily take any deeper interest in the implementation and adaptation of the systems. In Zaire, for example, one CAD was sold without any software.

Kaplinsky (1983) suggested that CAD systems are likely to diffuse in precisely those sectors where export growth was high in the 1970s and where LDCs planned to specialize in the 1980s. Unequal diffusion between developed and developing countries could erode the international competitive advantage of LDC firms operating in markets where their competitors are using the systems.

This point is important, e.g., in the textile industry. CAD applications virtually eliminate the need for highly skilled graders and markers and allow substantial reductions in materials use. By 1982 more than 700 CAD systems had been sold, and nearly 50% of cloth produced in the USA came from firms using CAD systems. Less than 20 systems were sold in the developing countries, in the Asian NICs, and to domestic producers in Latin America.

18.4.3 Policy issues

Hoffman (1986) points out that policymakers, particularly in the poorer LDCs, frequently do not recognize the crucial role of technology in the development process. When policy statements attesting to its importance have been issued, they are often not backed by the political commitment. Existing institutions are often either ineffectual or concerned solely with science. In the latter case there seems to be an assumption that the development of technology will somehow automatically follow from that of science.

Many LDC governments are totally consumed with the short-term problems of trying to survive under conditions of severe resource constraints. Thus the development of technology policy has, understandably, received little attention.

Hoffman (1986) gives an example: the lack of foreign exchange restricts some governments' ability to import essential inputs, intermediates, and spare parts. Because of the lack of these inputs, capacity is underutilized and many plants are closed. These plants were established several years ago, and do not incorporate very sophisticated technologies. In many cases the spare parts and intermediates necessary to run the plant could have been produced locally. If the past policy had been directed toward the systematic development of local firms to supply parts and inputs, the effects of the crisis would arguably have been somewhat mitigated. Thus what seems to be a problem caused by lack of financial resources is due to the failure of government and managers responsible for industrial development to accumulate human and technological resources.

Government policies toward technology transfer usually focus on increasing the productive capacity as cheaply and quickly as possible. Little effort is put into acquiring technological capacity along with the productive capacity. Recipient firms obtain the hardware and train some operators, but they rarely acquire the underlying know-how and expertise required to improve and adapt the imported techniques. Many problems arise as a result.

Performance efficiency of LDCs often declines over time, whereas in developed countries performance efficiency normally increases. The difference between the two situations is caused almost entirely by the lack of an indigenous, in-plant technical change capacity in LDCs. By not striving to maximize the learning component of the transfer process, LDCs are missing enormous opportunities to develop technical change-related capabilities not only to improve the efficiency of existing plants, but to participate in designing and running the plant and equipment.

However, the governments of less developed countries have adopted a great variety of technology-related policies in their efforts to industrialize. It is hard to draw any simple conclusions on the actual development and the prevailing policy. Only in a few LDCs have the policies been clearly formulated and documented. This seems to be particularly true regarding the introduction and diffusion of new technology. The few more coherent policy efforts concentrate on the development of the electronics sector and to a lesser extent on the use of computers and the information sector. Singapore, Malaysia, and India have made some efforts here (Hoffman, 1986).

Only in some of the most successful NICs – South Korea, Taiwan, Singapore, and a few Latin American countries – have governments formulated general long-range programs to support technological development.

18.4.4   Institutional framework

Technological development can by no means be reduced to policy measures or be explained purely by governmental behavior. It is a question of the whole setting, interaction between different actors within the national economy and between the domestic agents and the conditions set by the international framework. Or, as formulated by Dosi *et al.* (1986), the diffusion of technology is "the outcome of evolutionary processes whereby the interaction between agents induces changing incentives, selection mechanism, and learning processes." Interpreting the technological development and evaluating the future trends in a country is a complex task. It should include a detailed investigation of the structural features in the economy and the behavior of the involved actors as well as a study on the changes in the international environment. This includes:

- A survey on the economic and industrial development so far: the performance of different industrial sectors, the overall structures of branches and companies (even manufacturing plants), the structure of the labor force, and the changes in these structures over time.
- The relations between labor market parties and their institutional or legal regulation. This is a very central issue because it involves the development trends of both labor costs and domestic demand. A main argument against the future development prospects of LDCs (even the foremost NICs) has been that the development of domestic demand and life-styles in these countries lags behind the industrial development. On the other hand, rising purchasing power means rising labor costs. However, the diffusion of flexible automation seems to be diminishing the

influence of labor costs anyway. In this situation the danger of losing old relative advantage may not be comparable to the advantages of modernizing life-styles in line with the modernization of production structures. In fact, many NICs seem to be already moving in this direction.

- The organizations and scale of technological capabilities. According to Dosi *et al.* (1986), the rate of diffusion of new technologies is quite directly comparable with the average technological capabilities of the country. High-technological capabilities also mean the ability to evaluate the properties of the technologies, and use them effectively and possibly even improve them. To evaluate a country's technological capability it is necessary to survey its scientific and technological infrastructure as well as its industrial R&D in manufacturing corporations. Even the education system producing scientists and engineers should be reviewed.

18.4.5 Implementing automation at the company level

Companies thinking about implementing automation technologies into the manufacturing processes face a several initial difficulties even in developed countries. The implementation of computerized systems is a difficult task, particularly when interfaces with the rest of the organization are numerous: the more complex the systems, the more experience and skills needed.

In LDCs, where support from software houses, computer experts and hardware suppliers is scarce, the implementation problems multiply (Meredith, 1987). The problems include several factors:

- Internal skills. Advanced manufacturing systems are highly complex, frequently computerized, and tremendously expensive. Both extensive experience and up-to-date education in such systems are required for successful implementation. The barrier of insufficient internal skills is obviously much higher in LDCs, where experience in these systems is practically nonexistent and up-to-date skills are scarce. Consequently, the implementation of large systems can be practically impossible for domestic LDC firms. It is perhaps only the largest companies in the most developed NICs, manufacturing NC machines, robots, and other automated capital goods, that have the capability to install whole FMSs in one effort.
- Multiplicity of implementation paths. Implementing factory automation entails further difficulties because of the multitude of apparent potential paths available to the firm. The decision of which path to take depends

on the situation of the implementor and the characteristics of the technology in question; how mature the technology to be implemented is; how much experience the firm has; who the supplier of automation apparatus and control systems is and what the relationships between suppliers and adopters are. The likely success of the chosen automation path depends also on the external surroundings of the firm: the institutional framework and public policy in the country or region where the firm is operating.

- Limiting or multiplying synergy. A further complication is the need to use a building-block approach to gain what is often referred to as "synergy": that is, adopting one system of technology at an early stage can limit a firm's options at a later stage. The chosen path of implementation can also have consequences on future options.

- Incremental skill building. The technological skill and experience gained by the staff must be considered as well: mistakes are expensive. Moving slowly and deliberately will often pay dividends in the end. The power of these technologies lies primarily in their ability to be integrated into large, complex systems that automatically interrelate with minimal human intervention; therefore, the staff must be able to identify problems or benefits when the system is extended.

- Support infrastructure within firms. These technologies require an infrastructure of supporting policies, systems, and procedures considerably different from what exists in most firms today. For example, no longer will castings with sand or blowholes be acceptable in raw materials as they were in the past, when workers, trying to expedite an order, welded the defective spots instead of returning the castings to the vendor's foundry. But even more extensive changes will be required. Maintenance is expected to assume a much more important role; direct labor hours will be less appropriate for allocating factory overhead; middle management and support staff will shrink in numbers; and quality-control personnel will have different responsibilities.

- Support infrastructure within firm networks. In developed countries flexible automation is increasingly part of modernization. The process of introducing flexible automation should in most cases comprehend the whole production chain – or "filiere" (c.f. Maillat, 1982), if we think of all the side and support functions needed as well. These filieres should include support agents needed to adopt modern manufacturing technology. Even subcontractors are often tightly bound within the process of comprehensive technological change.

In LDCs advanced technologies are usually either imported as turnkey packages or manufactured, installed, adopted, and used within one large corporation – (c.f. Baark, 1986; Fransman, 1986a, 1986b). This is true even in the most advanced NICs (see Westphal *et al.*, 1985).

There are four basic stages in the implementation process:

(1) Strategic Planning: The formulation of the strategic business plan of the firm and the selection of technologies on the basis of firm's competitive strategy – not just immediate financial return.

(2) Project Planning: The planning of project structure for the implementation process.

(3) Installation: The acquiring and allocating of project resources for the task elements.

(4) Integration: The inclusion of automation in the ongoing manufacturing activities of the firm.

Meredith (1987) stresses the importance of an assessment analysis, which should be performed before or during stage 1. The task of the analysis is to determine whether the firm can automate efficiently. There are two major sets of aspects involved – the technological and the organizational.

The organizational aspects deal with the "personality" of the firm, its culture, operations, values, and so on. The major aspects are top management involvement; employee anticipation; history of change (past changes in the workplace should have been smooth rather than troublesome); normal functioning (the daily operations of the firm should normally be carried out in a planned, thoughtful way rather than in a crisis mode).

The technical aspects should be assessed at the top and middle level of the firm. The top level should include regular evaluations of the factors required to succeed in the firm's markets. The firm should know what the competitive task is for that market, how it changes, and when and why it will change again. At the middle level, an important consideration is the current level of automation in the firm. A company experienced in automation is much more likely to be successful implementing another automation project. The proper functioning of the managerial system is another factor at this level. This includes both the formal and informal systems, the computerized as well as noncomputerized systems.

Need for a thoroughly planned process of implementation may be even more urgent in an LDC firm than in a firm operating in a developed country. It is not, however, self-evident that the process should always follow the guidelines formulated by Meredith (1987). The national and regional features may influence the firm and add new paths to the process. These aspects are the ones that should be studied and compared.

18.4.6 Main problems of flexible automation in LDCs

The possibilities of LDCs obtaining comparative advantage in CIM are seen in two opposing ways. The pessimistic view rests on the enormous gulf between the levels of economic, technological, political, and social developments in the North and the South. Developed countries can be seen embarking on a new and accelerating path of expansion because of their capacity to adjust quickly to the rigorous demands of new technology. The constraints under which most LDCs operate suggest that they could be left behind.

The optimistic view stems from a perception that the current period is one of tremendous flux and flexibility as well as of crisis. The new production system is still very much in its infancy. The old rules of the game are being thrown away and a search for new solutions involving a much broader range of participants is under way. LDCs should continue to seek to reform the international institutional systems, particularly those associated with technology transfer. They must undertake fundamental internal reforms. The relative malleability of the current situation may afford them better prospects to do so than at any other time in the last 30 years.

South–South trade is one possibility. It has grown quite considerably in recent years, and is important in certain sectors. The environmental conditions, infrastructure, and price factor relations are somewhat similar in LDCs. For this reason products successfully designed by local producers for local markets in one LDC could find export markets in other LDCs, particularly in the same region. These advantages have to be more systematically cultivated. The best prospects seem, however, to be valid mostly for the NICs and some countries following them. The situation is considerably worse, and the options remain ill-defined, for the poorer countries.

New policies are needed. A clear and explicit set of active innovation policies that require a broad range of inevitably expensive and controversial interventions on the part of governments is needed. There is also a need for fundamental changes in the social and institutional relations which govern the actions and interactions of groups both within the economy and at the international level.

The need for deeper structural changes seems to be quite acute in many LDCs. Major innovations usually concentrate on key sectors and give rise to problems of structural adjustment between sectors. The process of diffusion of innovations is a cyclical phenomenon that starts slowly but moves into a rapid growth phase. This swarming process has extremely powerful multiplier effects on capital goods, components, and downstream innovations, which give rise to expansionary effects on the whole economy that can lead to long upward swings in growth, output, investment, and employment.

However, the ability of different countries to succeed will obviously vary. Following such a path constitutes a fundamental normative challenge to the very nature of development in LDCs.

References

Baark, E., 1986, "Information Infrastructures in India and China," in E. Baark and A. Jameson, eds., *Technological Development in China, India, and Japan: Cross-Cultural Perspectives*, Macmillan Press, London, UK.

Ballance, R.H., and H.W. Sinclair, 1983, *Collapse and Survival: Industrial Strategies in a Changing World*, Allen and Unwin, London, UK.

Bruno, M., 1970, "Development Policy and Dynamic Comparative Advantage," in R. Vernon, ed., *The Technology Factor in International Trade*, National Bureau of Economic Research, New York, NY.

Chudnovsky, 1986, "The Entry into the Design and Production of Complex Capital Goods: The Experience of Brazil, India, and South Korea," in F. Fransman, ed., *Machinery and Economic Development*, Macmillan, London, UK.

Cline, W.R., 1982, *Reciprocity: A New Approach to World Trade Policy*, Institute for International Economics, Washington, DC.

Dosi, G., L. Orsenigo, and G. Silverberg, 1986, *Innovation, Diversity and Diffusion: A Self-Organization Model*, SPRU, University of Sussex, Sussex, UK.

Erber, F.S., 1986, "Capital Goods and Economic Development: Brazil," in F. Fransman, ed., *Machinery and Economic Development*, Macmillan, London, UK.

Fransman, F., ed., 1986a, *Machinery and Economic Development*, Macmillan, London, UK.

Fransman, F., ed., 1986b, *Technology and Economic Development*, Wheatsheaf Books Ltd., Brighton, UK.

Grunwald, J., 1985, "The Assembly Industry in Mexico," in J. Grunwald, K. Flamm, and K. Flamm, *The Global Factory: Foreign Assembly in International Trade*, Brookings Institution, Washington, DC.

Harris, N., 1987, *The End of the Third World: Newly Industrializing Countries and the Decline of an Ideology*, I.B. Tauris and Co. Ltd., London, UK.

Hoffman, K., 1986, "The Impact and Policy Implications of Microelectronics," in *The Management of Technological Change*, Commonwealth Economic Papers No. 21, London, UK.

Jacobsson, S., 1986, *Electronics and Industrial Policy: The Case of Computer Controlled Lathes*, Allen and Unwin, London, UK.

Kaplinsky, R., 1983, "Computer Aided Design: Electronics and the Technological Gap between DCs and LDCs," in S. Jacobsson and Sigurdson, eds., *Technological Trends and Challenges in Electronics*, Research Policy Institute, University of Lund, Sweden.

Kozma, F., 1982, *Economic Integration and Economic Strategy*, Martinus Nijhoff, The Hague, Netherlands.

Maillat, D., ed., 1982, *Technology: A Key Factor for Regional Development*, Gower, Brookfield, VT.

Majchrzak, A., and V. Nieva, 1984, *CAD/CAM Adoption and Training*, National Science Foundation, Division of Industrial Science, Washington, DC.

Meredith, J., 1987, "Implementing the Automated Factory," *Journal of Manufacturing Systems* 6(1).

Ohmae, K., 1985, *Triad Power: The Coming Shape of Global Competition*, Free Press, New York, NY.

Roobeek, A., and M.R. Abbing, 1986, The International Implications of Computer Integrated Manufacturing, CIM: How CIM is Transforming the Mass-production Concept with Global Sourcing into a Flexible Production Concept With Regional, Triad-sourcing, Unpublished report.

UNIDO, 1986a, *International Comparative Advantage in Manufacturing, Changing Profiles of Resources and Trade*, Vienna, Austria.

UNIDO, 1986b, *Industry and Development, Global Report*, Vienna, Austria.

Westphal, L.E., L. Kim, and C.J. Dahlman, 1985, Reflections on the Republic of Korea's Acquisition of Technological Capability, in N. Rosenberg and C. Frischtak, eds., *International Technology Transfer: Concepts, Measures, and Comparisons*, Praeger, New York, NY.

Woronoff, J., 1986, *Asia's "Miracle" Economies*, M.E. Sharpe, Inc., New York and London.

Chapter 19

Computer Integrated Manufacturing in a World of Regions

Duane Butcher

In the 1980s the world entered a period of serious international economic disequilibrium. The most obvious examples of this disequilibrium were the US trade deficit and the Japanese and West German surpluses. In the course of economic events, imbalances such as these are inevitably self-correcting. Unfortunately, correction normally requires a downward revision of the terms of trade and relative national income of at least one party. Now there is evidence that new manufacturing technology, itself partially responsible for the disequilibrium, will generate self-correcting forces and reduce the economic dislocation of the adjustment process.

The technological changes implicit in computer integrated manufacturing are far-reaching and inevitable. Flexible manufacturing allows firms to shift from a strategy of selling production to one of producing to order. CIM also allows firms to increase business speed, the new manufacturing watchword. However, CIM also requires and facilitates organizational changes between and within firms which will have an important impact on the world economy. New production and inventory management techniques create a mutual interdependence of firms within supply chains. Simultaneously, the importance of geographic proximity of all parts of the chain increases. Japan's lead in the application of technology in this context explains much of its current competitive edge.

Because of the rapidity of product development, the relative transferability of manufacturing technology, the importance of close-knit supply chains, and the need to meet differentiated demand, firms will find it more and more important to situate production inside lucrative markets. The need to defend or develop market share is now far more important than the desire to avoid import barriers, or to exploit low wages, as the primary motivation for foreign direct investment. As a result, intraregional trade will replace a substantial portion of international trade in manufactured items. The capital inflows, which in one sense are necessary to offset trade deficits (as in the US case), are partly in the form of the investment which will equalize manufacturing capability (CIM) and reduce imbalances in trade. Capital flows, ownership conditions, intellectual property rights, technology transfer, and services will be the international economic issues of the future.

This chapter focuses on how organizational change, abetted by technological developments, is producing a "world of regions" and briefly suggests some of the implications that policymakers and existing international institutions should address.

19.1   A View of CIM

CIM is a "veritable cornucopia of costs and benefits, tradeoffs and uncertainties for the firm" (Schoenberger, 1989b). It is a constantly evolving, ever-changing set of solutions to an exceptionally complex matrix of choices between manufacturing technologies and organizational setups. CIM incorporates and enables – but is not defined by – automation. Individual solutions may vary from hard automation through a variety of flexible automation systems to manual operation.

Firms exist in many different time frames, designing the next generation of product, producing for existing needs and desires, converting production to meet evolving demand, and planning further changes (Stalk and Hout, 1990). To function in such a world, all components of firms must be able to access, create, and share data. All functions are interdependent, and all components must be interconnected in real time in a dynamic and continuous manner. CIM simply becomes the application of a common, integrated base for all the firm's functions, from design to purchasing to manufacturing, even to feedback from the consumer or distributor. The IIASA CIM Project describes CIM as: "the application of computers and microelectronics to supplement, and partly replace, low-level human decision making in manufacturing" (Ayres, 1989a). Gunn says that much of "world class manufacturing" is creating new policies, procedures, and practices to deal with issues of staff and career planning, education and training, measurement and reward, communications, and needs for new incentives. In many cases, the "existing body of corporate knowledge has to be 'creatively destroyed' and constructively replaced," as total restructuring takes place (Gunn, 1988). It is clear that if CIM is to operate successfully the human element must be incorporated, and every operation in the chain of production must be subjected to scrutiny. First comes effective management; second, integration and computerization. Automation can and does come at many times during the process of integration (Ettlie, 1988b).

For the purposes of this chapter, the following description will suffice: CIM is the on-line, real-time, computerized integration of the complete operations of a manufacturing firm. CIM includes, but is not limited to, the integration of the following computer assisted automation technologies: just-in-time inventory management (JIT), manufacturing resources planning (MRP II), total quality control (TQC), group technology (GT), computer assisted design (CAD), computer assisted engineering (CAE), computer assisted manufacturing (CAM), CAD/CAM, flexible manufacturing systems (FMS), computer numerically controlled machine tools (CNC), machining centers (MC), industrial robotics (IR), design for manufacturability (DFM), automated guided vehicles (AGV), computer assisted process planning (CAPP), vision systems (VS), and automated storage and retrieval systems (AS/RS).

19.2 The Shift in Global Competition

CIM technology began to emerge as a real possibility in the 1970s. At that time the USA was still the only large market for which exporters could prepare exclusively. The market was characterized by superficial product differentiation in styling and heavy emphasis on marketing, especially in the auto industry. Products were broadly standardized and mass produced, output was relatively stable, within cyclical variations, and the real price of goods had declined steadily during the 1950s and 1960s. The life cycle of products varied from 5 to as many as 13 years. Neither the USA nor Japan could, as exporters, focus on the small, heterogeneous markets of the European Community (EC), where widely differing import barriers (tariff and nontariff), strong company loyalty, and design continuity were the rule.

Only in Japan was intense value competition the rule. Little brand loyalty was visible, and new products had a huge impact on sales. As a result, frequent new models and rapid response to changes in the market became common in the bellwether markets for automobiles and consumer electronics. It is no coincidence that the Fordist model of production, with its increasing division of labor and functional delinking of skills via wide separation of the elements of production, was most appropriate to the US market. Nor is it surprising that the metamorphosis of the Fordist model began in Japan. Japanese success with new production technology enabled it to move from virtual insignificance in the 1950s to become one of the top three exporters in the 1970s.

In the 1970s, as trade barriers came down, intense competition began to emerge, based first on access to the US market and much later on the emerging EC. In the automobile sector, the number of firms capable of competing on a comparable basis internationally jumped from no more than 5 or 6 in the early 1950s to 13 or 14 in the early 1980s and to 15 or perhaps even 20 today. Demanding and sophisticated customers gained access to information about competing goods, and demonstrated ever-decreasing brand loyalty – a willingness to discern. The major global markets have become buyers' markets akin to the Japanese mode: strong rivalry for customers, a need for responsiveness, a demand for excellence, rapid product innovation, and after-sales service have became critical to commercial sales (Clark, 1989b).

Simultaneously, technological change has accelerated the production of newer and more sophisticated products. The auto industry saw 120 new models introduced between 1983 and 1988, as sales of individual models dropped. In the Japanese auto market, the most competitive of all, the number of models went from 37 in 1970, with 2.4 million units sold, to 56 models in 1985, but only 3.1 million in total sales. Between 1970 and the mid-1980s the diversity of technology in engines and drive trains in the USA increased by a factor of seven (Clark, 1989b). Video recorders, personal computers, and digital disks appeared in the consumer electronics market. Philips, which produced about 100 different color TV models in 1972, put 500 on the market in 1988. It produced 10 types of disk players in 1982, but 150 in 1988 (Kumpe et al., 1988). In 1988, NEC alone introduced 20 new PC models in Japan.

In this process, time has emerged as the dominant competitive factor (Stalk and Hout, 1990). The old paradigm for manufacturers might have been: the most value for the lowest cost. Modern manufacturers must add: the most value for the lowest cost in the least amount of time. IBM Japan calls this competitive factor "business speed." One study estimated that each day of delay in introducing a $10,000 auto to market would cost a firm $1 million, not including future lost market share (Clark *et al.*, 1987). Toyota and Honda claim to have reduced the development time for new cars to two years. Philips has reduced product life cycles for audiovisual products to one year, in some cases even less. Production cycles also have shrunk. Motorola reportedly has cut the time it takes to build and ship electronic pagers from three weeks to two hours. Panasonic now puts bicycles from Japan in the hands of its US customers within three weeks of order (including six days for air freight). Everex Computers claims it can manufacture and ship PCs and peripherals within 48 hours (Case, 1990).

Time-based production also reduces one key uncertainty of doing business: forecasting demand. The further into the future a firm must forecast demand, the greater the probability of error in over- or underproduction. Production to order, rather than sale of production, is the new imperative. Although price may never have been as important a factor in market decisions as pure theoreticians had assumed, in today's affluent, fragmented markets non-price elements of competition are at least as important as price.

19.3 Changing Interfirm Relationships

The emergence of unstable, highly differentiated, rapidly changing markets; of new forms and sources of competition; of shrinking product life cycles; and of CIM technologies has directly influenced the direction of organizational changes in production. In the 1950s and 1960s large firms producing finished goods could afford arms-length, often adversarial, relations with their suppliers. Taylorist–Fordist production techniques permitted and promoted spatial separation of production and scattering of functions and skills. The common practice in the USA was to maintain large inventory buffers, to maximize the number of suppliers producing each component, and to negotiate fiercely (at fairly frequent intervals) over price.

Using CIM, short runs and a larger mix of product types on the assembly line are becoming economically feasible. CIM is making it possible for firms to resolve the Abernathy productivity dilemma between the need to cut manufacturing costs by maximizing standardization and the pressure to introduce new and improved products more and more rapidly. Simultaneously, the complexity of machinery and tighter interdependency of machines render the production process more fragile. Techniques of JIT inventory and TQC demand exact timing and coordination of the flow of parts up and down the production chain. Smaller production runs and frequent model changes demand flexibility, on-time delivery, and often drastic quality improvements from parts suppliers. Such concepts as design for manufacturing, in which parts are designed from conception to be most easily assembled in a final product, cannot succeed unless each firm in the chain is fully involved. Effective organization, usually at the behest or initiative of a lead customer or a supply-chain leader, means better, more timely information is available to all elements up and down the supply chain. The time it takes to do normal work is compressed. Efficiency is most enhanced where networks of firms are linked in geographic proximity and firms interact in a face-to-face manner on a daily basis. Benefits to be gained increase in direct proportion to the fraction of unique parts (complexity) in the final product (Helper, 1989b).

The result is a pronounced shift in the interface through which firms interact with each other. In the Toyota system the final producer of the finished good develops a long-term, mutual dependency with relatively few vendors, as compared with the old adversarial system. Often only one firm is the source of a particular component or family of parts. Final assemblers make long-term commitments to purchase supplier output. Buyers involve themselves in great detail in the operations of supplier firms. Close interaction takes places between engineering and design personnel of buyers and sellers. Buyers advise on production techniques, and may even invest in supplier firms to assure the most flexible, highest quality output. Buyers also may finance raw material and equipment costs, and even share the cost of excess inventory due to order changes. Typically, manufacturers also must review the extent of vertical integration: whether to make parts (Stalk and Hout, 1990; Schoenberger, 1989b).

With greater knowledge of supplier firms, buyers can negotiate on price; this aspect is quite different from the adversarial model. Longer-term contracts with less rebidding are more common, and purchasing typically may be accomplished directly from the shop floor. Reciprocally, suppliers often are involved in the design process of the buyer's final product, and may be included in the production-planning process. In effect, suppliers absorb some of the fluctuations in final demand by doing much of the labor-intensive work, and by shortening the life cycle of parts – at labor costs less than those of larger buyers.

Supplier firms often are heavy users of CIM technology, and may be large firms in their own right. Many engage in unique research and development. First-tier suppliers with significant engineering and manufacturing capability have begun to take responsibility for groups of parts, subsystems, or modules which greatly facilitate the assembly process when incorporated into computer controlled production planning. The first-tier firms may in turn organize a network of subcontractors and lower-tier firms. Japanese auto companies typically now deal directly with 200 to 300 first-tier suppliers (Clark, 1989c). These supplier–buyer relationships do not end the chain. To stay competitive modern firms must include retail outlets, distribution chains, and even the consumer in their information network (Japan Machinery Federation, 1988).

Whether termed computerized production and inventory control systems, manufacturers resource planning II, or JIT, the implication for manufacturing is the same (Senker, 1986). Modern production techniques require a careful reassessment of the functioning of the production process from start to finish. A key element of that process is the relationship between final assembler and parts producer. The efficiency of the whole, not the parts, is key, and all the functions of supply networks are enhanced by direct, computer-to-computer electronic data links. The results: quality, fast and reliable delivery of defect-free products, and minimized manufacturing costs (Clark, 1989c).

19.4 Evidence of Change

The Japanese industry's management of supplier networks is one of its most significant advantages in worldwide competition. A 1989 survey found that 68% of Japanese subcontractors had never changed their buyer firm and that 53% had done business with the same buyer for 15 years or more (US Congress, Office of Technology Assessment, 1990). Clustering also seems to move offshore with Japanese firms (Helper, 1989b).

There is strong evidence that an irreversible change in supply relationships is taking place elsewhere around the world. Brodner *et al.* (1988) concluded that there was a clear trend against vertical integration and a substantial increase in the volume of subcontractors in industry. They cited a recent Dutch study which found that "fully 80 percent of 500 surveyed firms reported that their subcontracting had increased sharply in the past few years." Firms in Europe were planning to reduce the total number of subcontractors by 30% to 40%. Edquist reported similar movements in Volvo and other Swedish firms, and said that the majority of Scandinavian firms would be linked in JIT schemes by 1992 (Edquist and Jacobsson, 1988; Wandel and Hellberg, 1987).

In the Federal Republic of Germany the story is much the same. Doleschal (1989) reported strong and consistent efforts by buyers to reduce supplier networks and to reduce vertical integration, even though suppliers have played a critical part in the production of cars in Germany since the birth of the industry. He identified a variety of network ideas and strategies in the automobile supply chain. He also observed that five-year contracts with guarantees of minimum purchases have become the rule in the FRG. There also has been a considerable change in legal responsibilities and obligations of both suppliers and buyers. Production of parts is more closely tied to production plans, and the planning horizon is over six months in many cases. There is widespread use of electronic data networks, and the amount of knowledge buyers have of supplier-firm production methods and costs has increased substantially. Buyers often help in meeting supplier-firm investment costs, especially for the special needs of the buyer for quality assurance or documentation, and sometimes even provide automated data systems and instruments to offset their demand for improvement.

German firms also are moving toward a system of first-tier suppliers able to produce complete modules or groups of parts in JIT/TQC mode, and new relationships are springing up within the system between first- and second-tier firms independent of size. Doleschal believes new computer applications will continue to bring drastic changes within and between firms in the FRG in the next 10 years (Doleschal, 1989; Doleschal and Klonne, 1989).

Mendius and Weimer (1988) came to similar conclusions when they studied small- and medium-sized firms in Germany in 1988. They found an increase in research and development being performed by supplier firms, and confirmed extensive use of electronic data-processing networks and communications systems. They also found a substantial increase in the degree of integration between auto manufacturers and retailers.

In the USA also the trend toward closer supplier–buyer relationships is under way. Helper's 1989 survey of US auto suppliers showed 92% of supplier firms provide their statistical process control charts to their customers, as opposed to only 16% in 1985. The number of suppliers who provide breakdowns of the cost of production also jumped substantially. The average contract length between US buyers and suppliers almost doubled in five years ending in 1989, from 1.2 to an average of 2.3 years. Before 1980 it was rare for buyers to contract for more than one year. By 1985, 70% of Ford's contracts were multiyear.

The number of suppliers who participate in the design of their products also has increased. Many suppliers now even see themselves as taking the lead in providing technical assistance to their customers. All but 4% of those surveyed had made or received plant visits for that purpose. The average number of suppliers competing with another firm providing the same product fell from 2.0 to 1.5 over the last five years, evidence of a tightening of supplier networks. (Compare, for example, the early 1980s, when automakers often had six to eight firms supplying a single part and dozens competing for a general class of component.) Helper also observed far less commitment by buyers to in-house component production. Wholly owned component plants have been closed in several cases, and all the major automakers have established programs to evaluate in-house parts and components in a competitive manner. Significantly, for JIT proponents, the number of suppliers delivering in lots that would last from 8 to 30 days fell from 44% to 16% (Helper, 1989a).

Ford reduced its suppliers from 5,000 in 1982 to 2,300 in July 1987. Xerox has reduced its supplier base from more than 5,000 companies to only 370. It has included suppliers in the design process, and begun providing performance specifications to suppliers instead of the old detailed blueprints to encourage supplier design (Burt, 1989).

Shapiro (1985) documented a growing awareness of problems with the adversarial model. He found substantial changes in purchasing departments, which often had been so insular that they exacerbated adversarial relationships with other firms (or even their own). He confirmed a substantial increase in supplier input into design and specification of parts, and found that the focus of relationships had shifted away from lower prices for parts toward lower total manufacturing costs. These and other studies also have confirmed the move toward first-tier suppliers of subassemblies and modular components. US retailers also have begun to realign themselves into super dealer chains with substantial purchasing and pricing leverage against manufacturers (Kumpe *et al.*, 1988).

US industry appears to lag behind Japanese and parts of European industries in improving supply-chain management. For example, Kelley and Brooks (1988) reported that only 3% of the firms they surveyed had received financial support from a customer, and only 9% had been asked by their customers to use modern, flexible automation technology in 1986. Helper (1989b) also found a considerable amount of skepticism still exists among suppliers about JIT and its effects. Many smaller firms suspect buyers will just shift responsibility for inventory to the supplier. Some observed changes may only be evidence of suppliers giving in to automakers' demands for higher quality and JIT (at low price), without creating better relationships. Nevertheless, the trend toward more tightly knit, cooperative supply networks in the USA, as around the world, is as irreversible as it is unavoidable – especially in the automobile industry (Kaplinsky, 1983; Helper, 1989c).

19.5   The Economic Value of JIT and Supply Networks

In the Toyota model, supplier–buyer proximity has become a *sine qua non* for success. Traditional agglomerations of industry seem to work best for supplier networks, and can become seedbeds of continued technological progress. Firms closer to developers of CIM technology, and those subject to the most demand from leading buyers, are most likely to adopt new production processes. In the automobile industry, processes such as body stamping, which once could be done at a central location and shipped to assembly plants, are being relocated at the assembly site.

Research has not, however, satisfactorily demonstrated either the optimum geographic size or the minimum critical mass of industry skills and expertise need for CIM. For example, although Mazda is in Michigan, GM and Nissan chose to locate new plants in Tennessee, while Honda and Mitsubishi chose Ohio. Each new location is within a one-day zone for delivery of parts from established industrial agglomerations; however, an increasing number of suppliers are now setting up operations near the new plants (Schoenberger, 1988). Wandel also found that 90% of chassis suppliers and 73% of body suppliers needed to be within 250 miles of the assembly plant to make JIT feasible at GM (Wandel and Hellberg, 1987). Prakke also observed increased regional clustering of firms in Europe, and that "production is being called back from distant countries and peripheral regions" (Brodner *et al.*, 1988).

Efficient supplier–buyer relationships become even more significant as vertical integration progressively is reversed. In 1980 the cost of materials, components, and services bought by US manufacturers averaged over 60%

of the cost of operations, up from 50% in 1960 and 40% in 1945 (Burt, 1989). Toyota creates only some 20% of the total value-added to each auto it assembles. Ford and General Motors probably account for about 50% and 70%, respectively (Kumpe *et al.*, 1988). In Germany, according to German Automobile Association (VDA) data, the percent of final production accounted for by auto assemblers has declined steadily from 41% of value in 1980 to 37% in 1987 and is nearing 30% or even 25% in 1990. Fully 70% of large electronics-producer costs may go into materials purchases. Xerox now purchases materials accounting for about 80% of total manufacturing costs (Case, 1990).

Perhaps the most dramatic evidence of the effect supply chains have on manufacturing has come from studies of the auto industry. Japanese automakers take an average of 13 to 14 months to produce major new tools and dies; in the USA the average lead time is almost 25 months. Clark (1989c) found the Japanese advantage over competitors was due to better planning of product content and to substantial differences in the organization and management of long-term supplier relationships. Overall, suppliers may account for one-third of labor-hour advantages, and perhaps a four- to five-month advantage in speed of delivery to the market. Cole and Yakashiji estimated that superior supplier relations gave the Japanese a $300 to $600 advantage per car in the early 1980s (Helper, 1990). Integrated problem solving between upstream and downstream designers in the supply network also was critical to the strength of the Japanese model in product development (Clark, 1989c). Polaroid reported savings of $27 million over a two-year period after it tightened its supplier network (Burt, 1989).

Helper's study found that US parts firms with effective buyer–supplier relationships had a notable improvement in performance, and exhibited more willingness to modernize production processes than firms outside such networks. For example, buyers use more CNCs as contracts lengthen. All firms with contracts of more than five years had adopted some CNC technology, but only 41% without a contract had done so. Results were the same for CAM, CAD/CAM, and other CIM technologies (Helper, 1989c). Shapiro (1985) concluded that the competitive position of the firm is highly positively correlated with the ability to manage suppliers. He found such relationships effectively encourage innovation, and that in-house talent is not enough to assure innovation. Failure to develop effective buyer–supplier relationships in effect hinders adoption of innovation into the final product.

The body of evidence about the quantitative impact of JIT/TQC-type systems, totally removed from networks (and apart from other CIM technologies), is equally convincing. Inventories have been reduced by 40% to 80% in such firms as Hewlett-Packard, Atlas Copco, Xerox, and Opcon (*Manufacturing Technology News*, 1990; Gunn, 1987a). Where interest rates are high, savings in some cases have exceeded direct labor costs of production. Electrolux reduced inventory over a four-year period from 25% to 18% of turnover. Each percentage point amounts to $164 million in interest saved alone. Additional savings from reduced employment and quality improvements were substantial.

Reductions in personnel costs from JIT application varied from 37% to more than 90%. Hewlett-Packard reported that labor costs declined by 37%, from 9% of total production costs to 2%, in its plants using JIT/TQC. Companies such as Harley-Davidson and GE report reductions in production time of up to 96%. These and other companies reported reductions in scrap and rework between a low of 23% and a high of 98% (Gunn, 1987a). Bradley reported a 67% drop in "total cost of quality." Zytec reduced production-cycle time by 96% (*Manufacturing Technology News*, 1990). Hewlett-Packard achieved reductions of 96%. Harley-Davidson and Texas Instruments reported that JIT/TQC increased productivity by 30% and 46%, respectively, while scrap went down by 23% and 60%. Hewlett-Packard reported JIT/TQC caused a 98% reduction in the number of nonconformities in parts (*Manufacturing Technology News*, 1990; Gunn, 1987a). Finch and Cox (1986) have shown that small firms can meet most of the requirements for successful JIT, and can achieve significant improvements in operations. In fact, supplier firms facing increasing demands from their customers may have no choice but to apply JIT and CIM themselves if they are to survive in tomorrow's world. This is confirmed by the IIASA FMS data base (Tchijov, 1989a, 1989b).

The conclusion is inescapable: efficient supplier networks are essential to improve inventory management and quality control. Combined with JIT/TQC or other systems, networks yield enormous economic benefits. Although close proximity may not be mandatory, no doubt propinquity is a decided advantage.

19.6   The Impact of CIM

The full impact of CIM technology will not be visible until a fully integrated firm emerges. Even then before-and-after measurements will be problematic. However, several reports are available on the results of applying different parts of the family of CIM technologies. (It is, of course, difficult to isolate the effect of separate technologies. For example, JIT may often have been used by the firms in the samples, and unit costs are particularly difficult to separate from savings in personnel.) A quick rehash of what has happened to flexibility and quality, as much as its impact on costs and operations, is in order.

Toyota reported recently that it had reduced design time by more than 50% using CAD/CAM (Suri *et al.*, 1989). Lockheed reported a reduction of 81% (Gunn, 1987a). Burt (1989) reported that new product development time and costs went down by 50%, using CIM technologies, and product lead time fell from 52 weeks to 18 weeks. Production time also dropped significantly, as setup and processing times were reduced, and the number of rejects was down by 93%. The service-call rate on GE's dishwasher production fell by 50% (Kaplan, 1986).

An ECE survey of 20 locations in Europe found 70% fewer machine tools were needed for the same volume of production after FMS was introduced (UN Economic Commission for Europe, 1986). IIASA reported FMS reduced the number of machines by an average of 78% for 84 cases, and work-in-progress by 77% for 64 cases (Tchijov, 1989a). Gunn (1987a) reported that large industrial firms achieved 200% to 300% increases in operating time of capital equipment using CIM. The McAlinden (1986) study found output per labor hour was up almost 140% after CNC/CAM was introduced. FMS caused an average increase in output of 490% in 61 cases (Tchijov, 1989a). McAlinden (1986) also reported that robotics resulted in productivity increases averaging almost 390%. Ebel and Ulrich (1987) report that the increase in output per draftsperson in Europe varies between 200% and 6,000%, and averages between 200% and 500%.

The impact on costs is still an important part of the total picture. A German aircraft manufacturer reported a 44% savings in personnel requirements from "CIM" as far back as 1980 (Schoenberger, 1988). In more recent years, reports on CIM found that labor-cost reductions of 5% to 20% are on the low side, and 90% to 95% reductions are not uncommon. Introduction of FMS brought an average reduction in unit costs of 37% in 55 cases in IIASA's world data bank in 1990, and a drop in labor costs of 77% in

137 cases (Tchijov, 1989a). Caterpillar and Yamazaki have reported more than 90% savings (*Manufacturing Technology News*, 1990; Mullens, 1986). CNC/CAM saved an average of 63% in unit labor costs in the 35 cases surveyed by McAlinden in 1986.

19.7   A New Paradigm for Trade and Investment

International trade theory has never explained satisfactorily how factors of production determine trade or how factors (especially technology) move internationally. The rigorous assumptions required in the theory of comparative advantage, or in later attempts at revising and modernizing pure trade theory, cannot encompass the existence of the multinational firm or explain competitive factors such as time, quality, design, after-sales service, and market share. Yet organizational and non-price factors are critical to investment and trade decisions. Location and geographical distance are equally important, but fall outside of traditional economics.

The new industrial division of labor (NIDL) model and its cousin the product cycle model hold that, as products and production processes mature and become standardized, and as competition increases, production will shift from advanced industrial (core) areas to developing countries to take advantage of lower-cost labor. (It was only a small additional step for some to conclude that a shift from manufacturing to services is a natural progression for a maturing economy, and that "manufacturing doesn't matter.")

Unfortunately, the NIDL's focus on factor markets and labor cost overlooks far more important technological changes in production – culminating in CIM (Schoenberger, 1987). For example, the NIDL ignores the fact that direct labor costs are an ever-decreasing part of total operations (no doubt already affected by various individual CIM technologies). One recent estimate put the ratio of material, overhead, and direct labor costs in a typical manufacturing firm at 50:40:10 (Edquist and Jacobsson, 1988). In electronics firms, direct labor costs appear to have dropped to as low as 2% of sales, and to be no more than 10% across a spectrum of industry (Kumpe *et al.*, 1988). Gunn (1987a) estimates that direct labor costs will vary between 0% to 6% of manufacturing cost in the future and are only 1% to 2% in many plants today. Although the Japan Management Association estimates direct labor costs at 10% to 14% of sales on average, analysts in Japan agree that costs are declining steadily, and that 3% to 5% of sales already are more accurate estimates among advanced producers in the electronics industry.

At Electrolux, direct labor cost of "white goods" is now only about 7% to 10% of total costs – less than distribution costs. Studies also show that assembly, though still labor intensive, is only 5% to 10% of typical production costs of automobiles, while parts-inventory costs alone may be as high as 20% (Altschuler *et al.*, 1984).

As the full benefits of CIM are felt, wages will decline even further as a percent of production costs, and become even less important in decisions regarding production sites. By contrast, transportation costs remain substantial, especially in low-wage areas. The author's survey of US-manufactured imports at the three-digit SITC level showed that the direct cost of freight and insurance alone averages more than 3% of the value of imports calculations. Wandel has reported that logistics, including the entire spectrum of transportation, handling, and distribution costs, contributes as much as 63% of value-added to final production (Wandel and Hellberg, 1987). A survey by the National Tooling and Machining Association found that "hidden costs" of making tools and dies offshore could add 5% to 15% to bid prices for shipping, 3% for additional communications, 5% to 10% for added inventory, and up to 35% for unanticipated design changes (Markides and Berg, 1988). According to a World Bank expert, lack of infrastructure can push the cost of exporting from many developing countries to as high as 30% of final costs. Despite computerized handling, transportation costs are likely to increase – especially when fuel prices climb upward. Transportation costs may exceed direct labor costs in a wide variety of industries.

The declining importance of labor costs in investment decisions was reflected in a study for the Japan Management Association (JMA). More than 50% of firms surveyed were adopting CIM to increase variety and handle small-batch production or to shorten lead times. Fully 39% wanted to forge stronger links between production and sales (Watanabe, unpublished). Reducing direct labor costs was a goal of fewer than 5%. A survey of Japanese automakers also found that firms installed industrial robots primarily for flexibility: to meet changing demand and frequent model changes. Labor saving was of secondary importance except for material handling (Japan Machinery Federation, 1988). Although many expected Japanese production to move offshore when the yen increased 30% in value against the dollar between 1986 and 1988, a recent Kaidanren survey of 214 Japanese manufacturers found that 54% said they would cope by upgrading products, adding greater value, and minimizing cost. The 29% who planned to shift production overseas said they would do so to supply export markets directly, not to export to Japan (Markides and Berg, 1988). One Japanese automaker reported that several strategies to import entry-level vehicles from Malaysia or India into Japan had evaporated.

For most firms, offshore manufacturing in low-wage regions is not a viable long-term strategy. At best it may be a short-term tactic to buy time to restore health at home. This is not a new development. Foreign direct investment (FDI) in manufacturing historically has gone into developed markets where labor costs are relatively high and markets strong. Between 1970 and 1984, US FDI in manufacturing soared from $31 billion to $86 billion, yet 76% of US manufacturing assets abroad were in Western Europe at the end of 1986. The share of US FDI destined for LDCs actually decreased somewhat between 1984 and 1988. As relative economic and financial strength shifted toward Europe and Japan, the USA became a prime recipient of investment, again reflecting market strength (Schoenberger, 1987; Survey of Current Business, 1990a, 1990b).

Although CIM reinforces arguments against investment in low-wage areas, and discredits the NIDL model, it reinforces incentives to invest as an alternative to exports. Knowledge of a market theoretically can be obtained by electronic linkage with market researchers or distributors, but speed of response can best be optimized by physical presence in the market. Long-distance service of a market cannot cope with timely identification and correction of defects, and requires large finished inventories to hedge against interrupted supply.

The attractiveness of direct investment increases with the level of competition in nonstandardized products and with the rapidity of change in the market. Where the market is large enough to support additional manufacturing or competition, investment is likely to take the place of exports. An increase in market size, such as the integration of the EC after 1992, adds to investment incentives. Regulations on creation of joint ventures, mergers, and other actions which affect the size, resource base, and technical scope of firms also will affect choice. If the value–weight ratio is high, and shipping costs low and not time sensitive, exports are less likely to be replaced by investment in the market. But even where strategic value is high or volume low, firms may choose joint ventures or cooperative arrangements such as R&D and production sharing, instead of exports.

Another important variable in the equation is the ready transferability of CIM technologies, the apparent freedom of supplier firms, and the ease with which buyers create supplier networks to meet their demands for JIT/TQC. Japanese supplier firms are successfully investing in the USA and UK where they compete for sales with domestic automakers (Rubenstein et al., 1979;

JMIF, 1989). The performance of the transplants in the USA also confirms that Japanese organizational techniques can be adapted and applied effectively elsewhere. The best Japanese plants produced a vehicle in 13.3 hours in 1989; the best transplants in North America took 19.9 hours. The best American and European firms took 17.7 and 21.8 hours, respectively. The average plant in Japan took 18.3 hours; the average transplant, 21.3; the average US-owned plant, 24.3; and the average in Europe, 33.3 hours. As to quality, Japanese firms averaged 36 manufacturing defects per 100 vehicles in 1989; transplants in North America had only 42; the best North American, 52; and the best European, 55 (Womack, 1989b).

CIM has introduced a new dimension to investment decisions. The new interfirm relationships reinforce market developments, making exports even less competitive with production directly in the market. JIT networks reward proximity, reinforce agglomeration benefits in industry, and alter business strategy; coproduction, joint development of products, and a host of other emerging relationships provide viable alternative vehicles for market entry. As labor costs approach or even fall below shipping costs, the value of physical presence in markets is further enhanced. The increasing dependence of industrial firms on service firms and the tendency to sell goods accompanied by service or software have the same effect. CIM increases the value to firms of a robust global manufacturing and distribution network which can meet the needs of different regions by adjusting sourcing, production, and distribution plans. Existing offshore plants will likely provide access to the markets in which they are located, rather than serving as export platforms to serve the industrialized core countries (Schoenberger, 1989c).

19.8 A New World of Regions

The world's markets already are forming into distinct regions bounded by fundamental economic and industrial factors, such as natural transportation barriers, relative shipping costs, optimum supply relationships, and cultural and ethnic characteristics of demand, market size, and income level. Governments also will be tempted to protect national firms and unions from the world outside the region, during periods of adjustment, so firms will have less to lose, but more to gain from seeking a presence in a strong regional market. The competitive advantage in acquiring the latest in CIM technology can only increase the likelihood of policy actions favoring investment – by region or by nation. Trade flows in manufactured goods between regions will dwindle, to be replaced by direct investment, with no loss of worldwide economic efficiency.

Economic reorientation, from east to west across oceans and from north to south within regions, already is visible. The European Community is set to establish the paramount supranational region when the internal market begins to merge in 1992. The 1988 USA/Canada Free Trade Agreement and overtures to Mexico create the presumption of an emerging North American market, as well. The Pacific Rim, though far behind the other two in political or economic integration, often is touted as another distinct region. As an illustration, EC auto production in 1989 was 15.7 million vehicles, as opposed to 15.3 million in the Asia-Pacific and 13.7 million in North America (JMIF, 1989). Though less affluent, the Soviet Union, China, India, and South America also seem likely regional entities. The key uncertainty in this scenario is Japan. Transport costs from Japan to North American will probably be less than those to the Pacific Rim, and Japan's current lead in production technology might well continue to offset transportation costs for the next 20 years. Japan could conceivably join the North American region or could remain on the boundary serving both regions.

For some goods R&D costs and economies of scale will be so large that the optimum region will be the globe. The best example is aerospace, where shipping costs are inapplicable, R&D costs huge, and demand far less differentiated than, say, for autos. Other products, such as common ball bearings producible using fixed, hard automation, will be produced in low-wage countries and traded much like commodities; that is, they also will supply the world. There is, of course, no reason that regions must be large enough to cross national boundaries. Sabel and others have identified a number of subnational regions: the "Third Italy," the "Second Denmark," the metalworking firms of Smaland in Sweden and Baden-Württemberg in the FRG, Silicon Valley and Route 128 in the USA, to name a few. These are, however, agglomerations of producers, and little research has been done to connect optimum agglomerations with market size. It is "easier to exemplify and typologize the flexibly specialized regional economies than rigorously to explain them" (Sabel in Hirst and Zeitlin, 1989).

19.9   The Global Firm

Fitting into this world of regions requires more of firms than just sales offices or production facilities in different regions. To maximize global after-tax

Table 19.1. Foreign direct investment (FDI) in the USA in billions of dollars.

	1983	1984	1985	1986	1987	1988	1989	Total
Manufacturing investment	3.1	3.1	12.1	16.8	19.7	36.1	28.3	119.2

Source: Survey of Current Business, 1990b.

profit, firms must optimize market and manufacturing expertise. Each firm must achieve a globally optimum configuration of supply-chain strategies and must integrate worldwide production choices in a consistent manner. Each firm will maximize design and engineering capabilities to provide satisfaction to its customers, but will find it increasingly difficult to cross regional barriers to meet cultural and social idiosyncrasies.

An increasing number of firms are undertaking to reach true global status, either through direct investment in the major markets of the globe or through arrangements and alliances. The activities of US multinationals have been well documented around the world since well before the appearance of *Le Defi Americain* in 1968. The current flow of FDI into the USA is graphic evidence of the emerging imperatives for physical presence in markets. Net inflow of FDI to the USA in 1989 reached a record $61.3 billion, and FDI in manufacturing went from only $3.1 million in 1983 to more than $36 billion in 1988 and $28 billion in 1989 (*Table 19.1*). The book value of foreign investor equity in US affiliates, only $8.2 billion in 1973, jumped to $91 billion in 1987. Japanese investment in US manufacturing went from only $1.1 billion in 1981 to $10.6 billion, second only to the UK, in 1988 (Schoenberger, 1988; Survey of Current Business, 1990).

British, French, and Japanese companies spent $7.9, $5.7, and $3.8 billion, respectively to purchase American companies in the first half of 1990, and Swedish companies were involved in deals totaling $8.5 billion during that period. Other large acquisitions were made by firms from Italy, Taiwan, and Bahrain. Total employment in European and Japanese subsidiaries in the USA now is over 3 million, and one report holds that more than 600 Japanese plants were operating on American soil in 1988 (Fuerbringer, 1990). Toyota, Honda, Nissan, Mazda, Mitsubishi, and even Suzuki/Isuzu and Fuji/Subaru have assembly plants in the USA. The major Japanese electronics firms also have invested large sums in a variety of consumer and intermediate goods plants. Nissan and Honda have announced commitments to design and engineer models in the USA, for the US market, by the mid- to late 1990s, and Honda claims that its North American autos will have more than 75% domestic content by 1991.

19.10 Diffusion

The speed at which CIM is penetrating manufacturing and the potential depth of penetration are critical to the speed of correction of international imbalances. Yet, any forecast must be made in the context of a constant evolution of the technology itself. Ayres (1989a) outlined the difficulties in estimating the rate of diffusion of CIM technology. He used robots, FMS, CAD/CAM, and vision systems, all automation technologies, as proxies for CIM. By the definition of this essay, however, CIM is not dependent on, nor does it require, full automation; effective, computer controlled integration of a shared data base can occur without "elimination of the human element" from the machining process. The organizational rearrangements between firms and the many process reorganizations within firms – the very essence of the JIT network–agglomeration hypothesis – can proceed without full automation or realization of, for example, artificial intelligence systems.

The Japanese Management Association (JMA) approached this problem by asking firms when they would accomplish "CIM." Although the question was somewhat ambiguous, the results provide the prospects for integration, rather than automation. JMA found that, as of 1989, more than 81% of Japanese firms were implementing CIM or planned to begin within two years. The figure was almost 80% for firms producing machinery, 86% for electronics and precision instruments, and 65% for transportation (auto, train, aerospace). Almost 35% of Japanese firms had an integrated production information system, and over 20% had in place a system integrating sales and production in 1989. More than 50% were planning to integrate sales, engineering, and production, but fewer than 1% said they had reached that target. Fully 70% of the firms surveyed expected to have CIM in place by 1992. In a similar approach, Peat Marwick Main & Co. found in 1988 that some 60% of US firms expected to have CIM in place within five years (Watanabe, unpublished). Schultz-Wild (1985) found more than one-fifth of West German firms were engaged in some form of computer integration in 1986–1987, concentrating on CAP/PPC, CAD/PPC, and PPC/CAM.

Estimates of the use of individual technologies are valuable as corroboration for the rate of adoption of CIM as a catchall term. (Improvements in CIM component technology also are a critical part of the observation that changing cost relationships discourage trade between regions.) Ayres (1989) estimated that CIM technologies probably had penetrated no more than 1% of maximum potential in discrete parts manufacturing in 1989. He suggested that 100% automation of inspection would be a practical reality by the year 2000. For two key indicators, software as a percent of total investment and the average utilization rate of CNCs, he predicted a substantial degree of automation could be reached by that time. Ranta (1989) estimated that flexible manufacturing systems per se are growing at an annual rate of about 40% (observed through 1988). By the year 2000 that rate would yield 1,800 to 2,000 systems in the world.

Japanese observers in MITI and JMA, and vendor firms such as IBM, NEC, and Toshiba, believe that virtually 100% of Japanese firms with more than 50 employees are using some form of computerized automation. One Japanese study, based on developments in robotics, concluded that automobile assembly lines could be between 35% and 75% automated by the year 2000 (Arai in Ayres, 1989c). The 1988 Survey of Manufacturing Technology by the US Bureau of the Census found that fully 70% of US firms with more than 500 employees have some linkage of CAD output to control of manufacturing machines or will have within five years. Of even more importance for interfirm relationships, over 72% of large firms had, or would have by 1993, some sort of intercompany computer network linking plants to suppliers or customers or both (Manufacturing Technology, 1988). A 1988 market study forecasts that 70% of all US machine tools will be operating in FMS configurations by 1997 (*FMS Magazine*, 1987).

By the year 1995 most large Japanese manufacturing firms will have reached the point of calling themselves CIM companies. A substantial portion of the integration, which is accelerating and reinforcing movements toward new interfirm relationships and locations will be in place in many firms in the industrialized world. Although Schultz-Wild's data may encourage some to be more cautious, by the year 2000 it is a good bet that fully one-half of manufacturing operations in the industrial world will be automated, and incorporated into a CIM configuration. Many substantial gains in quality, flexibility, and cost reduction will have been realized before that time.

19.11   Trade Flows

It is far too early for CIM to have had visible impact on international trade statistics. Nevertheless, it is possible to sketch some dimensions and directions. The automobile industry is a good place to begin. US imports of vehicles peaked at $87.9 billion in 1988, and were still $86 billion in 1989. The deficit on vehicles and parts accounted for more than 45% of the total merchandise trade deficit. Passenger car imports peaked at $48 billion in 1987, and were down to $44.56 billion in 1989. Car imports from Japan alone accounted for $20.2 billion in 1989 by census definition – almost 20% of the total deficit. These figures reflect three straight years of decline in passenger car imports, which the US Department of Commerce said was "directly attributable to the increase in Japanese production in the USA" (Survey of Current Business, 1990).

Japanese transplants produced about 795,000 vehicles in the USA in 1988, and over 1 million in 1989, according to the Japanese Motor Industry Federation (JMIF, 1989). Capacity will be about 2.5 million by 1992. Another 360,000 units will be produced by Japanese makers for US brand names. Womack (1989c) has estimated passenger car, van, and light truck capacity of transplants at 2.78 million by 1992. The Japan Machinery Federation (1988) predicts that total Japanese production in North America will reach 3 million by the year 2001; Japanese exports will sink to only 170,000 units. The German Association of Automobile Producers (VDA) predicts that Japan will account for 45% of the US market for automobiles by the year 2000, two-thirds of which will be produced in the USA. That would mean 5 million autos produced by Japan in the USA if consumption remains at about 16 million per year throughout the decade.

This chapter will settle for a middle-ground prognosis. Assume that American car manufacturers adopt CIM technology more rapidly and effectively than they have to date, that total sales stay at 16 million, and that Japanese producers settle for a lower market share than might actually be possible (to avoid adverse political reaction in the USA). It will then be entirely possible for Japanese auto production in the USA to reach 4 million units by the year 2000; Japanese imports will account for another 400,000. As *Table 19.2* indicates, the impact on the USA/Japanese trade balance will be substantial.

Japanese parts and engine exports to the USA have risen from $3.3 billion in 1985 to $11.2 billion in 1989, at least partly to provide inputs to transplant assemblers. In earlier years, trade analysts expected that at least

Table 19.2. US imports of Japanese passenger vehicles.

	1989[a]	1990[b]	1995[b]	2000[b]
Passenger cars (in billion US$)	20.2	18.0	12.0	6.0
Units (in million units)	2.1	1.8	1.0	0.4
Average unit value (in US$)	9,950	10,000	12,000	15,000
Japanese production in USA (in million units)	1.0	1.2	2.5	4.0
Total Japanese sales in USA (in million units)	3.1	3.0	3.5	4.4
Japanese parts/engines (in billion US$)	11.2	11.3	13.0	13.0
Passenger vehicle and parts imports (in million US$)	32.4	29.3	25.0	19.0

[a]Source: US Department of Commerce, Bureau of Census, Trade Data. Figures are based on census data, and are not identical to balance of payments data.
[b]Forecast.

30% of the value-added of offshore production would be imported. However, JIT/TQC configurations argue against that sort of relationship in the future. The major transplants in the USA claim local content already is approaching 75%. The Japan Automobile Parts Industry Association reported that 128 parts firms were manufacturing in the USA in early 1989 and another 24 had some sort of technology transfer arrangement with US firms; another 10 transplant parts firms were manufacturing in Canada (JMIF, 1989). According to another estimate, by early 1989, 232 Japanese plants were operating in the USA – either wholly or jointly owned – solely to supply parts to Japanese assemblers in the USA (Callahan, 1989). If the trend to JIT production for both transplant and US-owned plants continues imported content should go from about 30% in 1989 to 20% content in 1995. As indicated above, it will probably be feasible to import lower value items that have been made, for the most part, manually for many years, while CIM is being implemented, so 10% is used here as the base level for the year 2000.[1] It is, however, quite likely that the import of cheaper auto parts would come not from Japan but from the other Southeast Asian countries – or Japanese parts transplant firms in those areas. Thus, the US bilateral deficit with Japan would shift even more than these figures suggest.

In this conservative scenario, total imports of passenger cars and parts from Japan dwindle from $31 billion in 1989 to $25 billion in 1995 and to only $19 billion in 2000. Japanese production will replace at least another $2 billion worth of imports from Europe by 2000, as total Japanese sales in the USA increase from 3 million to almost 4.5 million vehicles. If European automakers turn their attention to Eastern Europe, if they fail to invest in the US market, they could lose a substantial portion of the $15.7 billion surplus they enjoyed with the USA in 1989. Japanese firms could, in that case, easily account for a total of 5 million vehicles sold in the USA, as the VDA predicts, and American firms could protect their market shares at the expense of European exports. The total impact of the reorientation of production location, by the more conservative estimate, will be a reduction in the US trade deficit of about $7 billion per year by 1995 and of $14 billion to $15 billion by 2000. If transplant production were to remain at 1 million units, and if the total in Japanese sales were achieved by export, the cost to the USA would be $18 billion in 1995 and $45 billion in 2000 (totally independent of adjustments in exchange rate, macroeconomic policy, or trade restrictions).

As noted above, Japanese production also will move toward the EC, where US automobile firms already have substantial presence. Japan exported 1.7 million autos to the EC in 1989 and sold 200,000 made locally. By 1992 Japanese transplant production will reach 820,000. The JMF (1988) also forecasts exports to the EC will dwindle to only about 140,000 by the end of the decade, as Japanese local production mounts.

Automobile trade was chosen as the prime example here because it was one of the biggest sources of trade friction in the 1970s and 1980s, and because vehicle production, sales, and maintenance are important parts of developed economies. The automobile industry also provides yet another piece of evidence that regionalization already has taken place to a large extent. US trade with the rest of North America in vehicles totaled $61.0 billion in 1989 – more than one-half of total US trade in vehicles and parts (Survey of Current Business, 1990). Car trade will not, of course, disappear. Imbalances in production and costs will persist. Those imbalances will, however, decrease as a more balanced "world of regions" with integrated production systems serving regional markets becomes the rule. There also is no automatic presumption that emerging industrial economies will be completely unable to export to developed countries. Likely low-wage producers of cheaper, mass-produced items for different regions might be Spain and Ireland for the EC and Mexico for North America. Mexico could merge very rapidly into the North American market if the Bush administration proposal for a free trade zone is accepted.

The US deficit on consumer durables was $35 billion in 1989, more than 30% of the merchandise trade deficit. Although no figures are available which allow comparisons similar to those of the automotive industry, electronics producers such as Toshiba, NEC, Matsushita, Electrolux, and Bosch already produce a variety of goods in the USA. Reports also have surfaced about Japanese decisions to relocate offshore production of consumer electronics back to Japan, and the possibility of doing so in the USA (Ohmae, 1985).

19.12 Implications

Accelerating product development, diversified markets, product differentiation, changing inter- and intrafirm relationships, and the impact of technology on cost and quality of production will produce a world economy as different from today as the globe is from a Mercator projection. For a wide spectrum of manufactured goods, FDI, motivated primarily by a desire to be present in valuable markets, will replace exports as the preferred vehicle of market access. The direction, makeup, and volume of world trade and FDI will be irrevocably altered.

The convergence of manufacturing technologies is leading most Western firms to adopt and adapt Japanese breakthroughs in factory management. The process is abetted and accelerated by Japanese FDI. That investment will help remove the massive flows of manufactured items, such as autos and consumer electronics, which have resulted in the large trade imbalances of the past 20 years.

Right now the flow of best-practice manufacturing technology in many cases equates to Japanese investment. Barriers to such investment already exist in the EC, and popular opposition has appeared in the USA. A long restructuring process lies ahead. International tensions easily could result. Some domestic firms in Europe and the USA will fail in the face of foreign competition, increasing pressures to ban foreign investments. Even more serious objections may arise when foreign investors seek to move into defense-related industries. Barriers to ownership of firms which control sensitive technology already exist in many nations. Xenophobia or economic chauvinism could prevent the spread of manufacturing efficiency and thus exacerbate problems and damage the world economic system more than do trade barriers. Nations also must review the implications of new interfirm relationships in terms of their effect on competition and antitrust/cartel regulations.

Regional concentration, including R&D, will be the most logical and efficient organizational pattern of production. New structures may be consistent with political structures, as in the EC, but may transcend national boundaries and cause adjustment problems elsewhere, as in the Pacific. Efforts to promote regional markets may very well require some sacrifice of economic sovereignty. Yet, integrating political mechanisms are lacking. No stable regional currencies or coordinated macroeconomic or environmental policies exist. International mechanisms to coordinate emerging regions and to solve interregional disputes also are lacking. Regional autarky could damage interregional trade and impair global efficiency in those areas where logical market size transcends regions. Where optimum regions are substantially smaller than national boundaries questions arise. Should flourishing regions help those in crisis? How will resources be allocated between regions? Where regions cross borders, as might be the case in an automobile region encompassing the Mexican border areas and the southwest United States, resource allocation and factor movement must be addressed.

Japan is seen by many as an economic Godzilla that will destroy all competition. Yet, many firms in Japan are not ready to compete on the world market (especially in areas such as chemicals and pharmaceuticals). An MIT/IMVP survey shows that even in the automobile industry the worst Japanese plants produce vehicles with some 98 defects per 100 vehicles, whereas the US average is 82. The best US auto firms – with 52 defects per 100 – could compete in Japan (Womack, 1989b). If Japan is unwilling to drop barriers to investment intolerable tensions could arise. This would be even more the case if barriers persist while other Japanese firms are expanding rapidly abroad (Kaslow, 1990).

Developing countries must study these forces carefully. The presumption is that industrial development based on low-wage export platforms will fail. The more appropriate strategy will be to focus on the domestic market and on attracting foreign technology to serve that market. Regional arrangements must be reexamined. Governments and assistance agencies also must rethink old conclusions about import substitution and about infant industries. Where a potential market exists there may well be reasons for following policies which promote local investment. Development should be based more on specialization and on the historical and cultural advantages than on the export platform model.

All governments must be concerned about access to emerging manufacturing technology. Electronic data networks will be a must, but will also introduce implicit problems of uncontrolled access. The question of haves and have nots may be as important for manufacturing technology as it is for basic science. Questions of intellectual property rights and software may well become more important than trade barriers.

In the absence of political forces which would unite emerging regions in a rational world system, firms themselves may supply the only cross-linking, postnational organization. Could it be that the firms will provide a binding horizontal, symmetrical substitute for global government? If company cultures prove stronger than national loyalty they could override regional differences and lead the way to the next stage of world organization.

19.13   Conclusion

This chapter has focused only on manufacturing. It ignores, for example, trade in the process industries (where computer automation has been in place for years) and trade in commodities. Nothing in this essay implies that CIM is a panacea for world problems, or that sound monetary and fiscal policy are not necessary to eliminate and prevent economic disequilibria. CIM is, however, a major new force with implications which go far beyond the provision of higher-quality goods at lower prices. The role of CIM, and the changes coming in its train, deserve far more study by both economic theoreticians and policymakers.

Notes

[1] The domestic content of US autos averaged 90% in 1986 (US Congress, Office of Technology Assessment, 1990, p. 134).

References

Altschuler, A., M. Anderson, D. Jones, D. Roos, and J. Womack, 1984, *The Future of Automobiles: A Report of MIT's International Auto Program*, MIT Press, Cambridge, MA.

Ayres, R.U., 1987, *Manufacturing and Human Labor as Information Processes*, RR-87-19, IIASA, Laxenburg, Austria.

Ayres, R.U., 1989a, *(1) Future Trends in Factory Automation, (2) Technology Forecasts for CIM*, RR-89-10, IIASA, Laxenburg, Austria.

Ayres, R.U., 1989b, "Impacts of Robotics on Manufacturing," *Technological Forecasting and Social Change* 35(2–3).

Ayres, R.U., 1989c, *US Competitiveness in Manufacturing*, RR-89-6, IIASA, Laxenburg, Austria.

Bessant, J., and B. Haywood, 1987, *Flexible Manufacturing Systems and the Small-to Medium-Sized Firm*, Innovation Research Group Occasional Paper No.2, Brighton, England.

Braunerhjelm, P., 1990, Svenska Industriforetagen for EG 1992, Industrins Utred-ningsinstitut och Overstyreslen for Civil Beredskap, Stockholm, Sweden.

Brock, W. *et al.*, 1990, *The Global Economy*, W.W. Norton, New York, NY.

Brodner, W. *et al*, 1988, FAST, Occasional Paper, Commission of the European Communities, February 1987, XII-145-87, Brussels, Belgium.

Burt, D.N., 1989, "Managing Suppliers Up to Speed," *Harvard Business Review* 7/8:127–135.

Callahan, 1989, *Automotive Industries* 2.

Case, J. 1990, "The Time Machine," *INC.*, June, 48–55

Clark, K.B., 1989a, "What Strategy Can Do for Technology," *Harvard Business Review* 11/12:94–120.

Clark, K.B., 1989b, *High Performance Product Development in the World Auto Industry*, Harvard Business School Working Paper 90-004, Harvard University, Cambridge, MA.

Clark, K.B., 1989c, "Project Scope and Project Performance: The Effect of Parts Strategy and Supplier Involvement on Product Development," *Management Science* 35(10):1247–1263.

Clark, K.B., and T. Fujimoto, 1989, "Overlapping Problem Solving in Product Development," in K. Ferdows, ed., *Managing International Manufacturing*, Elsevier Science Publishers, Amsterdam, Netherlands.

Clark, K.B., and R.H. Hayes, 1988, "Recapturing America's Manufacturing Heritage," *California Management R.*, Summer:9–33.

Clark, K.B. *et al.*, 1987, *Product Development in the World Auto Industry*, Brookings Papers on Economic Activity, 3/87, Washington, DC.

Clark, K.B. *et al.*, 1987, *Product Development in the World Auto Industry*, Brookings Paper on Economic Activity, Washington, DC.

Cohen, S., and J. Zysman, 1987a, *Manufacturing Matters*, Basic Books, New York, NY.

Cohen, M.A., and H.L. Lee, 1989, "Resource Deployment Analysis of Global Manufacturing and Distribution Networks," *Journal of Manufacturing Operation Management* 2:81–104.

Cohen, S. *et al.*, 1987b, "The Myth of a Post-Industrial Economy," *Technology Review* 2/3:55–62.

Cole, S., 1982, "The Microprocessor Revolution and the World Distribution of Income," *International Political Science Review* 3(4):434–454.

Cole, S., 1986, "The Global Impact of Information Technology," *World Development* 14(10/11):1277–1292.

Contreras, F. *et al.*, 1989, *The Bicycle Industry: International Implications of Modern Manufacturing Technologies and Practices*, Report 89-5, Department of Industrial Engineering, University of Wisconsin, Madison, WI.

Dahlman, C., 1989, *Impact of Technological Change on Industrial Prospects for the LDCs*, World Bank Industry and Energy Department Working Paper, Industry Series Paper 12, 6/89, World Bank, Washington, DC.

Dicken, P., 1986, *Global Shift: Industrial Change in a Turbulent World*, Harper and Row, London, UK.

Doleschal, R., 1990, *Die Automobil-Zulieferindustrie im Umbruch, Graue Reihe-Neue Folge 15*, Universität Gesamthochschule, Paderborn, Stitz and Betz, Dortmund, Germany.

Doleschal, R., and A. Klonne, eds., 1989, *Just-in-time-Konzepte und Betriebspolitik*, Universität Gesamthochschule, Paderborn, Stitz and Betz, Dortmund, Germany.

Ebel, K.-H., and E. Ulrich, 1987, "Some Workplace Effects of CAD and CAM," *International Labour Review* 126(3):351–370.

Edquist, C., and S. Jacobsson, 1988, *Flexible Automation*, Basil Blackwell, Oxford, UK.

Electrolux-Koncernen 1988, 1989, Electrolux, Stockholm, Sweden.

Eliasson, G., 1987, *Technological Competition and Trade in the Experimentally Organized Economy*, Industrial Institute for Economic and Social Research, Stockholm, Sweden.

Eliasson, G., 1988a, *The Firm as a Competent Team*, Working Paper 207, Industrial Institute for Economic and Social Research, Stockholm, Sweden.

Eliasson, G., 1988b, *The International Firm: A Vehicle for Overcoming Barriers to Trade and a Global Intelligence Organization Diffusing the Notion of a Nation*, Working Paper 201, Industrial Institute for Economic and Social Research, Stockholm, Sweden

Ernst, D., 1980, *New International Division of Labor, Technology, and Underdevelopment*, Campus, Frankfurt, Germany.

Ernst, D., 1985, "Automation and the Worldwide Restructuring of the Electronics Industry," *World Development* 13(3):333–352.

Esser, U., and G. Kemmner, 1989, *CIM: Mythen und Fakten der Computergesteuerten Produktion*, Management-Zeitschrift Industrielle Organisation, IO, nr.5, p. 81, Zurich, Switzerland.

Ettlie, J., 1988a, *Implementation Strategies for Discretionary Manufacturing*, University of Michigan, Ann Arbor, MI.

Ettlie, J., 1988b, *Taking Care of Manufacturing*, Jossey-Bass, San Francisco, CA.

Ettlie, J., 1989, "Organizational Integration and Process Innovation," Unpublished Paper, pp. 795–806.

Finch, B.J., and J.F. Cox, 1986, "An Examination of Just-In-Time Management on the Small Manufacturer: With an Illustration," *International Journal of Production Research* 24(2):329–342.

Fix-Sterz, J. *et al.*, 1987, *Flexible Manufacturing Systems and Cells in the Scope of New Production Systems in Germany*, FAST Paper 135, Commission of the European Communities, Brussels, Belgium.

Flaherty, M.T., 1986, "Coordinating International Manufacturing and Technology," in M. Porter *et al. Competition in Global Industries* Harvard Business School Press, Boston, MA.

FMS Magazine, 1987.

Fuerbringer, J., 1990, "Purchases by Japanese of US Concerns Climb," *New York Times*, July 17.

Gerstenberger, W. *et al.*, 1988, *Investitionen, Beschäftigung und Produktivität*, Ifo-Institut für Wirtschaftsforschung, e.V., Munich, Germany.

Grunwald, J., and K. Flamm, 1985, *The Global Factory*, Brookings Institution, Washington, DC.

Gunn, T.G., 1987a, *Manufacturing for Competitive Advantage*, Ballinger, Cambridge, MA.

Gunn, T.G., 1987b, "The Fallacy of Directly Pursuing Low Cost in Manufacturing," *Chief Executive* 42(11–12).

Gunn, T.G., 1987c, "US Global Players Win," *Managing Automation* 7.

Gunn, T.G., 1988, "People: The Primary Resource in World Class Manufacturing," *CIM Review*, Spring:6–9.

Helper, S., 1989a, *Supplier Relations at a Crossroads: Results of Survey Research in the US Automobile Industry*, School of Management Working Paper 89-26, Boston University, Boston, MA.

Helper, S., 1989b, *Strategy and Irreversibility in Supplier Relations: The Case of the US Automobile Industry*, School of Management Working Paper 89-22, Boston University, Boston, MA.

Helper, S., 1989c, "An Exit-Voice Analysis of Supplier Relations," Presented at the Second Annual International Conference on Socioeconomics, Washington, DC.

Helper, S., 1990, *Supplier Relations and Technical Change: Theory and Application to the US Automobile Industry*, Individual Paper, 1/90.

Hildebrandt, E., and R. Seltz, 1989, *Wandel betrieblicher Sozialverfassung durch systemische Kontrolle?* Wissenschaftzentrum Berlin für Sozialforschung, Edition Sigma Rainer, Bohn Verlag, Berlin, Germany.

Hirst, P., and J. Zeitlin, ed., 1989, *Reversing Industrial Decline? Industrial Structure and Policy in Britain and Her Competitors*, St. Martins Press, New York, NY.

Hoffman, K. *et al.*, 1985, "Microelectronics, International Competition and Development Strategies: The Unfavorable Issues – Editor's Introduction," *World Development* 13(3):263–272.

Hoffman, K. *et al.*, 1988, *Driving Force*, Westview, Boulder, CO.

Inaba, E.S., 1987, "The Present Status and Future Directions of Machine Factory Automation," *IM Report* 11:13–14.

Inaba, Y., 1989, "Japanese Automakers on the Offensive," *Tokyo Business Today* 5:14–19.

Ishitani, H. *et al.*, 1989, "Impacts of Robotics on Manufacturing," *Technological Forecasting and Social Change* 2/3(4):95.

Jacobsen, G., and J. Hillkirk, 1986, *Xerox: American Samurai*, Macmillan, New York, NY.

Jaikumar, R. 1986, "Postindustrial Manufacturing," *Harvard Business Review* 11/12:69–76.

Japan Machinery Federation (JMF), 1988, *Survey Research Report on the Development of New Manufacturing Systems*, Tokyo, Japan.

Japan Motor Industrial Federation Inc. (JMIF), 1989, *The Future of the Japanese Automotive Industry – A Report of the Consultative Committee on the Automobile Industry*, Japan Trade & Industry Publicity Inc., Tokyo, Japan.

Johnston, R. *et al.*, 1988, "Beyond Vertical Integration: The Rise of the Value-Adding Partnership," *Harvard Business Review* 7/8:94–101.

Jurgens, U. *et al.*, 1989, *Moderne Zeiten in der Automobilfabrik*, Springer-Verlag, Berlin, Heidelberg, New York.

Kaplan, R.S., 1986, "Must CIM be Justified by Faith Alone?," *Harvard Business Review* 34:87–95.

Kaplinsky, R., 1983, "Firm Size and Technical Change in a Dynamic Context," *Journal of Industrial Economics* 32(1):39–60.

Kaplinsky, R., 1984, *Automation, Technology & Society*, Longman, New York, NY.

Kaslow, A., 1990, "US Wary of Japanese Investments," *Christian Science Monitor* 7(26):8.

Kelley, M., and H. Brooks, 1988, *The State of Computerized Automation in US Manufacturing*, Research Paper, J. F. Kennedy School, Harvard University, Cambridge, MA.

Kelley, M. *et al.*, 1989, "From Breakthrough to Follow-through," *Issues in Science and Technology* 5(3).

Kellman, M. *et al.*, 1984, "The Nature of Japan's Comparative Advantage: 1965–80," *World Development* 12(4):433–438.

Kreinin, M.E., 1985, "US Trade and Possible Restrictions in High-Technology Products," *Journal of Policy Modeling* 7(1):69–105.

Kumpe, T. *et al.*, 1988, "Manufacturing: The New Case for Vertical Integration" *Harvard Business Review* 3/4:75–81.

Kutscher, R.E., and R.W. Riche, 1990, "Impact of Technology on Employment and the Workforce – The US Experience," Paper for the World Bank Seminar on Employment and Social Dimension of Economic Adjustment, 27–28 February, Washington, DC.

Lawrence, R.Z., 1983, "Is Trade Deindustrializing America? A Medium-Term Perspective," *Brookings Papers on Economic Activity* 1:129–157.

Lawrence, R.Z., 1989, The Issue of Competitiveness, Remarks for Conference on Trade Capital, Foreign Service Institute 1/30/89, Washington, DC.

Lay, G., and T. Michler, 1990, Stand und Aussichten der Fertigungsautomation in der Bundesrepublik Deutschland, Fraunhofer–Institut fur Systemtechnik und Innovationsforschungs Zentrum Karlsruhe, Germany.

Lay, G., and J. Wengel, 1989, Wirkungsanalyse der indirektspezifischen Forderung zur betrieblichen Anwendung von CAD/CAM–Systemen im Rahmen des Programms Fertigungstechnik 1984–1988, Kernforschungszentrum Karlsruhe GmbH, Germany.

Lehtinen, M., 1989, "Allen-Bradley's Methods for Competing," *Industrial Comput-ing* **8**(8):38–40.

Lemola, T., and R. Lovio, 1988, *Possibilities for a Small Country in High Technol-ogy Production – The Electronic Industry in Finland*, CP-88-3, IIASA, Laxen-burg, Austria.

Luria, D., 1989, Standards, Markets, & Economic Growth, Paper draft.

Magaziner, I. *et al.*, 1989, "Cold Competition: GE Wages the Refrigerator War," *Harvard Business Review* **3/4**:114–124.

Manufacturing Technology 1988, *Current Industrial Reports: 1989*, US Dept. of Commerce, Bureau of the Census, Washington, DC.

Manufacturing Technology News, 1990, Society of Manufacturing Engineering, Jan-uary, Dearborn, MI.

Markides, C.C., and N. Berg, 1988, "Manufacturing Offshore Is Bad Business," *Harvard Business Review* **9–10**:113–120.

McAlinden, S.P., 1989, *Programmable Automation, Labor Productivity and the Com-petitiveness of Midwestern Manufacturing*, Industrial Technology Institute, Ann Arbor, MI.

Mendius, H.G., and S. Weimer, 1988, "Improvement in Working Conditions in Small Firms with Close Links with Larger Firms," in *Proceedings of Conference on Partnership between Small and Large Firms held in Brussels*, 6/88, pp. 235–239, Belgium.

Merrifield, D.B., 1988, "FMS in the USA: The New Industrial Revolution," *Man-aging Automation* **9**:66–70.

Mody, A., 1989a, *Changing Firm Boundaries: Analysis of Technology – Sharing Alliances*, Industry and Energy Department Working Paper, Industry Series Paper 3, World Bank, Washington, DC.

Mody, A., 1989b, *Institutions & Dynamic Comparative Advantage: Electronics In-dustry in South Korea and Taiwan*, Industry and Energy Department Working Paper, Industry Series Paper 9, World Bank, Washington, DC.

Mody, A. *et al.*, 1989, *Emerging Patterns of International Competition in Selected Industrial Product Groups*, Industry and Energy Department Working Paper, Industry Series Paper 2, World Bank, Washington, DC.

Morris-Suzuki, T., 1988, *Beyond Computopia: Information, Automation and Democracy in Japan*, Kegan Paul, London, UK.

Mullens, P., 1986, "Four of Europe's Best," *Production* **10**:78–85.

Ohmae, K., 1985, *Triad Power: The Coming Shape of Global Competition*, Free Press, New York, NY.

Peters, H.J., 1989, *Seatrade, Logistics and Transport*, PPR Paper, World Bank, Washington, DC.

Peters, H.J., 1990, *India's Growing Conflict between Trade and Transport*, PPR Working Papers, World Bank, Washington, DC.

Piore, M., and L. Sabel, 1984, *The Second Industrial Divide*, Basic Books, New York, NY.

Porter, M.E. *et al.*, 1986, *Competition in Global Industries*, Harvard Business School Press, Cambridge, MA.

Preeg, E., 1989, *The American Challenge in World Trade*, The Center for Strategic and International Studies, Washington, DC.

Ranta, J., 1988a, "Impact Assessment of Automation Technology: Comments and Methodological Views," Paper for the Third IFAC/IFIP/IEA/IFORS Conference on Man-Machine Systems, 14–16 June, Laxenburg, Austria.

Ranta, J., 1988b, *Trends and Impacts of CIM*, WP-89-1, IIASA, Laxenburg, Austria.

Ranta, J., 1989, "The Impact of Electronics and Information Technology on the Future Trends and Applications of CIM Technologies," *Technological Forecasting and Social Change* 35(23):231–260.

Rolland, W.C., 1988, "Global Pressures: The Impact on US Manufacturing," *Managing Automation* 9:75.

Rubenstein, A.H. *et al.*, 1979, "Innovation Among Suppliers to Automobile Manufacturers: An Exploratory Study of Barriers and Faciliators," *R&D Management* 9(2):65–76.

Salzer, B., 1986, *Europas Mehr-Wert; Chance Fur die Zukunft: Gemeinsame Technologiepolitik*, Europa Union Verlag, Bonn, Germany.

Sanderson, S.W. *et al.*, 1986, "Impacts of Computer-Aided Manufacturing on Offshore Assembly & Future Manufacturing Locations," *Regional Studies* 21(2):131–142.

Schoenberger, E., 1986, "Competition, Competitive Strategy, and Industrial Change: The Case of Electronic Components," *Economic Geography* 62(4):321–333.

Schoenberger, E., 1988, "Technological & Organizational Change in Automobile Production: Spatial Implications," *Regional Studies* 21(3):199–214.

Schoenberger, E., 1989a, Foreign Manufacturing Investment in the US, *Commentary*, Fall:20–26.

Schoenberger, E., 1989b, "Some Dilemmas of Automation: Strategic and Operational Aspects of Technological Change in Production," *Economic Geography*:232–248.

Schoenberger, E., 1989c, "US Manufacturing Investments in Western Europe: Markets, Corporate Strategy and the Competitive Environment," Paper presented at the Annual Meetings of the Regional Science Association, July, Baltimore, MD.

Schulte–Hilling, 1989, *Situation & Prospects for Production Automation*, Part I, *International Comparison*, Research Paper 22/87 II, Cologne, Germany.

Schultz-Wild, R., 1985, "Introducing New Manufaturing Technology: Manpower Problems and Policies," *Human Systems Management* 5(85):231–243.

Schultz-Wild, R., 1989, "On the Threshhold of Computer-Integrated Manufacturing," *Computer Integrated Manufacturing Systems* 2(4):240–248.

Senker, P., 1986, *Towards the Automatic Factory? The Need For Training*, Springer-Verlag, Berlin, Heidelberg, New York.

Shaiken, H., 1987, *Automation and Global Production: Automobile Engine Production in Mexico, the US and Canada*, Center for US-Mexican Studies, University of California, San Diego, CA.

Shapira, P., 1988, "Industrial Extension: Learning from Experience," Paper prepared for the Assessment on Technology, Innovation, and US Trade, Office of Technology Assessment, US Congress, November, Washington, DC.

Shapiro, R.D., 1985, *Toward Effective Supplier Management: International Comparisons*, Harvard Business School, Division of Research, Cambridge, MA.

Simers, D. *et al.*, 1989, "Just-in-Time Techniques in Process Manufacturing Reduced Lead Time, Cost; Raise Productivity, Quality," *Industrial Engineering* 1:19–23.

Stalk, G., 1988, "Time: The Next Source of Competitive Advantage," *Harvard Business Review* 7–8:41–51.

Stalk, G., and T.M. Hout, 1990, *Competing Against Time, How Time-Based Competition is Reshaping Global Markets*, Free Press, New York, NY.

Steedman, H., 1989, "Productivity, Machinery and Automation," *National Institute of Economic Review* 5.

Stirling, C. *et al.*, 1985, *Under Pressure*, Westview, Boulder, CO.

Stoffel, J.M., 1989, "CE Reader Survey Reveals Latest Automation Trends," *Control Engineering* 1:51–53.

Stoffel, J.M., 1990, "CE Readers Project Bright Outlook for Controls," *Control Engineering* 1:56–57.

Stoll, H.W., 1986, "Design for Manufacture: An Overview," *Applied Mechanics Review* 39(9):1356–1364.

Suh, N.P., 1984, "The Future of the Factory," *Robotics and Computer-Integrated Manufacturing* 1(1):39–49.

Sullivan, L.P., 1987, "The Power of Taguchi Methods to Impact Change in US Companies," *Target*, Summer:18–22.

Suri, R. *et al.*, 1989, *International Implications of Modern Manufacturing Technologies and Practices: Some Initial Findings*, Technical Report No.89-9, Industrial Engineering, University of Wisconsin, Madison, WI.

Survey of Current Business, 1990a, US Department of Commerce, Bureau of Economic Analysis, March, Washington, DC.

Survey of Current Business, 1990b, US Department of Commerce, Bureau of Economic Analysis, May, Washington, DC.

Tani, A., 1987, *Future Penetration of Advanced Industrial Robots in the Japanese Manufacturing Industry: An Econometric Forecasting Model*, WP-87-95, IIASA, Laxenburg, Austria.

Tani, A., 1988a, *Distribution Models for Diffusion of Advanced Technology by Company Size*, WP-88-62, IIASA, Laxenburg, Austria.

Tani, A., 1988b, *Saturation Level of NC Machine-Tool Diffusion*, WP-88-78, IIASA, Laxenburg, Austria.

Tani, A., 1989, "International Comparisons of Industrial Robot Penetration," *Technological Forecasting and Social Change* 34:191–210.

Tchijov, I., 1989a, *FMS World Data Bank*, WP-89-33, IIASA, Laxenburg, Austria.

Tchijov, I., 1989b, *FMS in Use: An International Comparative Study*, WP-89-45, IIASA, Laxenburg, Austria.

Tchijov, I., 1989c, *Machining FMS: Tendencies of Development*, WP-89-51, IIASA, Laxenburg, Austria.

Tchijov, I., and R. Sheinin, 1988, *FMS Diffusion and Advantages*, WP-88-29, IIASA, Laxenburg, Austria.

Technecon, 1989, V.1, .1, Spring; V.1, .2, Fall; Industrial Technology Institute, Ann Arbor, MI.

Technological Advance & Organizational Innovation in the Engineering Industry, 1989, Industry and Energy Department Working Paper, Industry Series Paper 4, World Bank, Washington, DC.

Thurow, L., 1984, "Building A World-Class Economy," *The Future of Advanced Societies*, 11(12):16–29.

Treffpunkt CIM, 1989, Der Bundesminister für Forschung und Technologie, 9/89.

UN Economic Commission for Europe, 1986, *Recent Trends in Flexible Manufacturing*, United Nations, New York, NY.

US Congress, Office of Technology Assessment, 1988, *Paying the Bill: Manufacturing and America's Trade Deficit*, Washington, DC.

US Congress, Office of Technology Assessment, 1990, *Making Things Better: Competing in Manufacturing*, Washington, DC.

Van de Ven, 1986, "Central Problems in the Management of Innovation," *Management Science* 32(5):590–607.

Verband der Automobilindustrie, 1989a, *Das Auto International in Zahlen (International Auto Statistics)*, Druckerei Henrich, Frankfurt, Germany.

Verband der Automobilindustrie, 1989b, *Tatsachen und Zahlen aus der Kraftverkehrswirtschaft*, Bronners Druckerei Breidenstein GmbH, Frankfurt, Germany.

Wandel, S., and R. Hellberg, 1987, Transportation Consequences of New Logistics Technologies, 9/87.

Watanabe, K., "Innovation of Japanese Manufacturing Industries," Unpublished paper.

Watanabe, S., 1986, "Labor-saving Versus Work-amplifying Effects of Microelectronics," *International Labour Review* 125(3):243–259.

Watanbe, S., 1987, *Microeconomics, Automation, & Employment in the Automobile Industry*, John Wiley & Sons, New York, NY.

Womack, J.P., 1989a, *Strategies for a Post-National Motor Industry*, IMVP (International Motor Vehicle Program) International Policy Forum, 5/89, MIT, Cambridge, MA.

Womack, J.P., 1989b, *The Mexican Motor Industry Strategies for the 1990s*, IMVP International Policy Forum, 5/89, MIT, Cambridge, MA.

Womack, J.P., 1989c, "Seeking Mutual Gain: North American Responses to Mexican Liberalization of Its Motor Vehicle Industry," Paper presented at the 44th Annual Plenary Meeting of the Mexico-US Business Committee in Florida, 9 November, MIT Cambridge, MA.

World Class Manufacturing Operating Principles for the 1990s and Beyond, 1989, NCMS-88-MO-1A, National Center for Manufacturing Sciences, Ann Arbor, MI.

Wright, P.K., and D.A. Bourne, 1988, *Manufacturing Intelligence*, Addison-Wesley Publishing Co. Inc., Reading, MA.

Yamazaki, K., 1988, "The Forecast for Unmanned Factory Networking," *Metalworking Engineering and Marketing* 3:42–47.

Yeats, A.J., 1989, *Do Caribbean Exporters Pay Higher Freight Costs?* Discussion Paper 6, World Bank, Washington, DC.

Yuasa, M., 1989, "NEC's Global Production Strategy," *Tokyo Business Today* 5:28–31.

Epilog

Karl-H. Ebel

Policy Options

Although the diffusion of CIM is mainly market driven, it depends a good deal on government policies on industry, education and training, finance, commerce, defense and telecommunications procurements, aerospaces, and international trade. Governments usually monitor technological developments and may be involved in technology promotion and research and development depending on the social and economic system. At any rate, they set the framework and make political, economic, and strategic choices that will influence the direction and speed of technological development. They can counteract bottlenecks and constraints, promote a favorable climate for technology diffusion and transfer, encourage investment, support the training infrastructure, and provide direct assistance through credit schemes, tax concessions, advisory services, research grants, or pilot projects. Governments may also foster links between research institutions and industry as well as promote start-up enterprises in new technology fields. For a fuller discussion of such policy issues see Bessant (1989).

The difference in performance in CIM technologies between the United States and Japan has been partly traced to distinct government policies. In the United States defense procurement interests channeled research grants mainly into the development of highly sophisticated information technology and manufacturing processes for the production of complex military and space hardware. This specific technology has relatively limited civilian application. However, the United States has remained the leader in large advanced machinery and materials and in more sophisticated information technology applications. Japanese industrial policy concentrated on commercial applications of information technology and less-specialized NC and CNC machinery production. Such machines, which are in the lower price range, are within the reach of small- and medium-sized enterprises. Incentives were provided to firms for installing this equipment. As a result, Japan has become the market leader in these fields and is reaping considerable benefits (Bar *et al.*, 1989).

To preserve or enhance the competitive position of the industries of their countries in world markets, governments have usually opted to promote promising new technologies, and this includes CIM. Various measures and schemes addressing problems encountered in different societies are being applied. Most are designed to enhance the technological capability of industry and to develop the required human resources. This is not the place to review such measures or to assess their efficacy. Here it is sufficient to note that government action has been supportive in the worldwide technology race.

CIM: An Opportunity for Developing Countries?

While enterprises in industrialized countries struggle with the implementation of the CIM concept, developing countries are concerned about the widening technology gap between rich and poor nations. Would a transfer of CIM technologies help to close this gap? The answer depends on numerous factors. There can be no hard and fast rule.

Frequently the example is cited of the newly industrializing countries (NICs) or threshold countries such as Brazil, India, South Korea, Malaysia, Mexico, and Singapore, where some diffusion of such relatively mature technologies as CNC, CAD, and robotics is taking place. Governments usually favor or support such investments. It is argued that these countries are benefiting from the skill-saving nature of these technologies and that there appears to be a much wider potential for their use, particularly as regards export-oriented industries (Edquist and Jacobsson, 1988).

But is such a course advisable, realistic, and practicable for most developing countries? Obviously, this depends primarily on the constraints and the pattern of factor endowment that these countries face, among which a chronic capital shortage and a cheap, abundant, but low-skilled, labor supply are the most outstanding. Moreover, the infrastructural support for CIM (e.g., adequate supplies of energy, raw materials, components and spare parts, reliable transport systems, easy access to high-technology markets, competent engineering consultancy services, and research and training institutions) is lacking. Adequate maintenance of hardware and software can rarely be guaranteed. In addition, an adequate production volume and a large enough market to amortize the high investment needed for CIM must be ensured. All this implies that there is generally not sufficient demand for sustained development in the field of high technology.

These are formidable obstacles to the introduction of advanced manufacturing technologies. At any rate, whatever industrialization goals are pursued, there can be no justification for squandering scarce capital on costly CIM projects destined to become white elephants. Turnkey operations proposed by eager vendors are the most likely to suffer that fate: they tend to function badly and become rapidly obsolete. It should be borne in mind that the investment required in complex CIM projects could easily be in the range of US $400,000 to US $600,000 per work place.

Entry barriers to CIM are high, and there are many pitfalls. However, developing countries are confronted with the predicament that in many industrial activities, particularly in the capital goods sector, their competitive advantage of cheap labor is declining. Cheap labor is no substitute for the productivity and quality improvements achieved by highly skilled workers operating flexible automation systems, or for faster delivery and service and the rate of product innovation achieved with advanced technology. If developing countries are to gain a foothold in competitive international markets for more complex products and do not want to be completely outdistanced, they will need to develop some local capabilities in this field. This entails difficult choices.

The most promising approach would appear to be to start with the relatively mature components of the CIM technology, i.e., CNC and CAD systems and microprocessor monitoring systems, but without going right into the complexities of networking. Personal computer systems for maintaining records and programs may also bring great benefits. These systems would have to be incorporated in the manufacture of products for which a market could be found or developed, possibly in cooperation with enterprises from industrialized countries. They would have to fit into a strategic framework allowing a step-by-step adoption of more advanced technologies as experience is accumulated and learning proceeds. The adaptation of hardware and software to local conditions must be assured. Haphazard wholesale transfer of advanced technology usually spells failure.

Certain organizational innovations induced by CIM can be implemented without costly hardware and software. They concern essentially the logical and efficient organization of the production and information flow, changes in managerial responsibility, and training of the work force. It has been proved that the productivity increases to be obtained through such methods can be spectacular (Bessant and Rush, 1988). In 1989 the United Nations Industrial Development Organization (UNIDO) launched a project on planning and programming the diffusion of industrial automation technologies in the capital goods industries of developing countries to help them formulate policies to assimilate such technologies. It is hoped that this project will provide more realistic insight into the possible scope of the application of advanced technologies which, after all, have been conceived in the societal context of the highly industrialized countries; and even here they have often failed to reach their objectives. They can hardly be grafted on to industries without adequate preparation.

Outlook

Chances of reconciling divergent views on CIM are slim. Many of its advantages or faults are in the eye of the beholder. However, it is definitely not the panacea for all problems encountered in production that some seem to see in it. The promised land of total manufacturing integration is still far away, although an increasing number of enterprises appear to be engaged in an evolutionary process toward it.

By any standard, the introduction of CIM is a risky undertaking. If it is to be successful the firm's manufacturing organization and product range will have to be reviewed and rationalized. The pace of transition will depend on the knowledge, qualifications, and abilities of the staff. The neglect of the human factor, the absence of systematic training and personnel planning, and the maintenance of ossified and obsolete organizational patterns are fatal to the implementation of CIM.

CIM is certainly not a miraculous solution for enterprises falling victim to market forces; however, if used as a strategic means to seize new opportunities, it may help them win back lost terrain. Enterprises are well advised to avoid great technology leaps that are liable to fail and to build up their technological capabilities systematically but gradually.

CIM is a leading-edge technology, and its introduction requires long-term strategies, much research and development, and, possibly, forgoing immediate financial benefits. The most essential element in such a strategy is the preparation of the work force for the impending changes. This requires consultation at all levels and a systematic training effort. To neglect staff training is inevitably very costly in machine downtime and scrap production.

All experience gained so far favors a cautious and incremental approach in order not to overstretch the assimilating and learning capacity of the work force with the negative results which that implies. The fact remains that the trend toward more manufacturing integration is bound to continue, and scientific advances will continue to offer solutions to outstanding technical problems. CIM is most likely to fail where it tries to supplant essential human qualities. The subjugation of people to machines and technical systems is proving more and more counterproductive. Instead, a type of work organization is needed that enables and motivates people to use their theoretical and empirical knowledge and skills in mastering and using advanced means of production. CIM will only be as good as the people in charge of it.

This engages the social responsibility of the system designers; they should make the limits and the capabilities of technical systems explicit and design them as a tool to service the people operating them. A residual role for humans in the manufacturing process is dehumanizing. CIM should result in an organization that makes the best possible use of people's knowledge, skills, capabilities, and talents, which are vastly underutilized. Moreover, system designers should not lose sight of the objective of creating humane working conditions and improving the quality of working life. The exposure of operators to excessive stress has become a major topic of study on advanced manufacturing systems. More research needs to be directed to these objectives.

Are we heading in the wrong direction? Evidence suggests that the difficulties and complexities of introducing CIM on a large scale were initially underestimated. The technocentric approach aiming at the "unmanned factory" is now questioned for good reason: so far it has failed to produce the expected results. Its social and economic costs in terms of worker alienation, absenteeism, and production losses have often been disregarded. This is having a sobering effect on the unconditional technocrats. There is probably not just one type of "factory of the future" but many alternative solutions to manufacturing problems. In fact, from the technical point of view there is a proliferation of choices.

There is little evidence of dramatic negative effects on employment. CIM will do away with unskilled jobs, but this will happen gradually since its diffusion is slow. At any rate, it will have only a marginal effect on overall employment levels in the foreseeable future. Moreover, job reduction is seldom a rationale for the use of CIM technologies because direct labor costs in modern manufacturing are in any case relatively low when compared with capital costs; such technologies are more likely to be used to economize on scarce skilled labor and overcome labor shortages. In particular, CIM opens up opportunities for an optimal use of capital equipment by making operating times of factories independent of the working time of the work force for whom a wide array of flexible working schedules can be devised.

Will CIM really spell the end of Taylorism? It is definitely too early to pronounce its methods dead and buried. Taylorism will continue to subsist in mass production alongside dedicated automation and machinery, and so will the corresponding hierarchical management structures. However, mass production and market dominance of mass-produced goods are declining in many manufacturing activities. The markets demand differentiated, diversified, and customized products, entailing a need for small-batch production. The flexible automation offered by CIM can do the job.

At times of accelerated technical change, there is a need not only for innovative responses from both management and workers' representatives, taking into account technological and economic imperatives, but also for new opportunities to be at hand to improve the working environment and conditions of work. In the continuing adjustment process unions may need to shed inherited organizational structures and philosophies. However, the onus of making new work organization effective is on management.

Integrated manufacturing systems are vulnerable to disruption. Running them efficiently and round the clock presupposes harmonious industrial relations, since work stoppages, go-slows, or other types of resistance stemming from demotivating working conditions can cause major losses. The success of CIM, therefore, presupposes mutual understanding and cooperation between management and the work force and its representatives. While the introduction of even well-designed CIM systems is bound to cause tensions, it also offers new opportunities for enhancing dialogue and breaking down barriers between the social partners – a chance not to be missed.

References

Bar, F. *et al.*, 1989, "The Evolution and Growth Potential of Electronics-based Technologies," *STI Review*, No. 5, OECD, Paris,.

Bessant, J., 1989, *Microelectronics and Change at Work*, ILO, Geneva, Switzerland.

Bessant, J., and H. Rush, 1988, *Integrated Manufacturing*, Technology Trends Series No. 8, UNIDO, Vienna, Austria.

Edquist, C., and S. Jacobsson, 1988, *Flexible Automation: The Global Diffusion of New Technology in the Engineering Industry*, Basil Blackwell, Oxford and New York.

List of Contributors

Oleg Adamovic
All-Union Research Institute for
  Economic Problems of Science and
  Technological Development
Moscow
USSR

Robert Ayres
Department of Engineering
  and Public Policy
Carnegie-Mellon University
Pittsburgh, PA
USA

Luigi Bodda
International Institute for
  Applied Systems Analysis
Laxenburg
Austria

Robert Boyer
CEPREMAP
Paris
France

Hans-Ulrich Brautzsch
"Bruno Leuschner"
Berlin
Germany

Duane Butcher
US Department of State
Foreign Service Institute
Arlington, VA
USA

Pavel Dimitrov
Sofia
Bulgaria

Rumen Dobrinsky
Institute of Social Management
Sofia
Bulgaria

Karl-H. Ebel
International Labour Office
Geneva
Switzerland

Peter Fleissner
Austrian Academy of Science
Vienna
Austria

Shunichi Furukawa
Institute of Developing
  Economies
Tokyo
Japan

Grazyna Juszcak
Institute of Econometrics
  and Statistics
University of Lodz
Lodz
Poland

Czeslaw Lipinsky
Institute of Econometrics
  and Statistics
University of Lodz
Lodz
Poland

Nadezda Mamysheva
All-Union Research Institute for
  Economic Problems of Science
  and Technology Development
Moscow
USSR

Shunsuke Mori
Department of Industrial Automation
Science University of Tokyo
Tokyo
Japan

Witold Orlowski
Institute of Econometrics
  and Statistics
University of Lodz
Lodz
Poland

Felix Peregudov
USSR State Committee
  on Education
Moscow
USSR

Sergei Perminov
All-Union Research Institute for
  Economic Problems of Science
  and Technology Development
Moscow
USSR

Mariusz Plich
Institute of Econometrics
  and Statistics
University of Lodz
Lodz
Poland

Wolfgang Polt
Austrian Academy of Science
Vienna
Austria

Iouri Solomentsev
USSR Academy of Sciences
Moscow
USSR

Lucja Tomaszewicz
Institute of Econometrics
  and Statistics
University of Lodz
Lodz
Poland

Kimio Uno
Faculty of Policy Management
Keio University
Fujisawa
Japan

Pentti Vuorinen
International Institute for
  Applied Systems Analysis
Laxenburg
Austria

Mitsuo Yamada
Faculty of Humanities and
  Social Sciences
Mie University
Tsu
Japan

Eduh Zuscovitch
Department of Economics
Ben-Gurion University
  of Negev
Beersheba
Israel

INDEX

512

516